E-Collaboration in Modern Organizations: Initiating and Managing Distributed Projects

Ned Kock
Texas A&M International University, USA

Information Science REFERENCE

INFORMATION SCIENCE REFERENCE

Hershey · New York

Acquisitions Editor: Kristin Klinger
Development Editor: Kristin Roth
Senior Managing Editor: Jennifer Neidig
Managing Editor: Sara Reed
Copy Editor: Shanelle Ramelb
Typesetter: Jamie Snavely
Cover Design: Lisa Tosheff
Printed at: Yurchak Printing Inc.

Published in the United States of America by
Information Science Reference (an imprint of IGI Global)
701 E. Chocolate Avenue, Suite 200
Hershey PA 17033
Tel: 717-533-8845
Fax: 717-533-8661
E-mail: cust@igi-global.com
Web site: http://www.igi-global.com

and in the United Kingdom by
Information Science Reference (an imprint of IGI Global)
3 Henrietta Street
Covent Garden
London WC2E 8LU
Tel: 44 20 7240 0856
Fax: 44 20 7379 0609
Web site: http://www.eurospanonline.com

Library of Congress Cataloging-in-Publication Data

E-collaboration in modern organizations : initiating and managing distributed projects / Ned Kock, editor.

p. cm.

Summary: "This book combines comprehensive research related to e-collaboration in modern organizations, emphasizing topics relevant to those involved in initiating and managing distributed projects. Providing authoritative content, it describes issues that have implications for distributed project management, implications surrounding the use of e-collaborative environments for distributed projects, and debate related to e-collaboration support for distributed project management"--Provided by publisher.

Includes bibliographical references and index.

ISBN-13: 978-1-59904-825-3 (hardcover : alk. paper)

ISBN-13: 978-1-59904-827-7 (ebook : alk. paper)

1. Project management. 2. Virtual work teams--Management. I. Kock, Ned F., 1964-

HD69.P75E36 2008

658.4'04--dc22

2007023866

British Cataloguing in Publication Data
A Cataloguing in Publication record for this book is available from the British Library.

Table of Contents

Detailed Table of Contents

This chapter reviews theoretical research on e-communication behavior, identifying two main types of theories: technological and social. Based on this review, it provides the rationale for the development of a new theory that is neither technological nor social. The new theory is based on evolution theory, whose foundations were laid out by Darwin. Three theoretical principles are developed from evolution theory: media naturalness, innate schema similarity, and learned schema variety. The chapter concludes by illustrating how the theoretical principles can be used as a basis for the development of a simple predictive model in the context of an online broker.

In this study, a methodology built on the conceptual foundation of metaphor research was used to comprehend and then interpret the Web presence of 15 nonprofit theatres that comprise the total regional theatre of southern New Jersey that exists on the Web. In order to add additional insight, earlier research findings from working with off-Broadway and regional theatre festivals were extended to analyze the Web presence of the theatres in southern New Jersey. The authors contribute to the literature by systematic and deep investigation of the strategic importance of the Web for nonprofit theatre groups in the southern New Jersey region. In addition, their use of the metaphor methodology in order to create a telling portrait of what transpires on the Web in relation to nonprofit organizations is also an original contribution. This work is meant to heighten the awareness of administrators to the rapidly accelerating need for the strategic use of e-collaboration. The authors propose that with the use of the Web, administrators can

Section III
Emerging Issues and Debate

This chapter develops an analytical framework for new forms of information warfare that may threaten commercial and government computing systems by using e-collaboration in new ways. The framework covers (a) a strategic model, (b) the strategic arena, (c) e-collaboration, and (d) ethics and law. The framework is then used to compare two recorded instances of major hacker wars that erupted in the shadow of kinetic conflicts. In both cases, the hacker war appears to have been a grassroots collaborative enterprise by loosely organized civilians, with neither government control nor permission. Collaborating across networks to coordinate their attacks, such hacker wars can attack both government and commercial computer networks without warning. The analysis shows how hacker wars demonstrate characteristics found in the frameworks, and that there are forms of e-collaboration that represent a potentially difficult new source of threat for globalized information systems.

A laboratory experiment was conducted with 42 student groups to evaluate the effects of transactional vs. transformational leadership styles and anonymity when groups supported by an electronic meeting system (EMS) discussed the ethical issue of copying copyrighted software. A confederate leader displayed either transformational or transactional behaviors. Transformational leaders motivate effort by raising the awareness of followers to make them aspire to higher order needs and values and developing them to fulfill their aspirations. Transactional leaders motivate effort by highlighting the contractual exchange involved in a relationship. Participants working with a transformational confederate were more likely to make arguments that challenged the copying of copyrighted software than those working with one who was more transactional. These arguments in turn caused groups exposed to such arguments to have greater deviation among their members in intentions to copy the software. Participants working with a transactional confederate were more likely to make arguments in favor of copying copyrighted software. These arguments in turn caused groups exposed to such arguments to have a greater mean of intention to copy the software. Implications for practice and future research on ethics and leadership are discussed.

Knowledge management approaches have been broadly considered to entail either a focus on organizing communities or a focus on the process of knowledge creation, sharing, and distribution. While these two approaches are not mutually exclusive and organizations may adopt aspects of both, the two approaches entail different challenges. Some organizational cultures might be more receptive to the community approach whereas others are more receptive to the process approach. Although culture has been widely cited as a challenge in knowledge management initiatives and many studies have considered the implications of organizational culture on knowledge sharing, few empirical studies address the influence of culture on the approach taken to knowledge management. Using a case-study approach to compare and contrast the cultures and knowledge management approaches of two organizations, the study suggests the ways in which organizational culture influences knowledge management initiatives as well as the evolution of knowledge management in organizations. Whereas in one organization, the knowledge management effort became little more than an information repository, in the second organization, the knowledge management effort evolved into a highly collaborative system fostering the formation of electronic communities.

Cohesion is regarded as something to strive for in virtual teams yet is difficult to attain. What happens, though, when cohesion is achieved; does cohesion, as assumed, enhance the virtual team? During a longitudinal participant observation study of a virtual software development team, a strange paradox was noted. A new software development methodology was introduced to the project, and the developers were initially committed to its use. Over time, the commitment gradually decreased to the stage where aspects of the new methodology were practically ignored. As the team was a virtual team, with group members rarely congregating as a whole for any length of time, it was hard to explain why this diminishing of commitment occurred. The remoteness and part-time participation of group members meant that the team deciding themselves to ignore aspects of the methodology was not a likely possibility. A review of existing research suggested that the concepts behind the diffusion of innovations (specifically software process innovations) may have a bearing. Although pertinent to the area of introducing new software development methodologies, diffusion theories did not provide a complete explanation for the decrease in commitment that was observed. The theory of competing commitments was applied, and it was discovered that one cause of the decreased commitment among team members was groupthink. Groupthink should not be a problem with virtual teams as there should be less cohesion—the lack of contact between members dictating the low level of cohesion. Further analysis showed that traditional peer groupthink was not the issue, but hierarchical groupthink influenced by the project manager had a large influence. These findings are in contrast to most expectations concerning cohesion and virtual teams, including the project management of virtual teams.

A framework to capture and manage distributed knowledge can address distributed knowledge management from creation to facilitation. Knowledge generation and dissipation need to be embedded in corporate processes. These processes need to have an underlying principle that eliminates the obstacles of collecting multiple knowledge perspectives within complex organizations. Moreover, extrinsic motivators, social-psychological forces, and organizational climate factors are believed to influence knowledge sharing. This study discusses a framework that provides a synergized view to collect, share, and manage the distributed corporate knowledge using organizational knowledge models and technology knowledge models. Structural, cognitive, relational, and technological factors derived from a synthesized literature review aid to formulate this framework. Using this framework, the role of peer-to-peer networks on distributed knowledge management in organizations is examined.

Preface

Back in the early 1990s, I was working as a consultant in ISO-9000 certification projects (International Organization for Standardization) for several organizations. One of the key components of those projects was the creation of rules and procedures, as well as the appropriate technological infrastructure, to allow groups of employees to address nonconformities in business processes. "Addressing nonconformities" is simply another way of saying that the employees were doing business process improvement. In business process improvement, usually a group of people analyzes a business process and solves the problems they discover in the process. Since the original ISO-9000 set of standards was based on total quality management ideas, we needed continuous and incremental process improvement across each organization, which meant many business process improvement groups addressing local quality-related issues.

Documenting the work of many business process improvement groups generates a lot of paperwork and is a tedious process. So I was looking into the idea of conducting business process improvement groups supported by simple e-mail-based e-collaboration systems. In that way, the e-collaboration systems themselves would automatically generate and archive the records of each group's work. That meant, also, that most of the communication in each group would take place electronically.

After running a few groups where most of the interaction took place through an electronic communication medium, it became clear that the key determinant of the success of those groups was the quality of their distributed (or virtual) project management. In other words, if the leader and/or facilitator of an e-collaboration-supported business process improvement group were to do the wrong things (e.g., not stick closely to a well-defined group process structure), then the group could easily derail and fail to accomplish its task. It seems that with groups that interact primarily through electronic media, it is particularly difficult to bring a group back to the right track after it goes on a tangent or gets into a polarized discussion; apparently, it is much more difficult than with face-to-face groups.

The importance of distributed project management has only increased since the early 1990s. Much research has been conducted in the area in the 1990s and early 2000s, culminating with a special issue on collaborative project management published in the *International Journal of e-Collaboration* in 2006. The special issue was published in two parts (Issues 3 and 4 of the journal's second volume) and was guest-edited by two outstanding researchers and colleagues: Jerry Fjermestad and Nicholas C. Romano, Jr. Several of the articles published in that special issue have been included in this book.

This book is a collection of chapters covering topics related to e-collaboration in modern organizations. Its emphasis is on topics relevant to those involved in initiating and managing distributed projects. The book is organized in three main parts: Section I, "Conceptual and Theoretical Issues"; Section II, "Distributed Project Management"; and Section III, "Emerging Issues and Debate." Each of these three parts contains five chapters. Most of the chapters in this book are revised versions of selected articles published in the *International Journal of e-Collaboration*. I have been serving as founding editor in chief of that journal since its first issue was published in 2005.

Section I of the book is made up of Chapters I to V and is dedicated to the discussion of conceptual and theoretical issues that have implications for distributed project management. Chapter I, by Kock and Garza, reviews theoretical research on e-communication and e-collaboration behavior and proposes a new theoretical framework based on human evolutionary ideas. In Chapter II, by Kendall, a methodology built on the conceptual foundation of metaphor research is used to review the Web presences of 15 nonprofit theatres in southern New Jersey. Chapter III, by Wong, Dow, Turel, and Serenko, describes an application of the American Customer Satisfaction Index to model the antecedents and consequences of customer satisfaction with e-mail systems. Chapter IV, by Blanchard, examines how a Listserv affects its members' sense of community with the sponsoring organization, challenging previous theories about the development of a sense of community and demonstrating the positive effects of Listserv membership. Chapter V, by Roussev, introduces a project management framework that arguably creates optimal conditions for the successful implementation of large software projects when those projects are outsourced either offshore or onshore.

Section II of the book, which comprises Chapters VI to X, addresses issues in connection with initiating and managing distributed projects in e-collaborative environments. Chapter VI, by Kock and Corner, describes an action research study of a computer-mediated business process redesign group conducted in a New Zealand university. In Chapter VII, Zigurs and Khazanchi propose a unified theoretical view of virtual project management and summarize data in connection with patterns of the effective management of virtual projects. Chapter VIII, by Becerra-Fernandez, Del Alto, and Stewart, presents a case study describing the development and use of the first Web-based collaborative and knowledge management platform deployed at NASA. In Chapter IX, Prikladnicki, Evaristo, Audy, and Yamaguti, discuss an integrated risk management process for distributed software development and report the results of an exploratory case study conducted in a software development center of a U.S. corporation's Brazilian subsidiary. Chapter X, by Schwaig, Gillam, and Leeds, identifies project management issues associated with offshore information technology outsourcing projects and provides specific recommendations for addressing those issues.

Section III of the book consists of Chapters XI to XV, and focuses on the discussion of emerging issues and debate related directly and indirectly to e-collaboration support for distributed project management. In Chapter XI, Baskerville develops an analytical framework for new forms of information warfare that may threaten commercial and government computing systems by using e-collaboration in new ways. Baskerville's engaging chapter is a revised version of an award-winning article, selected as the best article published in the *International Journal of e-Collaboration* in 2006. Chapter XII, by Kahai and Avolio, discusses a laboratory experiment involving 42 student groups and evaluates the effects of transactional vs. transformational leadership styles and anonymity when groups supported by an electronic meeting system discussed the ethical issue of copying copyrighted software. Chapter XIII, by Leidner, Alavi, and Kayworth, uses a case-study approach to compare and contrast the cultures and knowledge management approaches of two organizations; the study suggests the ways in which organizational culture influences knowledge management initiatives. Chapter XIV, by McAvoy and Butler, presents a longitudinal participant observation study of a virtual software development team where a strange paradox was noted: A new software development methodology was introduced to the project, and the developers were initially committed to its use; over time, however, the commitment gradually decreased to the stage where aspects of the new methodology were practically ignored. In Chapter XV, Vaidyanathan discusses a framework that provides a synergistic approach to collect, share, and manage distributed corporate knowledge.

The contributing authors are among the most accomplished researchers in the world today in the areas of e-collaboration and distributed (or virtual) project management. I am most grateful for their hard work in connection with the development of this book, which has been a great pleasure to edit together with my colleagues at IGI Global. The blend of conceptual, theoretical, and applied chapters found here makes me confident that this book will serve both academics and practitioners very well. I hope that the book will stimulate further research on distributed project management and related e-collaboration issues, and help project managers successfully lead virtual projects enabled by e-collaboration technologies.

Acknowledgment

No book project can be completed successfully without the support of a dedicated editorial team. I would like to thank the team at IGI Global for their excellent editorial support for this book project. Special thanks go to Jan Travers, Kristin Roth, and Corrina Chandler. Many thanks are also due to Meg Stocking, Angela Sweigart, and their journal editorial team for their support in the development of several issues of the *International Journal of e-Collaboration* (IJeC). Since most of the chapters in this book are revised versions of *IJeC* articles, many of the chapters published here would not exist without Meg's and Angela's keen involvement with the journal.

Several years ago, my wife and I had been discussing the idea of living in a part of the United States that had a strong Latin influence. Both of us have been raised in Brazil, a country that shares many cultural characteristics with other Latin American countries. So, I joined Texas A&M International University, located in the south Texas city of Laredo, near the U.S. border with Mexico. We moved to Laredo with our four kids, and have been having a great time since.

My output in terms of books has certainly gone up since I moved to Laredo. The other day I joked with a friend that I wanted to become the Stephen King of information systems! Maybe it has been the warm weather, but I think one of the reasons for my increased productivity is the high recognition and support given by the university's administration to faculty scholarship. Special recognition in that respect should be given to Ray Keck, the university's long-term president; Dan Jones, provost; and Jacky So, dean of the College of Business and Economics.

Another reason for my productivity is the support and encouragement of a wonderful group of colleagues with whom I have been sharing the third floor of Pellegrino Hall on the university's beautiful campus. They make up the recently created Division of International Business and Technology Studies, which I have had the pleasure to serve since 2006 in the capacity of founding chair. My special thanks go to Jacques Verville, Jackie Mayfield, Milton Mayfield, Ananda Mukherji, and Pedro Hurtado for the leadership roles that they have been playing in the division.

Last, but most important of all, I would like to thank my family for their love and support. This book is dedicated to them.

Section I
Conceptual and Theoretical Issues

Chapter I
The Ape that Used E-Mail: An Evolutionary Perspective on E-Communication Behavior

Ned Kock
Texas A&M International University, USA

Vanessa Garza
Texas A&M International University, USA

ABSTRACT

This chapter reviews theoretical research on e-communication behavior, identifying two main types of theories: technological and social. Based on this review, it provides the rationale for the development of a new theory that is neither technological nor social. The new theory is based on evolution theory, whose foundations were laid out by Darwin. Three theoretical principles are developed from evolution theory: media naturalness, innate schema similarity, and learned schema variety. The chapter concludes by illustrating how the theoretical principles can be used as a basis for the development of a simple predictive model in the context of an online broker.

INTRODUCTION

Given the title of this chapter, it is prudent to begin it with a clarification. This chapter is not about a chimpanzee or gorilla that used e-mail. It is about a much more modest (no chimpanzee or gorilla has ever been shown to have been able to speak intelligently, much less send and receive e-mail) yet important topic, namely the multimillion-year development of our biological apparatus for communication and how it affects electronic communication (e-communication) behavior.

Defining E-Communication and E-Communication Behavior

The *e* in *e-communication* stands for *electronic*, so the term e-communication refers to, essentially, any form of computer-mediated communication plus more traditional forms of electronic communication, such as telephone communication (since the telephone is also an electronic device). The term e-communication includes computer-mediated communication over the Internet as well as over other computer network infrastructures, thus also including computer-mediated communication that takes place through group decision-support systems and local-area-network-based communication tools.

E-communication behavior refers to the behavior of users toward e-communication technologies. For example, it has been shown that individuals in groups engaged in knowledge-intensive tasks and interacting primarily over e-mail tend to take 5 to 15 times longer on average to prepare and make individual contributions (i.e., electronic postings) than if they were interacting face to face only (Kock, 1998, 1999, 2005). In this case, what could be called "decreased contribution speed," with contribution speed being measured in words per minute, is a component of e-communication behavior, or the behavior of the users toward e-mail. Behavior toward e-communication tools is described in two main ways: (a) by contrasting the behavior of people using e-communication tools with the behavior of people in the absence of e-communication tools (i.e., interacting face to face), and (b) by contrasting the behavior of people using e-communication tools that incorporate different elements (e.g., asynchronous vs. synchronous electronic conferencing).

E-communication has its roots in the 1960s, when the first e-mail systems emerged, largely running on mainframe computers. In those early days, only a tiny minority, largely made up of people who spent their working days in front of a computer screen, used computers for communication (Sproull & Kiesler, 1991). For the majority of people, face-to-face (FtF) conversation, telephone calls, and paper-based documents were the communication media of choice.

The interconnection of mainframes, followed by desktop computers, through networks and the Internet, has changed the above picture dramatically, making e-communication an alternative choice for many business and social interactions. Significant technological innovations made this choice even more attractive, such as the group sense fostered by features of computer conferencing systems, the synchronicity and facilitation features of group decision-support systems, and virtual-presence features of video-enhanced media spaces, which were made feasible by extensive applied research as well as the advent of cheaper technology and increasing bandwidth and connectivity.

Increased use of e-communication media has led to demands from business and government for a better understanding of e-communication behavior. These demands were met by intensive empirical and theoretical research in the 1980s and 1990s, which have led to the development of several theories that can be classified into two main groups—technological and social—theories that have been essentially in a tug-of-war for many years. Technological theories have been traditionally deterministic in the sense that they have tried to provide a basis for predicting e-communication behavior based on a finite number of variables. Social theories have in many cases been developed to overcome limitations of technological theories, claiming that technological theories are oversimplified. Nevertheless, social theories have been, more often than not, unable to provide a useful basis on which to predict e-communication behavior. While technological theories provide a simplified view of e-communication, usually focusing on communication media and collaborative tasks as predictive factors, social theories try to understand e-communication as a social and very complex phenomenon, and often end up being more explanatory and descriptive than predictive.

Rather than joining the tug-of-war between technological and social theories, perhaps a desirable alternative line of action would be to try to think outside the box and devise a theoretical model that is neither purely technological nor social. This is the choice made in this chapter, which uses evolution theory (i.e., a biological theory) to explain e-communication behavior. In doing so, it provides a logical basis on which the implicit assumption made by technological theories that individual behavior is uniform and predictable can be understood. However, instead of arguing that individual behavior is uniform, this chapter shows that there are biological influences that induce individuals to present similar behavior. Moreover, these biological influences are isolated from social influences, the latter also being seen as strong and equally important. This isolation provides a basis for a unified understanding of e-communication behavior as a combination of biological and social influences, as well as an understanding of constraints posed by communication media and collaborative tasks.

This chapter is organized as follows. In the "E-Communication Behavior Theories" section, a brief theoretical review of research on e-communication behavior is conducted. The review identifies two main types of theories, technological and social, and provides the rationale for the development of a new theory that is neither technological nor social. The next section, "The Evolution of our Biological Apparatus for Communication," discusses how communication evolved over millions of years in the human species; this discussion is focused on traits that are relevant for the understanding of behavior toward e-communication technologies. The section "Key Theoretical Principles" follows with principles developed from evolution theory that have a direct application to our understanding of e-communication behavior. The section "Using the Theoretical Principles to Predict the Behavior of Customers of an Online Broker" illustrates how the theoretical principles can be used as a basis for the development of a simple predictive model tying media naturalness to customer satisfaction and revenues at an online broker. Finally, the conclusion summarizes the main points of the chapter and suggests future research directions.

E-COMMUNICATION BEHAVIOR THEORIES

Technological theories of e-communication place particular emphasis on the fit between task and medium as a determinant of the communication process and outcomes. That is, the foci of these theories are the communication medium and the task being accomplished through it. Examples of technological theories are the media richness theory (Daft & Lengel, 1986; Daft, Lengel, & Trevino, 1987; Lengel & Daft, 1988), the gains and losses model (Alavi, 1994; Nunamaker, Dennis, Valacich, Vogel, & George, 1991), and the task-technology fit theory proposed by Zigurs and Buckland (1998).

Among technological theories, perhaps the best known is the media richness theory (Daft & Lengel, 1986; Daft et al., 1987; Lengel & Daft, 1988), which has been quite influential among e-communication tool developers and researchers (Jackson & Purcell, 1997; Kock, 1998; Kock, Lynn, Dow, & Akgün, 2006; Lee, 1994; Markus, 1994) even though it was developed before the advent of most of the e-communication tools in use today. The media richness theory argues that rational individuals predictably favor the use of specific communication media to accomplish certain tasks. The media richness theory classifies different communication media according to a richness scale that features FtF interaction at the top and printed documents at the bottom, with e-communication media somewhere in between (Lee, 1994; Markus, 1994). A key hypothesis of the media richness theory is that rich media are more appropriate to support equivocal communication (which is likely to occur in complex tasks) than lean media, and that aggregate data about rational individual media choices would consistently support this hypothesis.

Social theories of e-communication place emphasis on the role of the social environment and socially constructed information-processing schemas (i.e., mental information-processing structures) in defining behavior toward e-communication technology. Examples of these theories are the social influence model (Fulk, Schmitz, & Steinfield, 1990), critical mass theory (Markus, 1990), adaptive structuration theory (DeSanctis & Poole, 1994; DeSanctis, Poole, Dickson, & Jackson, 1993; Poole & DeSanctis, 1990), and the technology metastructuration model (Orlikowski, Yates, Okamura, & Fujimoto, 1995).

A historical analysis of theoretical developments in e-communication suggests that many social theories have been developed to fill a gap arguably left by previous technological theories. An influential study conducted at a large risk-management service provider by Markus (1994) illustrates this link. The study builds on the social influence model and shows that social influences can shape individual behavior toward e-communication media in ways that are inconsistent with the media richness theory's predictions (Daft et al., 1987). Markus convincingly questioned the rigidity of the richness scale proposed by the media richness theory by showing that social pressures can change some attributes of e-communication media seen as static by the theory. For example, the study showed that pressure from senior managers on their subordinates to reply promptly to e-mail sent to them increased feedback immediacy, a feature of the e-mail medium that the media richness theory claimed to be static; therefore, e-mail was shifted up from its relative position on the media richness scale.

By focusing on the communication medium and task, technological theories implicitly assume that behavior and outcomes are determined only by those factors and not by specific characteristics of the individuals interacting through the communication medium. That is, technological theories implicitly assume that, for each specific combination of communication medium and task, individuals will behave in very similar and predictable ways. This has probably been the main target of criticism from social theoreticians, who argue that social characteristics of individuals interacting through a communication medium to perform a task and characteristics of the environment surrounding those individuals are as important, if not more, to determine behavior and task outcomes as are medium and task characteristics.

There are pros and cons to the debate regarding technological vs. social theories. On the positive side, it has provided plenty of fuel for research; technological as well as social theorists have published many research papers, often with new and interesting points. On the negative side, it has led to a generalized perception that, since all theories that incorporate strong predictive elements are attacked and their flaws uncovered, e-communication behavior is to a large extent unpredictable (DeSanctis et al., 1993; Postmes, Spears, & Lea, 1998; Trevino, Webster, & Stein, 2000; Zigurs, Buckland, Connolly, & Wilson, 1999).

Joining the tug-of-war between technological and social theories is not necessarily a bad thing and is probably better than ignoring theory altogether. However, if we try to think outside the box (i.e., from a perspective that is neither technological nor social) about this tug-of-war and its underlying assumptions, we may be able to make significant progress toward a unified theory of e-communication. One key assumption made by the proponents of technological theories is that individual behavior is not only predictable but also uniform. Social theories argue otherwise, based on the assumption that a complex web of social factors and interactions influences individual behavior. Perhaps the solution to the problem is to try to isolate those influences that are uniform across individuals from those that are not. While social theories have already established that social influences are not uniform, varying depending on social background and culture (whose diversity today is not only celebrated, but also encouraged), technological theories have failed to identify the source of uniform influences implicit in their hypotheses. This source, this chapter argues, is our biological communication apparatus.

THE EVOLUTION OF OUR BIOLOGICAL APPARATUS FOR COMMUNICATION

One of the fundamental premises of evolution theory, whose foundations were laid out by Darwin (1859/1966), is that all living organisms evolved from one common ancestor through a process that follows a few simple laws (Boaz & Almquist, 1997; Campbell, 1992; Dawkins, 1986, 1990; Dozier, 1992; Gould, 1977; Isaac, 1993; McCrone, 1991; Wills, 1993; Wilson, 1998).

- **The Inheritance Law:** Offspring inherit a large proportion of their parents' biological characteristics through their genes. The similarity between the combined genetic code of the parents and that of the offspring is very high.
- **The Mutation Law:** When members of a species generate offspring, natural genetic mutations occur that lead the offspring to develop biological characteristics that are different from those of the parents. These genetic mutations are usually incremental and arbitrary.
- **The Natural Selection Law:** Those offspring whose new biological characteristics give them an edge for survival and mating over the others are the most likely to pass the genes responsible for those biological characteristics to their own offspring.

According to evolution theory, the human species also evolved, according to the laws listed above, over millions of years, which suggests that it shares certain biological characteristics with all living beings, particularly those closer to it in evolutionary terms such as the great apes (e.g., gorillas and chimpanzees). The evolutionary pace set by the evolution laws is usually very slow (Boaz & Almquist, 1997; Lorenz, 1983), leading to the development of physical, behavioral, and cognitive traits over long periods of time. These periods may span thousands or millions of years and are contingent on breeding speed and mortality rates. In the case of the human species, this process is not believed to have led to significant physical and cognitive modifications in the last 100,000 years (Campbell, 1992; Dawkins, 1986; Dozier, 1992; Wills, 1993; Wilson, 1998).

During the vast majority of the evolutionary processes that led to the human species, human beings and their ancestors communicated FtF. Research evidence suggests that facial expressions and simple sounds were extensively used for communication as early as 5 to 2 million years ago by Australopithecus afarensis and africanus, who were members of the australopithecine genus (Boaz & Almquist, 1997). This behavioral trait, also found in modern primates and many other mammals, has been refined over millions of years, leading to the appearance of the first rudimentary forms of speech and later complex speech (Isaac, 1993; Laitman, 1993). Only very late in the evolutionary process that led to the human species is there evidence of communication through pictorial representations, mostly in the form of cave paintings, which can be seen as early manifestations of written communication (Campbell, 1992).

The development of a sophisticated biological apparatus to communicate through facial expressions and sounds was an important element in our evolution. Such an apparatus includes a complex web of facial muscles, nerves, and specialized brain functions that, research shows, could not have been developed for any purpose other than communication (Lieberman, 1998). For example, while only a small subset of our facial muscles is used for chewing, a much larger number is used for expression of thoughts and feelings. Also, the development of a larynx located relatively low in the neck, a key morphological trait that differentiates human beings from their early ancestors, considerably increased the variety of sounds that we can generate yet, at the same time, significantly increased our chances of choking on ingested food and liquids (Laitman, 1993; Lieberman, 1998).

Based on the discussion above, it is reasonable to assume that sounds were not frequently used alone for communication (e.g., only sounds, without gestures and facial expressions). Also, asynchronous (i.e., time-disconnected) communication would have required some form of sound or symbol (e.g., a pictorial representation) storage artifact. Paintings (mostly in caves), which are probably the most rudimentary of such storage artifacts, appeared late in the human evolutionary cycle, after complex speech was developed. Sound storage artifacts appeared only much later, after civilization was well established. Therefore, one can conclude that synchronous FtF communication, with the use of discrete sounds (which later developed into complex speech) and visual cues, has been the predominant mode of communication used by human beings over millions of years of evolution, and that our biological communication apparatus has been optimized for it. Our brain, in particular, seems to have been structured to excel in FtF communication (Lieberman, 1998; McCrone, 1991; Wills, 1993) by allowing us, for example, to derive a wealth of accurate information and meaning from facial expressions and tone of voice even when they contradict what is being said (e.g., when a person is lying or speaking in a delirious way).

KEY THEORETICAL PRINCIPLES

The discussion above provides the basis for the development of principles that can help us understand e-communication behavior in ways that are significantly different from those presented by technological and social theories. In the discussion of the principles provided below, information-processing schemas (Bartlett, 1932, 1958; Cossete & Audet, 1992; Lord & Foti, 1986), which are mental information-processing structures, are referred to only as schemas. The goal of this chapter is to highlight the potential of evolution theory as a lens for understanding e-communication behavior, not to develop a detailed theory. Thus, we develop and discuss applications of only three principles even though many other principles could be developed based on a more in-depth review of evolution theory as it relates to the evolution of our biological communication apparatus.

- **The media naturalness principle:** Innate communication schemas bias an individual's perception of media naturalness. Media that incorporate all the elements of actual unencumbered FtF interaction (e.g., physical presence, ability to see and hear others, synchronicity) will be perceived as more natural for communication than other media, and therefore as requiring less individual cognitive effort (due to cognitive adaptation) to be used for communication than other media. The extent to which a communication medium incorporates actual FtF interaction elements defines its degree of naturalness.
- **The innate schema similarity principle:** Innate communication schemas are very similar across different individuals, therefore biasing media naturalness perceptions of different individuals in very similar ways.

- **The learned schema diversity principle:** Since the human brain is very adaptable, learned communication schemas (i.e., those that are not inherited, but acquired through interaction with the environment surrounding an individual, including the social environment) also influence media naturalness perceptions. However, this influence is not as uniform across different individuals as that of innate communication schemas. Therefore, learned communication schemas bias media naturalness perceptions of different individuals in different ways.

The three principles above illustrate both the power and the likely limitations of evolutionary-theory-based e-communication theories, and have been developed with that purpose in mind. The media naturalness principle and the innate schema similarity principle are complementary, whereas the learned schema variety principle highlights the fact that biology alone cannot explain the full complexity of e-communication behavior.

The media naturalness principle is a first step in the direction of developing an entirely new theory (or theories) that could replace the media richness theory and other similar technological theories. For example, the link between decreased naturalness and increased cognitive effort can be used to explain why people prefer to conduct certain collaborative tasks using media that incorporate elements of FtF communication like the ability to use gestures and tone of voice to aid communication as well as provide and obtain immediate feedback during the communication interchange (Daft et al., 1987). This link can also be used to explain why better outcomes can be generated through media of lower naturalness through compensatory adaptation (Kock, 1998; Kock et al., 2006) by providing a basis on which to hypothesize that low naturalness poses cognitive obstacles that individuals engaged in collaborative tasks may be able to overcome—a hypothesis that is incompatible with the media richness theory yet has been strongly supported by empirical evidence (Kock, 1998, 1999, 2005).

The innate schema similarity principle highlights one of the most important reasons why evolution theory should be used to explain and predict e-communication behavior: namely, its potential to explain innate influences that are common to all individuals, independently of cultural and social background. As the Internet makes the world smaller by bringing together people with completely different cultural and social backgrounds into virtual communities based on shared personal interests and common business purposes, it is important to develop a predictive theoretical model that can help us understand and anticipate, at a certain level and with perhaps a limited degree of certainty, behavioral traits of all of the members of those virtual communities. The innate schema similarity principle stresses the potential of evolution theory as a basis for the development of such a predictive theoretical model.

Finally, the learned schema diversity principle underscores evolution theory's inability to explain everything. That is, while it is important to be able to isolate the influence of innate schemas from that of learned schemas on e-communication behavior, which helps us apply the Cartesian "divide and conquer" method of scientific inquiry, e-communication behavior will always be the result of a combination of the innate and learned schema influences. For example, individuals who developed schemas (i.e., learned schemas) related to the use of e-mail by using it to perform equivocal tasks during many years would tend to see e-mail as a more natural medium to perform tasks of high complexity than other individuals who have not gone through the same experiential schema-development process. The influence of learned schemas, in this example, would partially suppress the influence of innate schemas, even though the innate schemas would still exist and have the same configuration as in other individuals.

Using the Theoretical Principles to Predict the Behavior of Customers of an Online Broker

As a way of illustrating the use of the three principles discussed above, let us consider the case of an online broker and its customer support process. This example is based on an action research study previously conducted by one of the authors involving one mutual fund management firm and two online brokers. The goal of this illustration is to provide an example of how the theoretical principles can be used as a basis for the development of a simple predictive model tying media naturalness to customer satisfaction and revenues. The predictive model refers to the generic customer support process performed by the online broker, and can be extended to most online brokerages.

The main goal of the customer support process is to help customers buy investment products, which usually include money market instruments, stocks, bonds, mutual funds, and derivatives. The customer support process is implemented by means of a set of short interactions (i.e., with a duration that goes from a few minutes to a few hours) between the customer and online broker through communication media.

For a customer to buy or sell an investment product through the online broker, he or she must engage in a set of interactions with the online broker using one or more media. Thus, it can be said that the customer and the online broker engage in a set of interactions using N different media that we may call M_1, M_2, $M_3 \ldots M_N$, where N is the number of different media used to complete the task. For example, for the task of buying shares of a mutual fund, a customer may first obtain basic information about mutual funds from the broker's Web site (which is primarily text based), then conduct a live text-based chat online with one of the broker's customer representatives to understand how mutual funds operate, then go back to the Web site to understand the steps involved in making the purchase, then call the broker and chat with a customer representative over the telephone to clarify some issues regarding the purchase transaction steps, and then, finally, perform the purchase online using the broker's Web site. In this example, three main media—the Web site, a live text-based chat system, and the telephone—were used in five communication interactions. Based on the media naturalness principle, it could be argued that these three media can be sorted in the following order of decreasing degree of naturalness: telephone, live text-based chat, and Web site. The telephone medium is the most natural of the three because it incorporates synchronicity and the ability to convey verbal cues, both present in the FtF medium. The live text-based-chat medium is the second most natural medium because it incorporates synchronicity, but not the ability to convey verbal cues. The Web site (which is primarily text based) is the least natural of the three because it does not incorporate either synchronicity or the ability to convey verbal cues.

The three principles, that is, media naturalness, innate schema uniformity, and learned schema variety, can be used to derive hypotheses related to the use of different media for the provision of customer support services by the online broker. These hypotheses refer to each of the customer support interactions. The hypothesis below is a direct application of the media naturalness principle.

H1: There is a negative causal link between the degree of the naturalness of the medium used for communication between the customer representative and customer, and the cognitive effort perceived by the customer in connection with the customer support interaction.

Previous computer-mediated communication studies that investigated perceived cognitive effort and satisfaction (Graetz, Boyle, Kimble, Thompson, & Garloch, 1998; Nunamaker et al.,

1991) provide the basis on which to hypothesize that increased perceived cognitive effort is likely to lead to increased customer dissatisfaction with the service provided by the online broker, and thus increased likelihood of using the services of a competing online broker. This is compounded by the low cost associated with opening an account at a competing online broker and the weak bond that ties providers and customers in the online brokerage industry (Spiro & Baig, 1999). This can be summarized in two hypotheses.

H2: There is a positive causal link between the cognitive effort perceived by the customer and his or her degree of dissatisfaction with the customer support provided by the online broker.

H3: There is a positive causal link between the degree of dissatisfaction experienced by the customer during a customer support interaction with the online broker and the likelihood that the customer will use the services of a competing online broker.

The innate schema uniformity principle suggests that hypotheses H1, H2, and H3 are likely to hold for the majority of customers since it is primarily based on the influence of innate schemas possessed by all the members of the human species. This could be expressed as a hypothesis (which would in fact be a metahypothesis as it refers to the other hypotheses) but will be omitted in this illustration for simplicity.

The learned schema variety principle, on the other hand, suggests that learned schemas are likely to also play a role in how customers perceive their communication interactions with the online broker by moderating the causal link established in hypothesis H1 between media naturalness and cognitive effort. That is, the less the customer knows about buying and selling investment products using the services provided by the online broker, the stronger the negative effect of media naturalness on cognitive effort. This leads us to hypothesis H4.

H4: The mismatch between the knowledge possessed by the customer and that required to perform the task (i.e., buy or sell an investment product) positively moderates the negative causal link between the degree of naturalness of the medium used for communication between the customer representative and customer, and the cognitive effort perceived by the customer in connection with the customer support interaction.

Given that the theoretical model above refers to each customer support interaction, it is implicitly assumed that each interaction will have an impact on the likelihood that the user will move to a competing online broker. Previous studies on the nature of customer-provider relationships in the financial service industry (Macdonald, 1995; Walkins, 1992) suggest that it is reasonable to also assume that the impact of each interaction will be both incremental and cumulative in most cases (exceptions would be, e.g., single interactions in which dissatisfaction would be extreme, leading the customer to leave the broker at once).

Since the four hypotheses presented above have each a clear negative form (i.e., a respective null hypothesis), they (as well as the causal model that summarizes them) can be tested and therefore disproved or shown to appropriately explain the results obtained from the study. This can be accomplished through positivist case and action research studies, and, less ideally, laboratory experiments or surveys of online brokers. Current trends in the use of communication media in the online brokerage industry provide support for the hypotheses above as they suggest a link between customer satisfaction and loyalty, and the use of communication media that incorporate elements of FtF interaction (e.g., synchronicity, the ability to see and hear the other party, etc.) to support customers (Dodson, 2000).

CONCLUSION

This chapter provides a brief theoretical review of research on e-communication behavior, which identifies two main types of theories—technological and social—and the rationale for the development of a new theory that is neither technological nor social. This new theory is based on evolution theory, whose foundations were laid out by Darwin, particularly as it relates to how communication evolved over millions of years in the human species. Finally, key theoretical principles are developed from evolution theory that have a direct application to our understanding of e-communication behavior and can be used in the development of a new theory of e-communication behavior. An application of these principles is provided through the development of a simple predictive model tying media naturalness to customer satisfaction and revenues in an online broker.

Why is it important to try to understand e-communication behavior based on evolution theory? Two main reasons can be given to answer this question. First, evolution theory provides a scientific basis on which to ground key hypotheses of technological theories. For example, the media richness theory proposes a richness scale but does not explain why different media are perceived according to that scale. This chapter shows that the richness scale proposed by the media richness theory is partly supported by evolution theory. If one assumes that some technological theories need to be refined and not replaced by a new theory, which may be the case, evolution theory can be very useful in their refinement and scientific grounding.

Second, evolution theory can be used as a basis for the development of an e-communication behavior theory that, while perhaps incorporating some of the hypotheses of certain technological theories, will be free from their previously identified flaws. Again, using the media richness theory as an example, the media naturalness principle replaces, with advantages, the static richness scale hypothesis of the media richness theory. As noted before, several studies succeeded in showing fatal flaws in this hypothesis by providing evidence, for example, that the media richness scale is not static (Markus, 1994), and that richness is not inherent in a communication medium and can vary depending on who is involved in the communication act (Lee, 1994). In fact, the static richness scale hypothesis proposed by the media richness theory has proven to be its main weakness as it opens the door, for instance, to the argument that a medium richer than FtF (e.g., a "super-rich" virtual-reality-based medium) will be even better than FtF for tasks involving intense communication. What is argued here, in contrast, is that any non-FtF medium will be perceived as less natural than FtF. One of the reasons for this is because our biological apparatus for communication has been optimized by Darwinian evolution for FtF communication. This implies that the users of a super-rich virtual-reality-based medium will also perceive it as less natural, thus requiring more cognitive effort for communication (because, e.g., it may induce perceived information overload) than the FtF medium.

The widespread use of e-communication technologies today is only matched by the increasing uncertainty about the effects of these technologies on humans. Given this, the search for reliable theories that help us predict e-communication behavior is strongly warranted. This chapter is a first step in that search and a small one toward the final goal of e-communication researchers, which is to develop a grand theory that can be used to explain e-communication behavior in its full complexity. Evolution theory alone will never lead us to such a grand theory, but it will be instrumental in its development.

ACKNOWLEDGMENT

We would like to thank Amotz Zahavi, Geoffrey Miller, and Donald Hantula for comments and suggestions regarding the connection between evolutionary biology and evolutionary psychology principles and some of the ideas proposed in this chapter. An earlier version of this chapter was published as an article in Volume 5 of the journal *Communications of the Association for Information Systems* (AIS); we thank the journal's publisher for permission to use that article as a basis for this chapter.

REFERENCES

Alavi, M. (1994). Computer-mediated collaborative learning: An empirical evaluation. *MIS Quarterly, 18*(2), 159-174.

Bagozzi, R. P. (1980). *Causal models in marketing.* New York: John Wiley & Sons.

Bartlett, F. (1932). *Remembering: A study in experimental and social psychology.* Cambridge, MA: Cambridge University Press.

Bartlett, F. (1958). *Thinking: An experimental and social study.* New York: Basic Books.

Boaz, N. T., & Almquist, A. J. (1997). *Biological anthropology: A synthetic approach to human evolution.* Upper Saddle River, NJ: Prentice Hall.

Campbell, B. C. (1992). *Humankind emerging.* New York: HarperCollins.

Cossete, P., & Audet, M. (1992). Mapping of an idiosyncratic schema. *Journal of Management Studies, 29*(3), 325-348.

Daft, R. L., & Lengel, R. H. (1986). Organizational information requirements, media richness and structural design. *Management Science, 32*(5), 554-571.

Daft, R. L., Lengel, R. H., & Trevino, L. K. (1987). Message equivocality, media selection, and manager performance: Implications for information systems. *MIS Quarterly, 11*(3), 355-366.

Darwin, C. (1966). *On the origin of species.* Cambridge, MA: Harvard University Press. (Original work published 1859)

Davis, J. A. (1985). *The logic of causal order.* London: Sage.

Dawkins, R. (1986). *The blind watchmaker.* New York: W.W. Norton & Company.

Dawkins, R. (1990). *The selfish gene.* Oxford, UK: Oxford University Press.

DeSanctis, G., & Poole, M. S. (1994). Capturing the complexity in advanced technology use: Adaptive structuration theory. *Organization Science, 5*(2), 121-147.

DeSanctis, G., Poole, M. S., Dickson, G. W., & Jackson, B. M. (1993). Interpretive analysis of team use of group technologies. *Journal of Organizational Computing, 3*(1), 1-29.

Dodson, J. (2000, September 18). It's time to chat with your customers. *Internetweek.* Retrieved from http://www.internetwk.com/ebizapps/col091800.htm

Dozier, R. W., Jr. (1992). *Codes of evolution.* New York: Crown Publishers.

Fulk, J., Schmitz, J., & Steinfield, C. W. (1990). A social influence model of technology use. In J. Fulk & C. Steinfield (Eds.), *Organizations and communication technology* (pp. 117-140). Newbury Park, CA: Sage.

Gould, S. J. (1977). *Ever since Darwin: Reflections in natural history.* New York: W.W. Norton & Company.

Graetz, K. A., Boyle, E. S., Kimble, C. E., Thompson, P., & Garloch, J. L. (1998). Information sharing in face-to-face, teleconferencing, and electronic chat groups. *Small Group Research, 29*(6), 714-743.

Isaac, G. L. (1993). Aspects of human evolution. In R. L. Ciochon & J. G. Fleagle (Eds.), *The human evolution source book* (pp. 263-273). Englewood Cliffs, NJ: Prentice Hall.

Jackson, M. H., & Purcell, D. (1997). Politics and media richness in World Wide Web representations of the former Yugoslavia. *Geographical Review, 87*(2), 219-239.

Kock, N. (1998). Can communication medium limitations foster better group outcomes? An action research study. *Information & Management, 34*(5), 295-305.

Kock, N. (1999). *Process improvement and organizational learning: The role of collaboration technologies.* Hershey, PA: Idea Group Publishing.

Kock, N. (2005). *Business process improvement through e-collaboration: Knowledge sharing through the use of virtual groups.* Hershey, PA: Idea Group Publishing.

Kock, N., Lynn, G. S., Dow, K. E., & Akgün, A. E. (2006). Team adaptation to electronic communication media: Evidence of compensatory adaptation in new product development teams. *European Journal of Information Systems, 15*(3), 331-341.

Laitman, J. (1993). The anatomy of human speech. In R. L. Ciochon & J. G. Fleagle (Eds.), *The human evolution source book* (pp. 56-60). Englewood Cliffs, NJ: Prentice Hall.

Lee, A. S. (1994). Electronic mail as a medium for rich communication: An empirical investigation using hermeneutic interpretation. *MIS Quarterly, 18*(2), 143-157.

Lengel, R. H., & Daft, R. L. (1988). The selection of communication media as an executive skill. *Academy of Management Executive, 2*(3), 225-232.

Lieberman, P. (1998). *Eve spoke: Human language and human evolution.* New York: W.W. Norton & Company.

Lord, R. G., & Foti, R. J. (1986). Schema theories, information processing and organizational behaviour. In H. P. Sims & D. A. Gioia (Eds.), *The thinking organization* (pp. 20-48). San Francisco: Jossey-Bass.

Lorenz, K. (1983). *The waning of humaneness.* Boston: Little, Brown & Co.

Macdonald, J. (1995). Quality and the financial service sector. *Managing Service Quality, 5*(1), 43-46.

Markus, M. L. (1990). Toward a critical mass theory of interactive media. In J. Fulk & C. Steinfield (Eds.), *Organizations and communication technology* (pp. 194-218). Newbury Park, CA: Sage.

Markus, M. L. (1994). Electronic mail as the medium of managerial choice. *Organization Science, 5*(4), 502-527.

McCrone, J. (1991). *The ape that spoke: Language and the evolution of the human mind.* New York: William Morrow and Company.

Nunamaker, J. F., Dennis, A. R., Valacich, J. S., Vogel, D. R., & George, J. F. (1991). Electronic meeting systems to support group work. *Communications of ACM, 34*(7), 40-61.

Orlikowski, W. J., Yates, J., Okamura, K., & Fujimoto, M. (1995). Shaping electronic communication: The metastructuring of technology in the context of use. *Organization Science, 6*(4), 423-444.

Poole, M. S., & DeSanctis, G. (1990). Understanding the use of group decision support systems: The theory of adaptive structuration. In J. Fulk & C. Steinfield (Eds.), *Organizations and communication technology* (pp. 173-193). Newbury Park, CA: Sage.

Postmes, T., Spears, R., & Lea, M. (1998). Breaching or building social boundaries? Side-effects of computer-mediated communications. *Communication Research, 25*(6), 689-715.

Rice, R. E. (1992). Task analyzability, use of new media, and effectiveness: A multi-site exploration of media richness. *Organization Science, 3*(4), 475-500.

Spiro, L. N., & Baig, E. C. (1999, February 22). Who needs a broker? *Business Week, 3617*, 113.

Sproull, L., & Kiesler, S. (1991). Computers, networks and work. *Scientific American, 265*(3), 84-91.

Trevino, L. K., Webster, J., & Stein, E. W. (2000). Making connections: Complementary influences on communication media choices, attitudes, and use. *Organization Science, 11*(2), 163-182.

Walkins, J. (1992). Information systems: The UK retail financial services sector. *Marketing Intelligence & Planning, 10*(6), 13-18.

Webster, J., & Trevino, L. K. (1995). Rational and social theories as complementary explanations of communication media choices: Two policy-capturing studies. *Academy of Management Journal, 38*(6), 1544-1573.

Wills, C. (1993). *The runaway brain: The evolution of human uniqueness.* New York: Basic Books.

Wilson, E. O. (1998). *Consilience: The unity of knowledge.* New York: Alfred A. Knopf.

Zigurs, I., & Buckland, B. K. (1998). A theory of task/technology fit and group support systems effectiveness. *MIS Quarterly, 22*(3), 313-334.

Zigurs, I., Buckland, B. K., Connolly, J. R., & Wilson, E. V. (1999). A test of task-technology fit theory for group support systems. *Database for Advances in Information Systems, 30*(3), 34-50.

Chapter II
Metaphors for E-Collaboration:
A Study of Nonprofit Theatre Web Presence

Julie E. Kendall
Rutgers University, USA

ABSTRACT

What constitutes regional commerce? What creates and enhances a regional identity? In the United States, regions can be quite large and may even cover geographical territory from several surrounding counties or states. They are larger than any one individual company, shopping street, or district. Regional cooperation of commercial businesses is often manifested through special events, cooperative advertising with coordinated signage, extended opening hours, and special discounts that contribute to building a sense of community, and which eventually develop a sense of region. The political and environmental exigencies for the creation and expansion of regions have meant an increase in the popularity and importance of regions and a subsequent movement to enhance and differentiate their identities. We now see the rise of regional governments, water authorities, and educational institutions among many others. One little-explored idea has been the use of e-collaboration to forge, reinforce, and sustain a regional identity via the virtual world. Although geographical separation of many miles might dictate that bricks-and-mortar theatres cannot easily collaborate physically (i.e., they cannot share costumes, props, ushers, and so on), the possibility of e-collaboration opens potential opportunities for attracting wider audiences, reaching and ultimately casting fresh talent, and building reciprocal audiences who possess a passion for the arts and who have the means and desire to travel to attend performances throughout the geographical region. In this study, a methodology built on the conceptual foundation of metaphor research was used to comprehend and then interpret the Web presence of 15 nonprofit theatres that comprise the total regional theatre of southern New Jersey that exists on the Web. In order to add additional insight, our earlier research findings from working with off-Broadway and regional theatre festivals were extended to analyze the Web presence of the theatres in southern New

Jersey. We contribute to the literature by systematic and deep investigation of the strategic importance of the Web for nonprofit theatre groups in the southern New Jersey region. In addition, our use of the metaphor methodology in order to create a telling portrait of what transpires on the Web in relation to nonprofit organizations is also an original contribution. Our work is meant to heighten the awareness of administrators to the rapidly accelerating need for the strategic use of e-collaboration. We propose that with the use of the Web, administrators can move toward creating a regional theatre Web presence for South Jersey, one which would make use of an evolutionary metaphor. To this end, we suggest the use of an organism metaphor. Through the creation of reciprocal hyperlinks, theatres can be supported in improving their practice of colocation on the Web, wherein they will be taking strides to cooperate as a regional theatre community.

INTRODUCTION

Web presence is the perception of influence and organizational identity that organizations attempt to create in their customers and Web site visitors. Used as a strategic positioning instrument, Web presence goes well beyond the basic graphics, text, and hyperlinks that are the building blocks of a Web site. The Web presence of the organization should have a positive effect on its operations, meaning that these operations should be made easier, faster, and more efficient and effective. An organization's competitiveness is expected to be sharpened through a strong Web presence so that it will gain additional market share, expand operations into new markets, and attract additional customers (Abuhamdieh, Kendall, & Kendall, 2000, 2002, 2007).

Developing an organizational Web presence is expected to enhance the organization's adaptation and growth by enhancing its relationship with its customers. This will make the organization more alert to its customers' needs and expectations. This translates into a good outlook for an organization's growth prospects.

Nonprofit organizations traditionally lag behind commercial enterprises in their approach to implementing integrated information technology, particularly in the area of developing a strategic IT plan. This occurs for a number of reasons, but they include the lack of expertise and knowledge concerning the importance of information technology to an organization (specifically in the performing arts community), the lack of a predictable source of funding for endeavors that are earmarked as exclusively for IT development, and reticence to include IT as a funding priority when grant requests are made. Additionally, many funding agencies specifically will not grant requests for standard items required to build IT infrastructure, such as computers, software, and expertise to develop information systems and IT policy. Because of these concerns, oftentimes information technology enhancements are not even broached (Te'eni & Kendall, 2004).

I chose to look at an exhaustive list of regional theatres that have a Web presence in southern New Jersey (widely known to residents as South Jersey). South Jersey is made up of eight counties and has a population of about 2.3 million people (U.S. Census Bureau, 2003). This fairly densely populated region can support a number of different theatre groups, but the theatres are spread over the region.

As surprising as it may seem, creative arts have a critical part to play in the economy of New Jersey and in the economy of South Jersey as well.

A report published in 2000 that examined the economic impact of arts funding in New Jersey a decade ago found that $18 million in NJSCA funding resulted in $1 billion of annual economic activity in New Jersey. The findings showed that during that time there was over $546 million in direct spending by arts groups and over $474 million in ancillary spending by visitors. This generated $27 million in tax revenue, over 17 million audience members including 4.5 million schoolchildren, and over 11,000 jobs involving over 47,000 artists and over 700 arts groups (*The Arts Mean Business: A Study of Economic Activity, 2000. NJSCA/ArtPRIDE*, 2000).

Mindful of the potential economic and cultural impact of regional performing arts groups in New Jersey, I set about to systematically examine how theatres currently use their presence on the Web, and to recommend how they can improve their service by changing their Web metaphors and how they can cooperate and collaborate with other theatres in the region.

METHODOLOGICAL APPROACH

In beginning my research, I set as one of my standards to be inclusive rather than exclusive in identifying theatres to study. I subsequently identified 15 nonprofit regional theatres that had developed and maintained a Web presence. I conducted original research seeking to identify nonprofit theatres located in southern New Jersey using several different search engines including Copernic Agent Professional and Google. Copernic uses more than 1,000 search engines, consults multiple search engines at once, combines their results, removes duplicates, and keeps the best information gathered from the search engines it queries. Membership listings in arts alliance organizations including the Artpride New Jersey, Discover Jersey Arts Online, the New Jersey Theatre Alliance, New Jersey Theatre League, and the South Jersey Cultural Alliance.

When I began to determine which metaphor was in play on each theatre Web site, I used two different approaches. One approach was to read and interpret the language used on the Web pages in order to determine the presence or absence of predominant metaphors that I had already identified in organizational users of other information systems and in systems development methodologies (J. E. Kendall & Kendall, 1994).

Another way to observe what metaphor was foremost in serving as an organizing concept on a theatre's Web site was to assess the visual and multimedia elements of the Web page to understand the content, design, and technical elements of the pages. I examined the theatre Web sites using eight dimensions provided as guidelines for Web designers (K. E. Kendall & Kendall, 2005). The eight dimensions include (a) overall appearance (layout and composition), (b) use of graphics, (c) use of color, and the use of sound or video, (d) use of new technology and products, (e) content, (f) navigability (internal and external links), (g) site management, and (h) communication with users. By assessing both the language employed and the technical elements adopted, I was able to determine what metaphor was representative of each theatre Web site.

Metaphors shape the way we think. By using words and visual aspects that people understand and believe in to make linkages with the new and unfamiliar, a metaphor enables the person who hears it to envision a new reality. Metaphors assert that one object actually is another (Weaver, 1967). Invoking a metaphor means opening the door for a listener to use all previous associations in approaching a new subject.

Researchers have argued the power of metaphors to shape our very reality (Duncan, 1968; Graber, 1976; Lakoff & Johnson, 1980). Ott (1989, pp. 29-30) notes, "Metaphors help organization members put meaning into things they experience and realize apparent contradictions and paradoxes they encounter. Metaphors help organizations tie their parts together into meaningful wholes." For example, saying, "Time is money," calls into play many interrelated ideas, including that time is a means of payment, that it perhaps can be spent like money, and that things like time off can be earned or banked. The user is informed of a variety of possible courses of action because of the entailment aspect of the original metaphor.

In the past 20 years or so, organizational researchers have recognized the importance of metaphors. Morgan (1986, pp. 12-13) in his pioneering work goes even further by concluding, "By using different metaphors to understand the complex and paradoxical character of organizational life, we are able to manage and design organizations that we may have not thought possible before."

Metaphor analysis was first used in the study of management information systems by J. E. Kendall and Kendall (1993), who built on the seminal work of Clancy (1989) concerning business metaphors. Clancy identified six metaphors commonly found in organizations: the journey, game, war, machine, organism, and society. Kendall and Kendall observed three other basic business metaphors: the jungle, zoo, and family, which pointed out weaknesses in the organization that Clancy did not identify in his work. The authors proceeded to map the attributes of the nine metaphors to the development of different types of information systems and then hypothesized the likelihood of success given the organizational climate and the presence of particular metaphors. Further information about metaphors in systems development can also be found in J. E. Kendall and Kendall (1994).

CLASSIFYING THE WEB SITES OF SOUTH JERSEY THEATRES

The Web sites seemed to cluster around seven of the nine metaphors introduced earlier. Of the 15 Web presences evaluated, three were classified as portraying a jungle metaphor, three evidenced a family metaphor, three called forth a journey meta-

Table 1. Organizing metaphors found on theatre Web sites and how they cluster

Web Presence Metaphor	Southern New Jersey Theatres
Jungle	Foundation Theatre of Burlington County College East Lynne Theater Company Department of Fine Arts of Rutgers-Camden
Family	Bridge Players Theater Company Mainstage Center for the Arts Sketch Club Players
Journey	Cumberland Players The Road Company Surflight Theatre
Machine	Haddonfield Plays and Players
Society	Burlington County Footlighters Fine & Performing Arts Center at Cumberland County College
Game	Puttin'on the Ritz Triple Threat Foundation for the Arts
Organism	Holly City Repertory Theatre

phor, one evoked a machine metaphor, two used the society metaphor, two used the game metaphor, and one represented an organism metaphor. None of the theatre Web presences were classified as exhibiting a strategic war metaphor or a chaotic zoo metaphor. Table 1 provides a summary of the metaphors and the southern New Jersey theatre Web sites that correspond to each metaphor.

Enacting the Jungle Metaphor on South Jersey Theatre Web Sites

The jungle metaphor is one that typically requires a guide to help secure a path out of the jungle. All three of the sites identified could be characterized by the sentiment, "Once we get through this season, maybe we'll have time to construct a real Web site."

The Foundation Theatre Company in residence at the Geraldine Clinton Little Theatre of the Burlington County College Web site contains the motto of the school, "We Can Get You There," displayed in the upper right-hand corner of the home page. Unfortunately, that slogan stands in ironic contrast to the search function provided, which does not help the visitor locate theatre productions. There is a hyperlink on the page to the calendar of events, which brings up a page of 12 radio buttons that the user can click on to designate that they are looking for a subject such as performing arts or student government events. The page features a phone number for the box office to get further information, but there is no way to book online or even to request information online. Neither is there any way for the organization to capture visitors' information when they visit the site.

Another theatre in the category of jungle metaphor is the East Lynne Theater Company in Cape May. The Web site is situated inside a shell of the sponsor NewJersey.com, which is an electronic consortium of New Jersey newspapers that display paid advertising. They offer free Web hosting and Web page set up for community groups. As a result, the pages appear in a generic frame, with links to the main site and a chance to build one's own site, so there is no specific design that conveys the mission or feeling of the theatre's Web presence. It is very difficult to locate this URL (uniform resource locator). The free frame provided by NJ.com is chaotic and confusing. Going to the "Links" page eventually brings up a series of 11 hyperlinks that refer to a grab bag of local tourism sites, and professional and amateur theatre Web sites. One link is broken, and another, when clicked on, brings up a page with the message "Being reconfigured." There is no clear organization; the page serves as more of a friendly list of theatres and other interesting Cape May tourist sites that someone has hastily put together. The hyperlinks do not appear to be diligently maintained or shaped in any particular way. They are overgrown with possibly irrelevant links.

The third Web site in the jungle metaphor category is that of the Department of Fine Arts of Rutgers-Camden. The maze of links is disorderly and confusing. Although there is a link on the page to the Rutgers Camden Center for the Arts, there are no reciprocal links to the theatre division of the fine arts department. In fact, clicking on a link to Rutgers lands the visitor in a central Rutgers University site, not the Rutgers-Camden site. Many synergies are being missed.

It is interesting to note that two of the three Web sites evidencing the jungle metaphor are schools (one a university, the other a community college). It is possible that these Web sites will always evince a metaphor chaotic in nature, even when they are developed more completely, simply due to their missions as schools that serve a diverse constituency.

Using the Family Metaphor on South Jersey Theatre Web Sites

Families tend to have more than one goal. Parents and children may love each other but may not see each other's point of view. The Web sites all communicate their family approach with the message of "We are a community; you are welcome to our house."

The Web site of the Bridge Players Theater Company is designed to use a graphic of theatre curtains as the frame. There is an original icon that is a drawing of the Tacony-Palmyra Bridge with the two masks of drama—comedy and tragedy—superimposed over the bridge. The Web site is enthusiastic but amateurish. It is clear that no voluntary help with the site is turned away; therefore, the results are somewhat uneven. Some of the Web pages are consistent in design, and others are not. Some things are almost touchingly naïve (such as the request for people not to use files that are posted openly) and are not password protected. The site includes a generic message board and a visitor guest book that permits linking to other people's URLs without a screening of the links for appropriateness to the overall Web presence. The security problems and the patched-together feel of the Web site certainly make it seem homespun. It is a friendly site and everyone is invited to participate.

The Mainstage Center for the Arts of Blackwood also clusters into the family metaphor. The Web site is lively and friendly, showing children's handprints on the main page, which includes links labeled "Theater," "Kids," "Harmony", "Music," "Dance," "Home," "Contact," "Mission History," and "Audition," as well as "Upcoming Events." The slogan reads, "Quality family entertainment at affordable prices!" There are nine different fonts used on the white background of the home page. There is also an icon showing an award from the *Courier-Post* newspaper as one of the "2000 Best of South Jersey" along with a sentence that reads, "Ranked One of the Best Arts and Entertainment Centers of South Jersey Five Years in a Row!" and a link to the *Courier-Post Online*; the site also links to the New Jersey State Council on the Arts and the Discover New Jersey Arts Web sites. In addition, this is the only Web site I examined that partners with Greatergood.com, which in turn partners with many e-commerce merchants, donating 15% of purchases from the participating e-commerce merchants to the Mainstage Center for the Arts. The mission stresses the positive benefits of engaging youth in theatre arts as a way to address increasing violence in society.

The third Web site that displays a family metaphor is the Sketch Club Players, Inc., of Woodbury (SCP). This Web site features a small logo of the company: the dramatic masks of tragedy and comedy intertwined, with the name of the company surrounding the masks in a circle. On the right-hand side are five clip-art images from popular clip-art packages, which give the Web site a familiar and homespun look. These are not hyperlinked, however. Everyone, regardless of their particular talent, is encouraged to participate. This particular family is very proud of their dwelling, stating, "We are proud to own our building, and we work hard to keep improving it." The site is hosted at Geocities.com, which is a commercial site that offers free Web hosting, or Web hosting for a nominal fee. Unfortunately, the design result is still somewhat amateurish. The link leading to the history of SCP ends with the comment "SCP hopes to keep supporting the performing arts in Gloucester County for many more years!" Their community involvement and their commitment to welcoming everyone is a very clear family message.

Employing the Journey Metaphor on South Jersey Theatre Web Sites

The journey metaphor invokes a goal and a struggle to reach the goal. The sentiment expressed overall by a theatre using this metaphor is "We try new things; some succeed, some don't. We continually discover what our patrons want; we will eventually get there."

The Cumberland Players performing at the Little Theatre in Vineland, Cumberland County, are clearly on a voyage of discovery. The journey metaphor is also emphasized by one of the reviews chosen for the Web site from the Atlantic City Press, which reads in part, "Instead of performing the same old overdone musicals and Neil Simon plays, the Cumberland Players are challenging area theatergoers with quality work." Although the box office is not online, the form page has been made quite functional. This is the only site in the group studied to include a navigational bar at the bottom of the printable ticket order form. The theatre is making full use of the navigational qualities of the Web in that way, and that plays nicely into its use of the journey metaphor as well. The site is well maintained and frequently updated, and appears eager to take visitors to new places. In response to a current arts funding crisis, the Web site has enabled two e-mail links, one to the New Jersey legislators and one to the governor of New Jersey, which permit people to request that the governor not suspend funding for the arts. The Cumberland Players "Links" page features the Discover NJ Arts alliance page as the first link, and has a series of links that do indeed permit the user to discover different arts alliances groups. The links encourage exploration, and descriptions of each link, as well as the icon to click on, are presented. This page is informative and leads visitors to check each link because of its vivid descriptions of what each link contains (much as a travel brochure might).

The Road Company of Williamstown in Gloucester County is another theatre group whose Web site evokes the journey metaphor. Even its name is evocative of a journey. The Web site is a mixture of clip art and animation, and features several links and a good deal of thought about where the company has been and where it is going. Clicking on the link "Theater Information" brings up the statement:

We're called The Road Company for a reason: we don't own a theater; we rent wherever we perform. For many years, that rented home was The Grand Theater in Williamstown. We performed over 30 shows in our years there. But with the current owners' decision to no longer use it for rentals; we once again have hit the road.

When I wrote to the group on another matter regarding an inappropriate link that was listed on the Web site (the owner of the domain had changed to a dorm Web-cam pay site), they were incredibly responsive and remedied the problem immediately by removing the link.

The third theatre using the journey metaphor is that of the Surflight Theatre of Beach Haven. On the left-hand side is a gray navigational bar with links such as "Home," "What's Playing," "Tickets," "About Surflight," "Our Programs," "Support Surflight," "Volunteer," and "Jobs at Surflight." The "Mailing List" link is displayed but not enabled; "Seating Chart" is enabled as is "Contact Us," however the "Site Map" is not a link. The home page shows two color photographs of two upcoming performances. Also included is a link to respond to the proposed arts funding budget cuts for New Jersey, and a number of icons for a variety of theatre alliances are shown. Clicking on the link "About Surflight" brings up a lovingly told chronicle of the historical journey of the theatre company thus far. Each attempt at something new is described; most, such as the summer children's breakfast or going not for profit, are described as great successes. There is a genuine feeling of discovery and exploration conveyed on this page.

Enacting the Machine Metaphor on One South Jersey Theatre Web Site

If you have ever heard anyone praise a well-organized enterprise by remarking that "It runs like a well-oiled machine," you understand what makes a machine metaphor. For the theatre enacting this type of metaphor, we can characterize the sentiment as "We designed a series that should apply to all; three musicals, and three classic plays each year."

The Web presence of the Haddonfield Plays and Players is the only one of the 15 theatre Web sites examined to evince a machine metaphor. The left-hand column is a list of 11 links headed by the ubiquitous hit counter numbering in the 16,000s, which is the number of visitors reportedly coming to the site since January 12, 2002. The link called "Database" brings up a page that claims to be a historical database that includes all the shows and full casts from past productions. There is a simple search function where the visitor can enter either a name or the name of a show, and bring up a separate screen where the information requested is displayed. The database and all other functions seem to work properly, but the traditionally warm, welcoming feeling of the other theatres employing the family metaphor is absent; absent also is the explorative spirit shown in the theatres with Web presences that I have characterized as having a journey metaphor. The machine metaphor as enacted here holds predictability and order, with several subscription series designed to apply to all.

The Society Metaphor and South Jersey Theatre Web Sites

Rules as a mainstay for the maintenance of the society is a strong feature of this metaphor. Theatres with this type of presence embody the sentiment "We do not allow exchanges or refunds. Payment for next year's series must be received no later than…" and so on.

On the Burlington County Footlighters Web site, clicking on "Order Tickets" brings up a page titled "Easy Internet Ticket Order Form," which leads the visitor to believe that this transaction can be completed online. However, upon further reading, it is clear that you need to print out the form and mail it back to the theatre. No e-mail or e-commerce transactions are permitted. Further, the theatre only takes checks; it does not have the ability to process credit cards. These rules are explicitly stated throughout. It mentions that the Web site's message board is moderated and thanks visitors for behaving in accordance with its rules. The link for auditions includes an agreement that the actors will follow the rules of the house regarding behavior before, during, and after the show. Included in the rules is the topic of paying dues. This preoccupation with the rules is highly consistent with the society metaphor, but this is the only Web site of the 15 that went to this extent to ensure compliance. Rules feature prominently on the Web site. The visitor gets the idea that the Footlighters have been in business a long time (since 1938), and one of the reasons for their longevity is that they spell out all of the rules, and they communicate them clearly. For instance, on the ticket order form, subscribers are instructed to print all information clearly, with an additional, wry comment that states, "We are not handwriting experts."

The second South Jersey theatre that embodies the society metaphor is the Fine & Performing Arts Center (FPAC) at Cumberland County College in Vineland. Rules are greatly in evidence on several of the pages. These include the "Rental Info" page, still within the frame of the county college, which gives a text version of the policies and procedures renters must follow in order to rent the Fine & Performing Arts Center. The frame gives the Web site an orderly look, and permits the visitor to orient easily to the information on the page and to navigate easily around the site. More rules are given on the "Stage Specs" page, which provides technical information necessary to set up a show within FPAC. This information is currently linked to the rental information and not hidden away from the casual browser who is looking at the site as an audience member or performer. The rules and technical data can be overwhelming to patrons who just want to find a good production for the weekend. The "Links" page features six links and two of them are to performing-arts or theatre-alliance-type sites, one is to another theatre group discussed in an upcoming section, and one is to a nearby symphony group.

Those sites do no appear to have reciprocal links with the Fine & Performing Arts Center.

The Game Metaphor and South Jersey Theatre Web Sites

In the game metaphor, the leader is often described as a coach, and the game is one of a competitive sport. The sentiment expressed by theatres possessing the game metaphor can be summed up as, "We strive to be better than the competition; we will continue to improve and win the hearts of theatregoers. We will win all of the awards." Two South Jersey theatres evoke this metaphor: Puttin' on the Ritz of Oaklyn, and Triple Threat of Cherry Hill.

Puttin' on the Ritz in Oaklyn is intending to win the big game, and all of the competitions leading up to it as well. Its slogan appears at the top of the home page—"Quality theatre, simply done"—and on each associated page there is a further definition of quality shown in maroon lettering at the top of the page. Tickets must be reserved via phone with a credit card or a subscription can be mailed on a form from the Web site with check or credit card payment. The box office is not online. Puttin' on the Ritz runs a children's theatre as well as the main theatre. The main theatre produces six shows a year, five of which are musicals from the past.

Triple Threat Foundation for the Arts of Cherry Hill has a competitive advantage in that they used online ticketing for about a year for most of their shows. The Web site resembles a corporate Web site and the links on the left-hand side are very revealing of the game orientation. They begin with the link, "News> TTFA takes 'Civil War' to competition for the NJ Theatre League Competition." This is the only theatre Web site to mention the competition though several of the regional theatres will compete in it. Another link brings up the news that three actors who performed at Triple Threat are now going to appear in the national touring company of *The Music Man*. The group is clearly proud of them, and considers them team members as well as winners. Triple Threat definitely has a goal to be professional in everything it does. In its children's performing-arts pages, the end of the page advertising classes states, "Triple Threat Students Succeed," which heads a boxed display of alumni names and those who have gone on to professional jobs. This is a very different arena to compete in, and except for the Puttin' on the Ritz Web presence, the other theatres do not aspire to this goal of professionalism, quality, or success in a bigger arena.

The Organism Metaphor and One South Jersey Theatre

The organism metaphor is a balanced metaphor that allows the entity to grow. In the realm of community theatre, the sentiment of the organism metaphor can be articulated as "We are a growing theatre; we are helping develop new plays and new talent."

The Holly City Repertory includes among its priorities on the "Mission" page producing and developing original works that have not been previously produced and works "produced outside of an educational institution." It also offers workshops and classes "to improve our company" and "to offer another creative outlet for local artists." Its mission statement concludes with the expression, "collaborating with any and all area artistic groups to create a nurturing, culturally rich atmosphere which will foster economic growth through tourism." The theatre's Web presence, including the graphics, layout, design, and content, is highly evocative of the organism metaphor. The Web site includes a link to the Riverfront Renaissance Center for the arts, which has an elaborate Web site featuring the cultural and community life of Millville, New Jersey, as well as many links to sites on the subject. It also has a link to the Glasstown Arts District, which has a reciprocal link to the Holly City Web site as well as to Cumberland County. Holly City Repertory's selection of plays to produce, events to sponsor, and calls for participation in the key activities of nurturing and growing through producing original works are all different ways to express the idea of cultivating the organism.

E-COLLABORATION AND NONPROFIT THEATRES

Examining the 15 Web sites systematically through the use of metaphors has brought many observations to the fore, and many insights have been realized. The following section examines the possibilities for e-collaboration of the theatres based on the metaphor currently evoked by their Web presence. E-collaboration has been broadly defined as the following: "Collaboration using electronic technologies among different individuals to accomplish a common task" (Kock, 2005).

This section focuses on how the metaphor is hindering or enabling the theatre, and expressly recommends changes in the Web presence metaphor being used that I believe will produce positive outcomes for the theatre in question.

The evocation of the jungle metaphor by three of the theatres can be interpreted to mean that the development of a Web presence is not a high priority or even a priority at all. If the theatres desire success in the future, it is imperative that they give thought, careful attention, and monetary resources to developing a Web presence that fulfills their missions. This responsibility can no longer be neglected or mischaracterized as an afterthought or an add-on. Ideally, as the organizations are experiencing strategic planning, they will also give thought to what their Web sites should contain, who should design them, and maybe even more critically, who will maintain and update them. If they are educational institutions trying to create a Web presence, they will always need to retain a broad constituency and thereby will always need to maintain a Web presence that appeals through a variety of metaphors. The jungle metaphor does not appear to be conducive to e-collaboration.

At first glance, one might argue that the family metaphor might encourage e-collaboration. A detailed analysis of the Web sites, however, shows that these organizations are quite protective of their families. They encourage their audiences to remain loyal and would also prefer that their actors continue acting with them, rather than pursuing opportunities with other theatre groups. In the end, these theatre families act as typical families do. Individual family members seek the companionship of members of other families, but entire families rarely have a reason or opportunity for collaboration with another entity. If the theatres evincing this metaphor desire to become more goal oriented, they will need to systematically evaluate their IT plans and possibly change the family metaphor that they are portraying via their Web presence. For now, the theatres have no overarching goals, but instead recognize that each member of the family has a life beyond the theatre group.

The journey metaphor highlights the fact that theatre productions are often a journey where obstacles are encountered and eventually overcome in order for the show to go on. When this metaphor appears on a theatre's Web site, the theatre shares its adventures with the patrons. The theatres must, as a matter of course, review whether the chaos of the journey remains appealing to their staff, patrons, and board members. The strong goal orientation of the journey metaphor makes it a good choice for Web presence for these South Jersey theatres. Their well-conceived Web sites are easily navigable and are excellent at depicting the spirit of exploration. Those on the journey do not necessarily seek to be rescued when storms develop; rather they would prefer to struggle through and eventually reach their intended destination on their own. These theatres show no interest in collaboration at the present time and may need to change their metaphor so that not everything they attempt involves a struggle.

There was only one theatre in my population of Web sites in the Southern New Jersey region with a Web site that evoked a well-oiled machine. The theatre has set a clear goal and exists in an orderly environment. The availability of a searchable database on the Web site in order to locate performers and productions is one dimension of the machine metaphor that is working well for it. However, a machine metaphor does not permit creativity and is essentially boring. Without creativity, the group employing a machine metaphor is in danger of becoming a de facto club rather than a community theatre. In general, the machine metaphor is not a positive one for regional theatres to pursue in their Web presence and certainly does not encourage cooperation.

There were two organizations evoking the society metaphor in the group of the Southern New Jersey theatres. Both of the organizations evoking the society metaphor are well established and highly rule based. Although societies try to cooperate, the societies involved can rarely agree on common goals and tend to be bound by the rules they set up. One may look to the United Nations in general or the Kyoto accords in particular as examples of the attempt to make societies cooperate with one another. In the final analysis, the society metaphor is not useful for facing changes that require the group to adopt a more collaborative position. I recommend that eventually those organizations evincing the society metaphor in their Web presences will need to rethink and redesign them to include a regionally cooperative metaphor.

There were two theatres with Web presences that evinced the game metaphor. The game metaphor as a Web presence can function in a number of successful ways for the theatre groups. In the cases of the two theatre groups here, the metaphor was supportive in fostering a competitive stance. The two groups are quite different. One is old, and the other is rather newly established. One has grown up with the Web, while the older one is just now recognizing the importance of a Web presence for maintaining a competitive position and is hiring professional help for the Web site's development and maintenance. Both organizations display Web sites that are visually and textually furthering their metaphor of the game. The team-building aspect of the game metaphor is a positive aspect of this Web presence. The game metaphor will allow these theatres to move to a regional cooperative rather than a competitive stance.

In an analogous situation, sports teams belong to leagues. The NFL (National Football League) negotiates contracts with television executives and shares the revenue among 32 different teams that comprise the league. This is merely one example of how a game metaphor can provide support for collaboration. Regional theatres can benefit from the same approach and consequently the game metaphor is flexible enough for them to embrace these strategic changes.

There was only one theatre that displayed the organism metaphor in the 15 southern New Jersey theatre Web sites that I studied. Developing a Web presence with an organism metaphor affords a theatre company a great deal of flexibility. It permits growth, change, and experimentation that other metaphors do not. The organism is a very positive metaphor for regional theatre Web presence. It allows the expression of creativity balanced by the requisite discipline to adapt and grow. It accommodates many different types of participants and an abundance of new ideas, and it can serve as a liberating metaphor for an organization.

Working together, the regional theatres can all be part of the same garden. They can allow one of the sponsoring agencies discussed earlier in the study to cultivate the garden. Once a theatre group recognizes that by using an organism metaphor it need no longer be constrained by its own history, past participants, or play choices, its ability to attract and retain participants will be exhilarating.

Based on my assessment of the 15 southern New Jersey theatre Web sites, I believe that I can identify which regional theatres are more likely to collaborate with others in terms of a strategic Web affiliation. The theatres have already developed Web sites that demonstrate that they are willing and able to be part of an umbrella organization. These theatres have existing Web sites that, under scrutiny, invoke metaphors of the game or organism. Less likely to use e-collaboration are those theatres that adopt a society, journey, or family metaphor. Even less likely to cooperate regionally are the three theatres that are still hacking their way out of the metaphoric jungle on their Web sites.

SUCCESSFUL WEB METAPHORS IN THE THEATRE INDUSTRY

Now that I have identified regional theatres that invoke the game and organism metaphors, the next step is to look for instances of successful leagues and gardens in the theatre industry.

In New York City, large professional theatres that are located within a defined neighborhood are referred to as Broadway theatres. The group that promotes Broadway shows is called The League of American Theatres and Producers, Inc. (http://www.LiveBroadway.com). The league began in 1930 as The League of New York Theatres, and its mission is to heighten awareness of and interest in Broadway theatre as well as to support the creation of more profitable theatrical productions. The league also negotiates collective bargaining agreements with all theatrical unions and guilds, maintains relevant research archives, and supports charitable efforts.

Although it is called a league, The League of American Theatres and Producers behaves more like a gardener. The emphasis is on cultivation, and the membership is so diverse that each of the members has its own way of nurturing shows. The league's membership consists of theatre owners and operators, Broadway and touring company producers, Telecharge and Ticketmaster, booking agents, hotels, and newspapers and magazines such as the *New York Times* and *The New Yorker*. Under the Live Broadway banner, the league brings to Broadway fans live plays as well as other theatrical events and services like the Tony (Antoinette Perry) Awards (together with The American Theatre Wing) and small annual shows such as *Broadway on Broadway* and *Broadway Under the Stars*.

Broadway's location possesses a distinct advantage in attracting theatregoers. All of the Broadway theatres are located in a relatively compact site, and the theatres benefit from the fact that theatregoers will notice a sign for a new play when leaving a show they have just seen. Marquees are powerful symbols, and the patron who notices the show a theatre is currently producing may decide to see that play or a future production.

We have long known that locating similar businesses together in clusters often helps the businesses perform better than those that are in dispersed locations. Recent research examines the geographical concentration of Internet industries (Kolka, 2002), hotelling (Huck, Muller, & Vriend, 2002), restaurant locations (Mariani, 2001), and the location of retail outlets (Netz & Taylor, 2002). For a complete look at location theory, readers should see the volume by Wesolowsky (2001). Many factors come into play when location decisions are made. Jacoby (2000) studied the location of e-commerce hot spots and pointed to the attitudes of store owners, customers, and their communities in making location decisions. Of interest in our study of regional theatre Web presence is that he also found that regions and the country could come into play in location decisions.

A more focused study asks the question about what exactly transpires within a cluster of similar businesses. Baptista (2000) looked at the clustering of technological firms in a region to find out whether innovations would diffuse faster because of their proximity to each other. While traditional location theory endorses the advantages to be found through locating similar businesses together in geographical clusters, the regional theatres of South Jersey cannot capitalize on building clusters of theatres into a theatre row (such as was recently accomplished in Manhattan on 42nd Street between 9th and 10th avenues, where a collection of five theatres are now available for rental). The theatres are geographically dispersed around eight separate counties. Indeed, many of the theatres were not built as theatres originally, so their individual locations within a community can vary widely from main street, a college campus, or an even an out-of-the-way residential neighborhood. Just heightening awareness of the existence of the theatre and then getting an audience to commit to attending performances in unknown locations is a challenge for all of the regional theatres.

E-commerce provides an answer for some of the location and publicity problems the theatres are experiencing. Just as with any business, theatres need to communicate with their audience the *who, what, when, where, why,* and *how* of their mission. In addition, theatres need to demonstrate their connection to the community and in particular their potential relevance in the lives of their audience members. This includes posting descriptions of their current offerings, creating offers that appeal to a variety of audience members such as weekend subscription packages and subscriptions to performances especially for children, and so on. The theatres need to communicate the locations of their theatres, the names and resumes of the current cast members, and information on how interested volunteers can support the companies through contributions of time or talent. In the past, newspaper advertisements were the most widely used vehicles for heightening awareness. In addition, theatres attempted to increase the number of series subscribers whiles also garnering repeat ticket sales from pleased audience members. Long-running shows that are seen by great numbers of theatregoers (shows such as *Phantom of the Opera* or *The Producers* come to mind) are able to rely on word of mouth for a continuous stream of eager theatre goers, but small, regional theatres often have very truncated production runs, often only over one Friday, Saturday, and Sunday. The expense of radio and television ads tend to put them out of the reach as communication tools of most regional nonprofit theatres.

A highly identifiable Web site with e-commerce features such as online ticketing and meaningful, organized links is an ideal place to promote current productions. Since patrons of theatres are educated and have sufficient disposable income to attend cultural events, they tend to have the means and desire to search for information on the Web. This turns out to be a rather one-sided relationship between the patrons who are online and the theatres putting on shows. While audience members may be online seeking local, live theatre productions, the regional theatres studied here do not possess the people, technical expertise, time, or budget to create a Web presence that will attract and retain people to the Web site.

From our study thus far, we have learned that the regional theatres that attempt a Web presence try to do so on their own separately from other theatres, without professional help or a template, starting with the basics. The volunteers who develop these initial Web sites create what it is in their power to do. Often, the question of a match between a theatre's Web site and its organizational mission is never even considered in this process.

It has been said that although modern transportation systems conquered the tyranny of distance, e-commerce can be credited with eliminating distance altogether. However, the process of helping audience members locate and then go to the Web site adds another dimension to distance. Even using the output from a familiar search engine such as Google, a theatregoer must still make informed choices about which of the listed links he or she will follow. The problem of electronically traveling to a different theatre Web site when one is already on one is that there is usually no information given about how to make this electronic journey. Location theorists tell us that there are distinct benefits that can accrue when similar businesses locate in proximity to each other. Making the e-commerce world of the theatre smaller and having sites "closer together" can have many beneficial effects. The way we characterize this internal and external navigational distance in e-commerce is through the term *e-distance*.

For the professional Broadway stage, we can see that theatres are connected in two important but different ways. One way is through their common ownership by a large, parent corporation. Eight Broadway theatres and 35 entertainment venues throughout the United States are owned by the Nederlander Producing Company of America. Its rival is the Shubert Organization, which is the parent of 17 Broadway theatres. Tickets for any of the productions at any of these theatres are sold by a common online ticket agency. The Shubert Organization owns Telecharge, and Ticketmaster is labeled as the world's largest ticket retailer by most sources. When an audience member arrives at these online ticket-agency Web sites, there is much to see and do. For instance, the sites provide a synopsis of the play and promote ready navigation among any of the current productions purveying tickets on the sites. For a small handling fee, audience members can purchase their tickets online at both sites. The entire theatre world comes alive on the Web on sites such as Playbill.com and Broadwayd.com, with daily postings of backstage gossip, show reviews, and photo opportunities at the fall of the first curtain.

Centralized online ticketing is not purely the province of Broadway shows. Some off-Broadway shows cooperate in a joint online box office called Ticket Central, where customers can buy tickets online for a dozen different plays. Clicking on a link called "In the Neighborhood" takes the Web site visitor to a neighborhood restaurant page, embellished with links to the Web sites of the restaurants listed. Even detailed parking instructions are presented. Part of the strategic alliance that was put together includes tie-ins for offering ticket holders discounts for parking and close-by restaurants if theatregoers display their evening's ticket stubs. The listing of the restaurants' links within the "In the Neighborhood" theme is very apt since the inculcation of community and a sense of helping each other is emphasized among all of the businesses present on this site.

Of the 15 regional theatres examined in South Jersey, only one of them features online ticket purchases for patrons. A few of the theatres do include links to local restaurants in their cities. The Holly City Repertory of Millville, New Jersey, is one of these.

The theatre alliance Web sites create links to multiple theatres in South Jersey, but the links are not often provided (by the theatres) in the reverse direction. When the links are displayed, they are presented in a hodgepodge, without apparent organization. If a patron is curious, he or she must click on the link to see where it leads. Some of the added links are no longer functional and some have changed destinations: the theatres' Web sites are not maintained sufficiently to catch errors, missing links, or inaccuracies.

CONCLUSION

In this period of uncertainty in funding prospects, the theatres in South Jersey need to join together. They can do this by e-collaboration that will permit them to link to one another on the Web or by setting up an umbrella organization that handles all ticketing and provides links to each of the regional theatres. In this way, theatres will be able to increase the level of interest among patrons. Rather than seeing themselves merely as members of a city or county arts community, the theatres can change their Web presence both rhetorically and visually to fashion a regional identity that can enable them to build knowledgeable audiences, obtain funding, and discover a broader range of acting talent.

Currently, the e-distance that separates the 15 regional nonprofit South Jersey theatres is greater than the true geographical distance among them. Several arts alliance groups, among them, Discover NJ Arts, New Jersey Theatre Alliance, New Jersey Theatre League, NJTheatre.com, and South Jersey Cultural Alliance, have created (as an overt manifestation of their expressed strategic mission to champion collaboration in the arts) Web presences that organize and display several of the New Jersey theatre links. However, the regional theatres do not collaborate with each other in promoting each others' productions. The New Jersey Theatre Alliance is currently doing trial runs of online ticket purchasing, but individual seats for requested performances are not available.

It is still a challenge for a theatregoer to locate a regional theatre production on the Web. Browsing a current schedule or bringing up detailed information about a play now showing means that the audience member must know all of the URLs for the theatres they would like to attend, or they must keep going back to one of the several theatre arts alliance Web sites for more links. The theatres present in southern New Jersey do not yet link directly to each other, so that means that the audience member becomes burdened with browsing.

There are only one or two instances where regional theatre Web sites are providing reciprocal links with other theatre companies. To make matters worse, some artistic directors jealously guard the resumes of the actors who work with the company from season to season, explicitly discouraging actors from working at another area theatre. The paramount goal then becomes to inculcate loyalty to a particular theatre company rather than perfecting the actor's craft, pooling area talent, sharing limited funding resources, or improving the overall quality of the South Jersey theatre experience. Actors who participate in regional nonprofit acting roles often feel quite comfortable with playing in any of the many regional theatres, taking their talents with them as they audition for yet another role. However, individual theatres have not adopted the perspective that permits them to visualize themselves as a collaborative community.

With those considerations aside for a moment, I recommend that the theatres follow through on their impulses toward creating a regional identity so that they are truly drawing audiences, talent, ideas, and funding from the region. This can be manifested metaphorically in their Web presence by rewriting content to widen the language to include more of a sense of community, or by adding more language reflective of a competitive game or a growing organism.

Specifically, the Web sites can be changed to ensure e-collaboration via inclusion of reciprocal links, especially for nearby theatres that have seasons that are clearly different. For example, a theatre season without musicals that presents several original dramas or an experimental theatre linking with another that does outstanding community outreach are prime candidates for this type of cooperation. Theatre Web sites can also be linked to other arts alliance Web sites. Theatre companies may also consider e-collaboration for the purpose of creating an exclusively regional Web site for southern New Jersey theatres that uses a different metaphor (perhaps the game or organism metaphor) that is able to reconceptualize individual theatre offerings as a lively and growing regional resource.

NOTE

This research was conducted as part of a fellow award made by the Walter Rand Institute of Public Affairs at Rutgers University, Camden. The author is grateful for the generous support and encouragement of the Rand Institute faculty and staff during the course of the fellowship. It was a fruitful and productive relationship.

REFERENCES

Abuhamdieh, A. H., Kendall, J. E., & Kendall, K. E. (2000). Beyond e-commerce: Evaluating the strategic impact of Web presence for nonprofit organizations. In M. Swink (Ed.), *Proceedings of the Decision Sciences Institute* (pp. 567-569). Atlanta: Decision Sciences Institute.

Abuhamdieh, A. H., Kendall, J. E., & Kendall, K. E. (2002). An evaluation of the Web presence of a nonprofit organization: Using the balanced scorecard approach in ecommerce. In R. Traunmuller (Ed.), *Information systems: The e-business challenge* (pp. 209-222). Boston: Kluwer Publishing.

Abuhamdieh, A. H., Kendall, J. E., & Kendall, K. E. (2007). E-commerce opportunities in the nonprofit sector: The case of New York Theatre Group. *International Journal of Cases on Electronic Commerce, 3*(1), 29-48.

The arts mean business: A study of economic activity, 2000. NJSCA/ArtPRIDE. (2000). NJ: NJ Foundation/Rutgers University Graduate School of Education.

The arts in New Jersey: A study of economic activity 1992-93. (n.d.). NJ: NJSCA/South Jersey Cultural Alliance/Eagleton Institute, Rutgers University.

Baptista, R. (2000). Do innovations diffuse faster within geographical clusters? *International Journal of Industrial Organization, 18*(3), 515.

Clancy, J. J. (1989). *The invisible powers: The language of business.* Lexington, MA: Lexington Books.

Duncan, H. D. (1968). *Symbols in society.* New York: Oxford University.

Graber, D. A. (1976). *Verbal behavior and politics.* Chicago: University of Illinois Press.

LiveBroadway.com. (2005). Retrieved January 2005 from http://www.liveBroadway.com

Huck, S., Muller, W., & Vriend, N. J. (2002). The East End, the West End, and King's Cross: On clustering in the four-player hotelling game. *Economic Inquiry, 40*(2), 231-241.

Jacoby, C. (2000). E-commerce hot spots. *Corporate Location, 3*, 18.

Kendall, J. E., & Kendall, K. E. (1993). Metaphors and methodologies: Living beyond the systems machine. *MIS Quarterly, 17*(2), 149-171.

Kendall, J. E., & Kendall, K. E. (1994). Metaphors and their meaning for information systems development. *European Journal of Information Systems, 3*(1), 37-47.

Kendall, K. E. (2000). Ecommerce aesthetics: Ecommerce for ecommerce's sake. *Information Resources Management Journal, 13*(3).

Kendall, K. E., & Kendall, J. E. (2005). *Systems analysis and design* (6th ed.). Upper Saddle River, NJ: Prentice Hall.

Kock, N. (2005). What is e-collaboration? *International Journal of e-Collaboration, 1*(1), i-vii.

Kolka, J. (2002). Silicon mountains/silicon molehills: Geographic concentration and convergence of Internet industries in the US. *Information Economics and Policy, 14*(2), 211.

Lakoff, G., & Johnson, M. (1980). *Metaphors we live by.* Chicago: The University of Chicago Press.

Mariani, J. (2001). The implications of your location. *Restaurant Hospitality, 85*(8), 20.

Morgan, G. (1986). *Images of organization.* Beverly Hills, CA: Sage Publications.

Netz, J. S., & Taylor, B. A. (2002). Maximum or minimum differentiation? Location patterns of retail outlets. *The Review of Economics and Statistics, 84*(1), 162.

Ott, J. S. (1989). *Organizational culture perspective.* Chicago: The Dorsey Press.

Te'eni, D., & Kendall, J. E. (2004). Internet commerce and fundraising. In D. Young (Ed.), *Effective economic decision-making by nonprofit organizations* (pp. 167-189). New York: The Foundation Center.

U.S. Census Bureau. (2003). Retrieved from http://eire.census.gov/popest/data/counties/tables/CO-EST2001-08/CO-EST2001-08-34.php

Weaver, R. M. (1967). *A rhetoric and handbook.* New York: Holt, Rinehart, and Winston.

Wesolowsky, G. O. (2001). Lectures on location theory. *INFOR, 39*(1), 124.

Chapter III
User Satisfaction with E-Collaborative Systems

Jeffrey Wong
University of Nevada, USA

Kevin Dow
Kent State University, USA

Ofir Turel
California State University, USA

Alexander Serenko
Lakehead University, Canada

ABSTRACT

E-mail is a critical component of most e-collaborative environments. This chapter describes an application of the American Customer Satisfaction Index (ACSI) framework to model the antecedents and consequences of customer satisfaction with e-mail systems. The ACSI framework is an established methodology in the marketing literature and appeared to be useful to assess the antecedents and consequences of individual satisfaction in many more circumstances than external customer purchases. We surveyed e-mail users to gather data to utilize in an ACSI model modified for e-mail systems. Our findings indicate that the ACSI model can yield useful insights into factors that contribute to and result from user satisfaction.

INTRODUCTION AND BACKGROUND

Information and communication technologies are the foundation for collaborative efforts that allow geographically dispersed individuals to work as a team (Kock, 2005). E-mail is one such ICT. The widespread use of e-mail technologies demonstrates its importance as a communication medium. For example, in 2003, 31 billion e-mail messages were sent daily worldwide with an average of 56 e-mails per e-mail address and 174 e-mails per person (Industry Canada, 2004). Furthermore, over 600 million people worldwide were using e-mail systems by the end of 2004 (Radicati Group Inc., 2004). The phenomenal number of users was brought about partially by the declining costs of computing, fees for long-distance communication (Sproull & Kiesler, 1991), and advances in computer and telecommunications technologies. Continuing advances in technology and the globalization of business will likely increase the widespread use of e-mail.

The prolific use of e-mail has also introduced some problematic situations. For example, user overload has resulted from the widespread use of e-mail as a communications medium (Sherwood, 2002). At the same time, the increasing number of unsolicited commercial e-mail messages (typically regarded as junk mail or spam) has exacerbated the information overload problem. Unwanted communication has been credited as being one of the most critical components of e-mail overload (Hinde, 2002).

The number of unsolicited messages is growing significantly and may soon account for more than half of all e-mail traffic (Krim, 2003). One explanation for this growth is that senders find it more expensive to target their e-mail messages to potential customers rather than simply send the same message to large distribution lists (Gopal, Walter, & Tripathi, 2001). Additionally, there are insufficient and ineffective antispam regulations in place. At the moment, only 26 countries have mandated antispam laws, leaving the opportunity for legal spamming activities in more than 100 other countries.

From the perspective of e-mail service providers, the negative effects of spam may hinder customer retention and acquisition. The magnitude of e-mail communication represents a potentially lucrative business opportunity for e-mail service providers. For them, the volume of e-mails and subscribers can translate into advertising dollars and in some cases, user fees. Accordingly, it is believed that spam can have crucial effects on both users' use of e-mail and on e-mail service providers' profitability. Our study seeks to contribute to the literature by studying the antecedents and consequences of user satisfaction with e-mail, and to explore the potential impact of spam on user satisfaction.

To survive in the competitive environment of technology services, it is essential to both attract and retain customers. User satisfaction is a critical factor in the determination of loyalty to a service provider (Reichheld, 2003). Customer turnover can be costly because of resources expended to replace customers lost, and the possibility of damage to an e-mail provider's reputation. Findings from empirical studies suggest that when customers perceive a firm is providing superior products or services, those firms enjoy higher financial returns than firms with less satisfied customers (Anderson & Fornell, 2000).

We employ the American Customer Satisfaction Index (ACSI) model developed in the 1990s (Fornell, Johnson, Anderson, Cha, & Bryant, 1996) in our study. The ACSI was created by leading researchers in customer satisfaction, Claes Fornell and Eugene Anderson, to measure overall customer satisfaction in a way that can be compared between companies or between industry segments. Customer satisfaction has increasingly become important as companies have become service oriented, and customers are faced with many alternative providers from which to choose services and products. The ACSI has become one of the most frequently utilized satisfaction measures in the marketing literature.

The value of customer satisfaction measured by the ACSI is derived from the premise that without satisfied customers, current and future revenue streams are jeopardized. The ACSI was designed to measure customer satisfaction in a standardized way that would provide insights into the consumer economy for companies, industry trade associations, and government agencies. The developers of the ACSI felt that their methods would help to meet the need for an overall measure of the consumption experience. The index measures the quality of goods and services as perceived by those who consume them, and is designed primarily to explain customer loyalty.

There are two principal concepts underlying the ACSI model. First, the constructs of the model represent different types of customer evaluations that can only be measured indirectly. This is why the ACSI uses a multiple-indicator approach that measures customer satisfaction as a latent variable. Second, the ACSI is built on a series of cause and effect relationships that allows the antecedents and consequences of overall user satisfaction to be examined. In the case of this chapter, we examine user satisfaction with e-mail systems. Figure 1 illustrates the modified ACSI model and relationships between variables used in our study. The goal of this model is to present a set of causal relationships among several quality-related constructs. These variables will be further discussed in the next section where we develop the relationships tested in this chapter.

The remainder of the chapter is structured as follows. The next section develops the relationships that are tested through the ACSI framework. The subsequent section outlines the research methodology and statistical results. The discussion, conclusion, and directions for future research are provided in the last section.

ACSI ELEMENTS AND FRAMEWORK

This study uses the ACSI framework to generate a score for e-mail services so that a comparison with other services may be made. This score, coupled with the causal model, depicts the perceptions, beliefs, and behaviors of e-mail users. As depicted in Figure 1, overall customer satisfaction has two antecedents and two consequences as proposed by Fornell et al. (1996). We have modified the model for this study by omitting the variables related to the perceived value construct because the e-mail services are generally free for any given Internet service provider.

The primary research questions addressed in this chapter result in five testable hypotheses. A secondary research question is also addressed.

RQ 1. What are the relationships among the antecedents and consequences of customer satisfaction with electronic mail?

Figure 1. The e-mail services customer satisfaction model (Adapted from the American Customer Satisfaction Model by Anderson and Fornell, 2000)

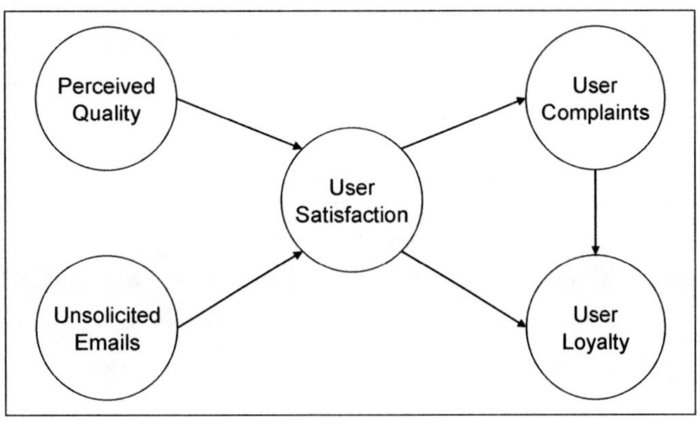

RQ 2. What is the level of customer satisfaction with electronic mail measured by the American Customer Satisfaction Index?

Satisfaction with e-mail is a users' reaction to their judgment of the state of fulfillment (Oliver, 1997). Users have a set of perceptions regarding their e-mail based on either previous experience or external influences. While using their e-mail system, these perceptions are compared with their actual experience, and a set of beliefs regarding the extent to which the e-mail service has met their expectations is developed. As a result, individual e-mail users adjust their perceptions accordingly.

The ACSI represents a customer-centered quality measurement system to evaluate the performance of individual companies, industries, economic sectors, and national economies. The index is calculated quarterly for many industries, including information technology. The ACSI is available for online news and information services, portals, search engines, online retailers, and telecommunications companies.

The contemporary literature presents a variety of models that measure the degree of customer satisfaction with products or services (Bansal, Irving, & Taylor, 2004). In our study, the ACSI was used because it demonstrates good psychometric capabilities and enables the generation of a standardized satisfaction score, which is comparable across industries and sectors.

The quality of an e-mail system positively influences the level of user satisfaction with it. It is derived from the degrees of personalization and reliability of the e-mail service. Further, the number of unsolicited e-mail messages, which is frequently cited as the major e-mail dissatisfaction reason, is posited to have a direct negative effect on satisfaction. Accordingly, the related hypotheses we test are as follows.

H1: Perceived quality will have a positive association with the level of customer satisfaction.

H2: The number of unsolicited spam messages received by a user is negatively associated with the level of customer satisfaction.

Prior research has found that satisfaction has a positive effect on both loyalty and retention (Bolton, 1998; Gerpott, Rams, & Schindler, 2001) and a negative effect on the number of customer complaints (Anderson & Fornell, 2000; Kim, Park, & Jeong, 2004). Marketing scholars have argued that in industries where some switching barriers exist, it is more valuable to examine loyalty than to measure retention. In high-switching-barrier industries (such as e-mail), customers that are dissatisfied might still remain with their service provider. In these industries, disloyal customers that remain with a company due to these high switching costs may ultimately have a negative effect on new customer acquisition as they share their opinions and beliefs with others (Reichheld, 2003).

In the case of e-mail systems, there are at least two potential switching barriers. First, for Web-based systems, one main barrier is the e-mail address (which is not portable to other service providers). In general, users find it inconvenient to regularly change their e-mail addresses because they have to locate all people and notify them about the change. Second, in the case of PC-based applications, the main barriers are the cost associated with the purchasing of new software, for example, Microsoft Outlook, and the setup efforts. It follows that the extent of user satisfaction is negatively associated with an individual's propensity to either formally or informally complain about services. At the same time, user satisfaction with e-mail positively influences the degree of loyalty to a specific e-mail system. User loyalty is a favorable attitude toward a specific e-mail system that leads to choosing the same system given a need for a new e-mail application. Accordingly, the related hypotheses we test are the following.

H3: The level of customer satisfaction is negatively associated with the frequency of user complaints.

H4: The level of customer satisfaction is positively associated with loyalty.

Following the ACSI framework, we test the relationship between the endogenous outcomes of overall satisfaction. There are no direct ways to test the efficacy of how an e-mail company deals with handling customer complaints. However, if a customer becomes satisfied with the way their complaints were handled, this treatment could engender loyalty. On the other hand, if the customer is dissatisfied with the manner in which complaints are addressed, the effect on loyalty would be expected to be negative. Therefore, we can only hypothesize that an association exists. Accordingly, the related hypothesis we test is the following.

H5: The frequency of user complaints is associated with loyalty.

METHOD AND DATA ANALYSIS

Survey Instrument and Construct Measures

A survey was used to sample participants from a population of e-mail users. All responses were kept anonymous and only limited demographic data were collected to determine the nature of the sample. This voluntary survey was administered to 200 undergraduate and graduate students. Two hundred surveys were administered and 186 surveys were returned. Eight instruments were either incomplete or partially complete and were excluded from data analysis. The final sample for analysis consisted of 178 usable responses.

The measures used in this study were adopted from previously validated instruments (Fornell et al., 1996) and also from existing literature to create measures for the five constructs used in this study to minimize the potential for measurement error. While it is acknowledged that all latent measures are imperfect, using existing and previously validated measures provides more of a consensus for appropriate representation of the constructs detailed in this chapter.

Perceived quality is simply an individual's judgment about the e-mail system and relates to the perception of quality based on how well the product fits personal requirements and whether expectations about reliability are met. Satisfaction is a measure of an individual's experience based on his or her evaluation of the e-mail system and can be thought of as the overall evaluation of the total consumption experience. Customer loyalty expresses an individual's intended behavior related to the e-mail system and it is operationalized with a single item used to reflect an individual's likelihood to remain loyal to the e-mail system. Customer complaints can be thought of the voicing of dissatisfaction and is measured with a single dichotomous item to reflect whether or not individual users have complained about their e-mail system. If subjects have complained, then they indicated to what extent have they have done so. Unsolicited e-mail is a categorical representation of the number of unsolicited e-mails received by the user on a daily basis.

Respondents to the survey indicated their age, sex, name, and type of the most frequently utilized e-mail system. In addition, actual e-mail usage details, such as the number of messages sent or received daily (excluding spam) and time spent working with an e-mail application were reported. The development of the research instrument followed an iterative process to ensure face validity of the survey. Specifically, a small group of information technology practitioners and academics was consulted and asked whether the items proposed in the instrument adequately measured the desired constructs. As a result of their feedback, several minor modifications were made.

Descriptive Statistics

The participants surveyed in this study represented a diverse sample. Their ages ranged from 18 to 50 with 51% of the participants male and 49% female. On average, respondents utilized their respective e-mail system for 5 years; the amount of time ranged from 3 months to 14 years. Table 1 outlines descriptive statistics for each of the items used in the survey. Table 2 outlines the Pearson correlations among the variables included in this study.

It should be noted that the correlation between perceived quality and satisfaction is relatively high. We believe, however, that this does not threaten the validity of the model for several reasons. First, the loadings of these items on the constructs to which they belong are higher than their cross-loadings. Second, these two constructs represent an independent and a dependent variable that are expected to be correlated. Third, other studies that utilized or adapted the American Customer Satisfaction Index framework also report high correlations between quality and satisfaction. All of these studies argue that there is a reasonable statistical support to treat quality and satisfaction as being distinctly different from each other. For example, researchers have consistently uncovered a high correlation between quality and satisfaction (Babakus, Bienstock, & Scotter, 2004; O'Loughlin & Coenders, 2004). Overall, it should be noted that even though some of the correlations in our model are fairly high, they are still within the norm, and as such are

Table 1. Descriptive statistics

CONSTRUCT	ITEM	N	MEAN	STD. DEVIATION	SKEWNESS	KURTOSIS
Quality	QUALITY 1	178	7.06	1.31	-0.83	1.47
	QUALITY 2	178	7.38	1.41	-0.52	0.05
	QUALITY 3	178	7.25	1.60	-0.56	0.30
Unsolicited E-Mails		177	2.29	1.61	1.06	0.28
ACSI	ACSI 1	178	7.28	1.43	-0.52	0.85
	ACSI 2	177	6.59	1.39	-0.43	-0.13
	ACSI 3	178	6.72	1.51	-0.43	-0.13
Loyalty		176	6.94	1.96	-0.60	0.13
Complaints		175	0.22	0.46	2.24	6.91

Table 2. Correlation matrix of variables in model

	Perceived Quality	Unsolicited E-mails	ACSI	Customer Complaints	Customer Loyalty
Perceived Quality	1.000				
Unsolicited E-mails	-0.114	1.000			
ACSI	0.828	-0.079	1.000		
Customer Complaints	-0.105	0.203	-0.187	1.000	
Customer Loyalty	0.531	0.021	0.573	-0.085	1.000

not inconsistent with the discriminant validity of our constructs.

The American Customer Satisfaction Index

To help understand the results of the survey, it is important to understand how the ACSI is calculated. The ACSI was calculated for e-mail systems (both in the aggregate and for each individual e-mail provider) based on the formula suggested by Anderson and Fornell (2000):

$$ACSI = \frac{\sum_{i=1}^{3} w_i \cdot \bar{x}_i - \sum_{i=1}^{3} w_i}{9 \cdot \sum_{i=1}^{3} w_i} \times 100, \qquad (1)$$

where w_i represents the weight of the ith item obtained from the outer model generated by PLS, and $\bar{x}_i$ represents the average of the ith item that loads on the ACSI construct.

Analysis

Prior to any analysis of the model, it is important to confirm the reliability and validity of measures used in this study. As such, we examine item reliability, convergent validity, and discriminant validity.

Item reliability. Item reliability indicates whether the indicators for a particular latent variable measure that latent variable only. Guidelines provided by Hair, Anderson, Tatham, and Black (1995) were used in determining the item reliability for each latent variable. Following their suggestion, only items with loading greater than or equal to 0.50 were retained. As can be seen in Table 3, each of the factor loadings are above the minimum threshold of 0.50. As such, all individual items were retained for the final structural model.

Convergent validity. Construct validity indicates the degree to which an observable variable represents the variable of interest, which may only be indirectly observed. It is measured with Cronbach's alpha using a popular rule of thumb at 0.70 (Fornell & Larcker, 1981). All values obtained exceed this value and therefore convergent validity has been satisfied.

Discriminant validity. Discriminant validity represents the extent to which measures of a given construct differ from measures of other constructs in the same model. Essentially, a latent construct should share more variance with its indicators than it shares with other latent constructs. To assess this, Fornell and Larker (1981) suggest the use of the average variance extracted: simply the average variance shared between a construct

Table 3. Loadings and cross-loadings for each item and construct

	Perceived Quality	ACSI	Unsolicited E-mail	Complaints	Loyalty
PQ1	0.863	0.702	-0.073	-0.132	0.402
PQ2	0.888	0.735	-0.069	-0.099	0.491
PQ3	0.841	0.708	-0.129	-0.042	0.484
ACSI1	0.818	0.864	-0.094	-0.160	0.543
ACSI2	0.547	0.778	-0.101	-0.156	0.345
ACSI3	0.698	0.893	-0.037	-0.163	0.541
JUNK	-0.104	-0.089	1.000	0.185	0.062
UC	-0.105	-0.187	0.185	1.000	-0.085
UL	0.531	0.537	0.062	-0.085	1.000

and its measures. The average variance extracted is obtained by the sum of the loading squared, divided by the number of items in the construct. This measure should be greater than the variance (squared correlation) shared between the latent construct and other latent constructs in the model. Each construct passed the test for discriminant validity, and the loadings and cross-loadings for each construct can be found in Table 3.

RESULTS

Structural Model

We employed partial least squares using the PLS Graph software package (Chin, 2001) to evaluate our model. Statistically significant levels of the estimated path coefficients were determined using the bootstrap procedure. The t-values we obtain are estimates of the bootstrap path coefficient divided by the standard error.

Figure 2 and Table 4 show the results of PLS analysis. The significant paths indicate that three of the five hypotheses are supported. Specifically, as hypothesized, the path coefficient from perceived quality to user satisfaction (0.829, $p<0.01$) provide support for H1, which stated that the quality of an e-mail application is positively associated with user satisfaction. The results do not support H2. This suggests that spam does not have a strong effect on user satisfaction. The path coefficient from user satisfaction to user complaints (-0.187, $p<0.05$) supports H3, which said user satisfaction

Figure 2. Results of statistical testing

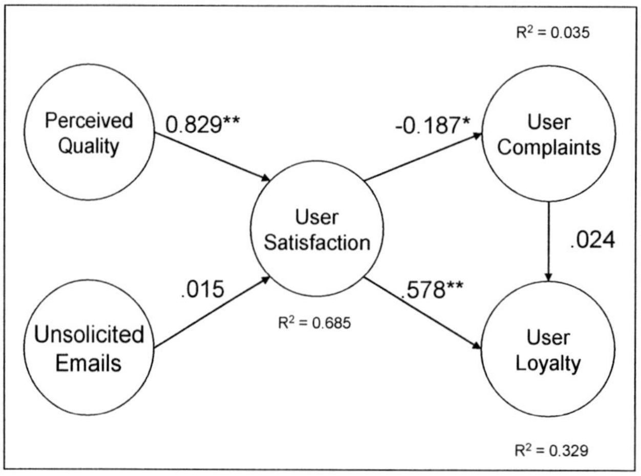

Table 4. Table of statistical testing

Hypothesis	Estimate	T-value	P-value	Supported?
H1: PQ → ACSI	0.829	30.682	<0.01	supported
H2: JUNK → ACSI	0.015	0.334	not significant	rejected
H3: ACSI → UC	-0.187	2.437	<0.05	supported
H4: ACSI → UL	0.587	8.394	<0.01	supported
H5: UC → UL	0.024	0.297	not significant	rejected

is negatively associated with complaints. H4, stating user satisfaction is positively associated with the degree of loyalty, is supported by the results (0.578, $p<0.01$). Finally, the path coefficient from user complaints to user loyalty was not statistically significant and does not support H5. In summary, most of the hypothesized linkages in the ACSI model were supported by our findings.

American Customer Satisfaction Index Results

An average ACSI score for all of the e-mail services was calculated to be 65.5, as seen in Table 5. This table is presented to get a sense of ACSI scores for other industries in North America for Quarter 4 of 2003. Table 5 further outlines the specific comparison of e-mail systems. As can be seen, the satisfaction score for e-mail service ranks just below that of the airline industry, and just above that of cable and satellite TV.

DISCUSSION

Our findings largely support the relationships between the elements of the ACSI model, suggesting that the ACSI is a feasible measurement of customer satisfaction related to e-mail usage. The degree of perceived quality of an e-mail application appears to strongly influence an individual's level of satisfaction with this e-mail system. A more satisfied e-mail user complains less about his or her e-mail experience and demonstrates a high degree of loyalty to a particular e-mail system.

Table 5. Current ACSI Scores for select industries in North America

Sector	ACSI
Search Engines	79
Personal Computers	77
Portals	76
Computer Software	74
News & Information	73
Fixed Line Telephone Service	70
Network/Cable TV News	69
Wireless Telephone Service	66
E-Mail Systems	65.5
Cable & Satellite TV	63

An interesting finding of our study is that unsolicited e-mail making it through the many spam filters of e-mail providers was not a significant determinant of customer satisfaction. This finding is surprising because the literature seems to indicate that spam is a real problem to e-mail users. We suspect that, in general, issues that are beyond the control of the technology do not seem to negatively impact the level of satisfaction with this e-collaborative technology. As such, users appear to be coping with these downside issues better than the experts would have predicted.

Alternatively, this finding may be partially explained by the motivational premises of attribution theory. This theory explains how people make causal explanations about events and describes the processes and behavioral outcomes of those rationalizations. According to attribution theory, people tend to give credit to themselves for events with positive outcomes and blame the external environment for events with a negative outcome.

When a spam e-mail message arrives, a person spends some time to determine the relevance of this message and to delete it. This is a negative outcome situation that e-mail users may blame on external factors, such as the quality of an e-mail system, a spammer, the lawmakers delaying antispam legislation, and so forth. Research in the area of human-computer interaction (HCI) suggests that people tend not to hold software responsible for mistakes, problems, or bugs because users expect computer applications to be generally unreliable (Lieberman, Rosenzweig, & Singh, 2001). Therefore, it is plausible that a recipient of spam attributes it to factors not associated with their e-mail application. For example, a spammer is considered directly responsible for the spam, not the e-mail service provider whose spam filters did not detect and remove the spam.

On the surface, the ACSI calculated for e-mail services appears to be relatively low compared with those of related products and services in other industries in North America. However, the proper way to use the ACSI score may be to compare scores from one period to the next instead of in a cross section among other goods and services. Alternatively, comparisons to similar services are more meaningful than industry or sector cross-sectional comparisons.

Finally, we found no evidence of a relationship between a user's tendency to complain about an e-mail system and his or her level of loyalty to the particular e-mail service. This may be due to high switching barriers that prevent people from moving from one e-mail system to another. Even though an e-mail service might be free of charge, the change of an e-mail address, time to install a new application, and software acquisition costs create high switching barriers for e-mail users. Therefore, even though people complain about their e-mail experience, they tend to stay with their current e-mail system. In addition, there are few substitutes for e-mail.

CONCLUSION

Satisfaction with e-mail systems is an important topic because of the significant role that e-mail plays in communication and e-collaboration. Our study employs the ACSI customer satisfaction model to offer insights into the antecedents and consequences of user satisfaction with e-mail systems.

We find evidence that perceived quality is a determinant of user satisfaction. In turn, user satisfaction is positively related to user loyalty, and negatively associated with user complaints. Our findings generally support the ACSI model of customer satisfaction. Surprisingly, unsolicited e-mail was not a statistically significant determinant of user satisfaction, which was an unanticipated finding based upon the literature discussing the potential effect of spam on e-mail users. Additionally, user complaints did not appear to have an effect on user loyalty, as some of the literature on customer loyalty might suggest. The findings of this chapter may be of value not only to e-mail users and e-mail providers, but also to researchers interested in other e-collaborative technologies.

Our study has its limitations. The sample consisted mostly of young Canadian students, which may not be generalizable to the entire e-mail user population. However, we feel that this population has strong familiarity with and dependence on e-mail, making it typical of regular e-mail users (Fallows, 2002). Ideally, it would have been preferable to survey a larger sample that draws from a cross section of individuals that includes people from business and governmental sectors and who are of heterogeneous geographical dispersion, using a wider variety of e-mail systems. Future research may employ a broader cross section of e-mail users to modify, validate, or expand on our findings.

REFERENCES

Anderson, E. W., & Fornell, C. (2000). Foundations of the American Customer Satisfaction Index. *Total Quality Management & Business Excellence, 11*(7), 869-882.

Babakus, E., Bienstock, C. C., & Scotter, J. R. V. (2004). Linking perceived quality and customer satisfaction to store traffic and revenue growth. *Decision Sciences, 35*(4), 713-737.

Bansal, H. S., Irving, P. G., & Taylor, S. F. (2004). A three-component model of customer commitment to service providers. *Journal of the Academy of Marketing Science, 32*(3), 234-250.

Bolton, R. N. (1998). A dynamic model of the duration of the customer's relationship with a continuous service provider: The role of satisfaction. *Marketing Science, 17*(1), 45-65.

Chin, W. W. (2001). *PLS-Graph user's guide, version 3.0.* Soft Modeling Inc.

Fornell, C., Johnson, M. D., Anderson, E. W., Cha, J., & Bryant, B. E. (1996). The American Customer Satisfaction Index: Nature, purpose, and findings. *Journal of Marketing, 60*(7), 7-18.

Fornell, C., & Larcker, D. F. (1981). Evaluating structural equation models with unobservable variables and measurement error. *Journal of Marketing Research, 18*(1), 39-50.

Gerpott, T. J., Rams, W., & Schindler, A. (2001). Customer retention, loyalty, and satisfaction in the German mobile cellular telecommunications market. *Telecommunications Policy, 25*(4), 249-269.

Gopal, R. D., Walter, Z., & Tripathi, A. K. (2001). Admediation: New horizons in effective email advertising. *Communications of the ACM, 44*(12), 91-96.

Hair, J. F., Jr., Anderson, R. E., Tatham, R. L., & Black, W. C. (1995). *Multivariate data analysis with readings* (4th ed.). Englewood Cliffs, NJ: Prentice Hall.

Hinde, S. (2002). Spam, scams, chains, hoaxes and other junk mail. *Computers & Security, 21*(7), 592-606.

Industry Canada. (2004). *The digital economy in Canada: Unsolicited commercial electronic mail (SPAM).* Retrieved July 30, 2004, from http://e-com.ic.gc.ca/epic/internet/inecic-ceac.nsf/en/h_gv00170e.html#stats

Kim, M. K., Park, M. C., & Jeong, D. H. (2004). The effects of customer satisfaction and switching barrier on customer loyalty in Korean mobile telecommunication services. *Telecommunication Policy, 28*(2), 145-159.

Kock, N. (2005). What is e-collaboration. *International Journal of e-Collaboration, 1*(1), i-vii.

Krim, J. (2003, March 13). Spam's cost to business escalates: Bulk e-mail threatens communication arteries. *The Washington Post*, p. A01.

Lieberman, H., Rosenzweig, E., & Singh, P. (2001). Aria: An agent for annotating and retrieving images. *IEEE Computer, 34*(7), 57-62.

Oliver, R. L. (1997). *Satisfaction: A behavioural perspective on consumer.* Boston: Irwin McGraw-Hill.

O'Loughlin, C., & Coenders, G. (2004). Estimation of the European customer satisfaction index: Maximum likelihood versus partial least squares. Application to postal services. *Total Quality Management & Business Excellence, 15*(9-10), 1231-1255.

Radicati Group Inc. (2004). *Market numbers summary update.* Palo Alto, CA.

Reichheld, F. F. (2003). The one number you need to grow. *Harvard Business Review*, 46-54.

Sherwood, K. D. (2002). *Overcome email overload with Microsoft Outlook 2000 and Outlook 2002: Get through your electronic mail faster.* Palo Alto, CA: World Wide Webfoot Press.

Sproull, L., & Kiesler, S. B. (1991). *Connections: New ways of working in the networked organization.* Cambridge, MA: The MIT Press.

Chapter IV
Organizational Sense of Community and Listserv Use:
Examining the Roles of Knowledge and Face-to-Face Interaction

Anita Blanchard
University of North Carolina at Charlotte, USA

ABSTRACT

This study examines how a Listserv affects its members' sense of community (SOC) with the sponsoring organization. It was expected that the Listserv would increase members' knowledge about and participation in the sponsoring organization department, which, in turn, would increase their SOC. The study examined Listserv members and nonmembers before and after implementation of the Listserv. As expected, Listserv membership increased knowledge and face-to-face activity, and knowledge and face-to-face activity increased sense of community. However, there was ironically no effect of Listserv membership on sense of community. These findings challenge previous theories about the development of sense of community while nonetheless demonstrating the positive effects of Listserv membership.

INTRODUCTION

Organizations can use one-way, information-dispersing Listservs to keep their members informed and connected. Listservs are group distribution e-mails in which members can conveniently send messages to one e-mail address, usually the Listserv name, instead of all of the individual members' e-mail addresses. Sometimes organizations assign members to Listservs, but often members choose to join a particular Listserv to stay informed about the organization or topics relevant to the organization.

Work organizations can use Listservs to keep employees updated on policies, announcing the entry or departure of key personnel, changes in benefits, and upcoming social events (e.g., the company picnic). Educational organizations can use Listservs to inform students about upcoming classes, research and internship opportunities, and extracurricular student activities (e.g., clubs). Social and professional organizations can use them to inform members of club-relevant announcements, involvement opportunities, and organize upcoming face-to-face (FtF) events. For example, alumni clubs can make announcements and promote viewing parties for athletic events; professional networking clubs can promote job opportunities and their monthly meetings.

These organizations may expect that this type of Listserv keeps Listserv members informed and active in the organization. However, how does the Listserv affect the Listserv members' greater attachment to the sponsoring organization? Researchers believe that electronic collaboration technologies such as e-mail and the Internet can increase members' attraction to and affiliation with their communication partners (Adams-Price & Chandler, 2000; Meier, 2000; Mesch & Levanon, 2004; Walther, 1996). Does this relationship extend to increasing affiliation with the larger organization?

This chapter will examine the relationship between an informational Listserv and organizational affiliation. Specifically, it will examine how a Listserv affects the amount of information members feel they have about the organization sponsoring the Listserv, the amount of face-to-face interaction members have with other members of the organization, and subsequently, a particular form of organizational affiliation: organizational sense of community (SOC). The next section examines the research on computer-mediated communication (CMC) and organizational sense of community.

Organizational Sense of Community

Sense of community has a long history within the community psychology literature. Sarason (1974) was one of the first researchers to identify that community members' feelings about each other and the community itself are important to the community's successful functioning. SOC leads to satisfaction with and commitment to the community, and is associated with involvement in community activities and problem-focused coping behavior (McMillan & Chavis, 1986).

McMillan and Chavis (1986) developed the SOC construct by defining it as an individual's feelings of membership, identity, belonging, and attachment with a group. Their descriptive framework of SOC has been widely accepted because of its theoretical base and its qualitative empirical support. This framework has four dimensions.

- **Feelings of membership:** Feelings of belonging to, and identifying with, the community.
- **Feelings of influence:** Feelings of having influence on, and being influenced by, the community.
- **Integration and fulfillment of needs:** Feelings of being supported by others in the community while also supporting them.
- **Shared emotional connection:** Feelings of relationships, shared history, and a "spirit" of community.

Recently, SOC has been gaining attention in a variety of groups and organizations including work organizations (Burroughs & Eby, 1998; Chavis & Pretty, 1999; Chipuer & Pretty, 1999; Clark, 2002; Obst & White, 2005), schools (Bess, Fisher, Sonn, & Bishop, 2002; Chipuer & Pretty; Royal & Rossi, 1999), and membership groups (Harris, 1999; Zaff & Devlin, 1998). SOC has even been reported in online groups (Blanchard & Markus, 2003; Foster, 2004; Koh & Kim, 2003; Roberts, Smith, & Pollock, 2002; Rodgers & Chen, 2005). In work organizations, SOC has been linked to positive outcomes including increased job satisfaction and organizational citizenship behavior

(Burroughs & Eby, 1998), and less work-family conflict (Clark, 2002).

SOC may be particularly relevant for organizations and groups using Listservs to disperse information and encourage participation. Certainly, there is a need for these organizations, particularly the membership organizations in which participants are not formally tied to the organization, to strengthen their participants' affiliation.

What contributes to organizational sense of community? Burroughs and Eby (1998) tested a framework of the antecedents and outcomes of their organizational SOC measure, which was based on McMillan and Chavis' (1986) measure. They hypothesized that employees' need for affiliation and tenure, the size of the work group, the number of friends one has in the group, transactional contracts (e.g., use of organizational programs and services), and relational contracts (i.e., employees' organizational commitment and the organizations' commitment to the employees) would lead to SOC. However, only relational contracts made a significant positive contribution.

Other researchers used their own SOC measures and tested factors contributing to them. Clark (2002) found that supportive supervision and the intrinsic value of the work lead to her measure of SOC. Royal and Rossi (1999) found that organizational variables (perceived orderliness of students and support for innovation) and time-related variables (employee tenure and time spent interacting with others) led to SOC (as they defined it) in a school. Schuster (1998) examined the processes of exchanging support that led to SOC (as she defined it) in a writers' group in an assisted care home for the elderly. García, Giuliani, and Wiesenfeld (1999) concluded that the community's history was an important factor in creating SOC. Zaff and Devlin (1998) found that the amount of interaction between community members and components of the physical environment led to SOC.

The findings from these previous studies can be grouped into two broad categories. The first is history or knowledge about the community. In a very complicated model, McMillan and Chavis (1986) predicted that knowledge about the community and its history is integral to SOC. Garcia et al. (1999) examined this relationship and found that knowledge about the community contributes to SOC. Thus, knowledge and information about the organization should increase SOC.

Can computer-mediated communication, then, increase knowledge? As Listservs are a communication medium, it seems quite reasonable that Listserv participants will increase their information about the sponsoring group or organization. Sproull and Kiesler (1991) have even shown that employees increased their attachment to the organization through participating in group e-mails like Listservs. This increased attachment was even more pronounced for employees who had been marginalized within the organization by working second shift (Huff, Sproull, & Kiesler, 1989). By reading informational e-mails sent to groups, members may increase their knowledge and information of the organization (Huff, Sproull, & Kiesler, 1989). Recent research has shown that information distributed online had a stronger effect on communication and participation than either elevation or print media (Nah, Veenstra, & Shah, 2006).

Two questions arise. About whom exactly does knowledge increase? With whom does SOC potentially increase? We argue that it is the organizational entity itself that sponsors the Listserv and subsequently sends out relevant information about the entity and its members. The sponsoring organizational entity could be the entire organization, an organizational department (e.g., a student's academic major), or a membership or social club. Indeed, equal access to information such as that dispersed through an organizational Listserv increases members identification with a target entity (Connaughton & Daly, 2004). Therefore, the first two hypotheses for this study are the following.

Hypothesis 1: Listserv membership is positively related to knowledge about the sponsoring organization.

Hypothesis 2: Knowledge about the sponsoring organization is positively related to SOC.

The second broad category of antecedents of SOC is the interaction, usually FtF, with other members of the community (Burroughs & Eby, 1998; Royal & Rossi, 1999; Zaff & Devlin, 1998). Clearly, members have to interact with each other to develop an SOC. There are quite likely intervening mechanisms between interacting and the development of SOC, but few researchers have addressed this. Possibilities include the exchange of support (Clark, 2002; Schuster, 1998) and organizational commitment (Burroughs & Eby, 1998).

Can the Listserv also increase FtF interaction between organizational members? If the sponsoring organization has opportunities for members to interact FtF, then potentially, yes. An information-dispersing Listserv can publicize involvement opportunities for its members. Examples include advertising about the company picnic, schools providing information about extracurricular activities, or membership organizations providing information about networking meetings. Indeed, recent research has found that members of a neighborhood Listserv are likely to know more people in their FtF communities (Mesch & Levanon, 2004). Thus, the next two hypotheses are as follows.

Hypothesis 3: Listserv membership is positively related to FtF activity in the sponsoring organization.

Hypothesis 4: FtF activity is positively related to SOC.

Although the effect of Listserv membership on increasing information and FtF activity and its subsequent effect on SOC are important, ultimately, what drives this intervention is being able to connect Listserv membership to increasing SOC. Therefore, we expect there to be a mediating relationship of information and FtF activity between Listserv members and SOC (see Figure 1). The final hypothesis addresses this issue.

Hypothesis 5: The relationship between Listserv membership and sense of community will be mediated by knowledge about the organization and FtF activity in the organizational department.

Figure 1. Proposed model of how a Listserv affects sense of community

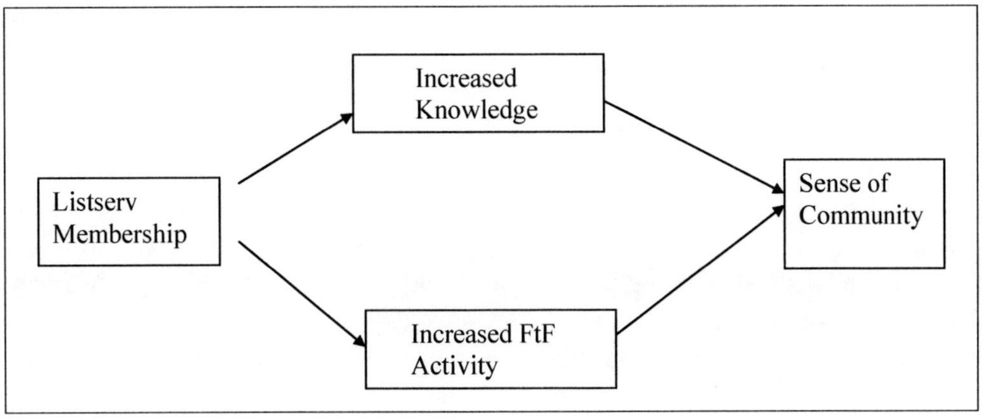

The next section describes the organizational context in which the study occurred.

Study Context

The Listserv used for this study was an informational Listserv implemented for the students of a large, urban southern California university. This Listserv was primarily developed to provide the primarily commuter student population with information about academic departments. Thus, in this study the sponsoring organization is the students' academic departments.

The communication on this Listserv tended to be one way; that is, professors and student leaders posted messages about administrative information, upcoming social events, and employment opportunities. The Listserv was used to advertise jobs, internships, research assistant positions, advisement information, club activities, and other information relevant for the students. Messages were posted at least once a week. Few discussions occurred on this Listserv. It can be described as an information-sharing organizational Listserv.

METHOD

Participants

Participants were advanced undergraduate students (i.e., juniors and seniors) of two academic departments within a large, urban university in southern California. The psychology (i.e., the test) department implemented a Listserv for its students while the sociology (i.e., the control) department did not. These two departments were chosen for comparison because of their similar size within the university and their similar social science orientation.

Data were collected at the beginning (before implementation of the Listserv) and the end of the academic school year (after implementation). During the first data collection, students were recruited during class periods with 327 psychology students and 239 sociology students participating. Although students could choose not to complete the surveys without penalty, nearly all of the students who were offered the survey participated. Thus, the difference in responses for the two departments reflects the difference in sizes between the two departments.

After the first round of data collection, participants in the psychology department were told about a new group e-mail system (i.e., the Listserv) that would provide useful information to them about upcoming events and news from the department. They were also instructed on how to join the Listserv.

During the second data collection, surveys were mailed to students' homes using addresses the students provided during the first round of data collection. Data were matched back to the first data collection resulting in a total of 266 usable surveys, a 47% response rate from the first data collection. The attrition between the first and second data collection is largely due to collecting data within the classroom in the first round and sending the surveys to participants' home addresses in the second. Even with a follow-up postcard, many participants did not return the second survey.

At this point, it was determined that not all of the psychology students who had the opportunity joined the Listserv. This meant that instead of two groups of participants (psychology test group and sociology control group), there were three groups of participants: psychology students who chose to join the Listserv (N=66), psychology students who did not choose to join the Listserv (N=112), and sociology students who did not get the opportunity to join the Listserv (N=88).

The demographic characteristics of the participants are similar across all three groups including age (mean=26.7, sd=8.0), years at the university (mean=3.58, sd=.54), days spent on campus (mean=3.56, sd=1.22), and minutes spent commuting (mean=31.31, sd=20.07). However, there were fewer women among the sociology participants (69%) as compared to the psychology students who were members of the Listserv (82%) and those who were not (89%).

Measures

Knowledge. Knowledge was measured by listing four statements about how much the students knew about the department. These four statements were "I generally know about things going on in this department," "I know about research opportunities," "I know about changes in classes," and "I know about upcoming student events in the department." These topics were typical of the information posted on the Listserv. Participants responded on a scale of 1 (*strongly disagree*) to 6 (*strongly agree*). A reliability analysis on these variables in the first data collection period produced a coefficient alpha of 0.82. The variables were combined into a scale of knowledge.

FtF activity. Face-to-face activity was measured by asking students about their membership in departmental student organizations (e.g., the Psi Chi psychology honors club), their participation in research and internship activities (e.g., a professor's research group), and their attendance of meetings within the department (see Appendix A). Responses ranged from 0 (*not a member*) to 5 (*very active member*) for the student organizations, and 0 (*none*) to 5 (5+) for the number of meetings attended. A reliability analysis on this scale in the first data collection period produced a coefficient alpha of 0.70, which is low but acceptable. The variables were combined into a scale of active participation.

Sense of community. Sense of community was measured using a modified version of the Sense of Community Index (Chipuer & Pretty, 1999). This index has been used in many organizations including educational organizations. Statements were modified to reflect participants' potential affiliation with their academic departments. Sample items include "I think this is a good department for me to be a student," "I feel at home in this department," and "It is important for me to be a student in this department." Responses ranged from 1 (*strongly disagree*) to 6 (*strongly agree*). A reliability analysis was also performed on this scale in the first data collection period, producing a coefficient alpha of 0.64.

This is a low reliability score, but it is considered typical of the Sense of Community Index (Chipuer & Pretty, 1999). This measure is widely considered to have a strong theoretical basis (Chipuer & Pretty; Long & Perkins, 2003; Obst, Smith, & Zinkieqicz, 2002; Obst & White, 2004). Several SOC researchers have conducted confirmatory factor analysis of this scale, which has yielded somewhat contradictory results (e.g., Long & Perkins, 2003; Obst & White, 2004). These researchers have found that although they can improve upon the overall Sense of Community Index, with which this study's low reliability score concurs, their subsequent exploratory factor analyses propose quite different subfactors of sense of community. Chipuer and Pretty argue that until these measurement issues are resolved, researchers should continue to use the combined Sense of Community Index so that comparisons can be made across studies. Therefore, the variables were combined into an overall SOC score.

RESULTS

Descriptive analyses of the data are presented in Tables 1 and 2.

Data were checked for analysis following the guidelines of Tabachnik and Fidell (2001). Because the FtF activity scale was highly skewed, it was transformed using an inverse function. This decreased problems with skewness and kurtosis so that it would not violate assumptions of normality and could be used in linear regression analyses. A consequence is that for this transformed scale, lower numbers represent more actual FtF activ-

Table 1. Descriptive analyses of study variables

	Preimplementation			Postimplementation		
	Psych With List	**Psych No List**	**Sociology**	**Psych With List**	**Psych No List**	**Sociology**
Knowledge	2.93 (.89)	2.93 (1.00)	2.94 (1.02)	3.67 (.79)	3.18 (1.17)	3.17 (1.00)
Inv. Activity	0.92 (.20)	0.94 (.16)	0.95 (.17)	0.80 (.26)	0.87 (.22)	0.95 (.16)
SOC	4.04 (.42)	4.01 (.42)	4.12 (.45)	4.26 (.58)	4.18 (.66)	4.22 (.59)

Note: There were 66 psychology Listserv members, 112 psychology nonmembers, and 88 sociology students. Inverse (inv.) activity is the inverse of the activity scale to deal with problems of skewness and kurtosis.

Table 2. Correlations between study variables

	1	2	3	4	5
1. Pre-Knowledge	(.82)				
2. Pre-Inv. Active	-0.19**	(.70)			
3. Pre-SOC	0.45***	-0.16**	(.64)		
4. Post-Knowledge	0.43***	0.00	0.33***		
5. Post-Inv. Active	-0.19*	0.61***	-0.09	-0.11	
6. Post-SOC	0.38**	-0.02	0.54***	0.59**	-0.15*

Note: Pre means preimplementation of the Listserv. Post means postimplementation of the Listserv. The numbers on the diagonal are the reliability coefficients from the first round of data collection. * $p<.05$, ** $p<.001$, *** $p<.001$

ity. Table 2 shows the correlations between the research variables.

A potential problem of this study's research design is its quasi-experimental nature; participants could not be randomly assigned to academic departments and thus preexisting differences could account for any changes that occur after Listserv implementation. To test for preexisting differences, a one-way ANOVA (analysis of variance) was conducted comparing the study's measures before Listserv implementation for the sociology students, psychology Listserv members, and psychology nonmembers. None of the preimplementation measures reported in Table 1 were significantly different between the three groups. Additionally, because gender was different for the sociology department as compared to the psychology department, further analyses were conducted to determine if gender was related to the study variables: It was not.

To test the hypotheses, we conducted path analyses using the following logic. First, we collected measures of all the variables of interest in both time periods. We expected that measures from the first data collection (e.g., SOC 1) would be reliable predictors of the same measures in the second round of data collection (e.g., SOC 2). If our hypotheses are supported, then the hypothesized variable would have a reliable relationship above and beyond the measures from the first data collection. We assessed whether the relationship was above and beyond by examining the beta weights from regression equations. Additionally, because we have three groups to compare (sociology, psychology without the list, and psychology with the list), we used contrasts to test the effects of using the Listserv. The two groups we contrasted were (a) the psychology Listserv members against the psychology nonmembers, and (b) the psychology Listserv members against the sociology students. Finally, the number of years at the university was used as a control variable. This is considered conceptually similar to the tenure data found to be related to SOC in previous research (Royal & Rossi, 1999), but was not an object of focus for this study.

The hypothesis testing strategy included (a) testing that the Listserv increased knowledge and FtF interaction as expected (Hypotheses 1 and 3), (b) testing that knowledge and FtF interaction increase SOC as expected (Hypotheses 2 and 4), and finally (c) that knowledge and FtF interaction mediate the relationship between Listserv membership and SOC.

To test Hypothesis 1, stating that the Listserv increased knowledge about the sponsoring organization, we examined whether Listserv membership increased knowledge above and beyond the original measure of knowledge. The results of this analysis show that Listserv membership is positively related to knowledge for the psychology Listserv members compared to the psychology non-Listserv members (β=.12, $p<.05$), and for the psychology members compared to sociology members (β=.12, $p<.05$; see Table 3). Therefore, Hypothesis 1 is supported. Participation in the Listserv reliably explained participants' knowledge about the department above and beyond their previous knowledge about the department.

To test Hypothesis 3, which said that the Listserv increased FtF activity in the sponsoring organization, we examined whether participation in the Listserv increased FtF activity above and beyond the original measure of FtF activity. The results in Table 4 show that although Listserv membership significantly explains FtF activity above and beyond previous FtF activity as compared to the sociology students (β=-.24, $p<.001$), it does not show these results when compared to the other psychology students (β=-.00, p=.99). Therefore, Hypothesis 3 is partially supported.

To test Hypotheses 2 and 4, which state that knowledge and FtF activity increase SOC, we conducted one hierarchical regression model.

Table 3. Testing Hypothesis 1: Listserv increases knowledge

Variable	B	SE B	β
Step 1			
Academic Year	.11	.12	.06
Step 2			
Academic Year	.00	.11	.00
Pre-Knowledge	.46	.08	.42**
Step 3			
Academic Year	.00	.11	.02
Pre-Knowledge	.45	.06	.42**
Psych List to No List	.15	.08	.12*
Psych List to Sociology	.17	.09	.12*

*Note: R^2 = .00 for Step 1. ΔR^2 = .17 for Step 2 ($p<.001$). ΔR^2 = .04 for Step 3 ($p<.001$). *$p<.05$, **$p<.001$*

Table 4. Testing Hypothesis 3: Listserv increases FtF activity

Variable	B	SE B	β
Step 1			
Academic Year	-.00	.03	-.13*
Step 2			
Academic Year	.00	.02	.01
Pre-Inv. Activity	.77	.07	.61**
Step 3			
Academic Year	.00	.02	-.03
Pre–Inv. Activity	.74	.07	.58**
Psych List to No List	-.00	.02	-.00
Psych List to Sociology	-.00	.02	-.24**

*Note: R^2 = .02 for Step 1. ΔR^2 = .35 for Step 2 ($p<.001$). ΔR^2 = .06 for Step 3 ($p<.001$). ** $p<.001$*

Table 5 shows the results of testing the effects of knowledge and FtF activity on SOC. Knowledge, FtF activity, and SOC from before Listserv implementation are entered in Step 2 as baseline measures. In Step 3, knowledge and FtF activity after implementation are entered to determine if they explain SOC beyond baseline. From the results, we can conclude that SOC at the beginning of the study strongly predicts SOC at the end of the study (β=.40, p<.001). However, both knowledge (β=.35, p<.001) and FtF activity (β=-.12, p=.06) at the end of the study are also related to SOC, although knowledge has a much stronger relationship than FtF activity. A total of 45% of

Table 5. Testing Hypotheses 2 and 4: Knowledge and FtF activity increase SOC

Variable	B	SE B	β
Step 1			
Academic Year	-.00	.07	-.03
Step 2			
Academic Year	-.01	.06	-.07
Pre-Knowledge	.11	.04	.19**
Pre-Activity	.15	.19	.05
Pre-SOC	.56	.07	.47***
Step 3			
Academic Year	-.01	.06	-.08
Pre-Knowledge	.00	.04	.07
Pre-Activity	.29	.22	.08
Pre-SOC	.48	.07	.40***
Post-Knowledge	.20	.03	.35***
Post-Activity	-.30	.17	-.12'

*Note: R^2 = .00 for Step 1. ΔR^2 = .33 for Step 2 ($p<.001$). ΔR^2 = .12 for Step 3 ($p<.001$). ' p=.06, * p<.05, ** p<.10, *** p<.001*

the variance of SOC is explained by this model as indicated by the adjusted values for R^2. Thus, Hypotheses 2 and 4 are supported.

To test Hypothesis 5, regarding knowledge and FtF interaction mediating the relationship between Listserv membership and SOC, we used Baron and Kenny's (1986) mediation testing strategy. This strategy includes testing that (a) Listserv membership is related to knowledge and FtF interaction, (b) Listserv membership is related to SOC, and (c) when knowledge and FtF interaction are entered in a hierarchical regression after Listserv membership, these two variables decrease the relationship of Listserv membership to SOC. Because Hypotheses 1 and 3 complete the first step of this strategy, the next step is to test that Listserv membership increased SOC above and beyond the original measure of SOC. The results of this analysis show that Listserv membership did not increase SOC (see Table 6). Therefore, Hypothesis 5 is not supported.

Because there is no direct relationship between Listserv membership and SOC, we cannot test Hypothesis 5 stating that this relationship is mediated by knowledge and FtF activity (Baron & Kenny, 1986). There can be no indirect (i.e., mediated) relationship if there is no direct one.

Exploratory Analysis

The results of these analyses are perplexing. Listserv membership increases knowledge about the sponsoring organization and FtF activity within it. Knowledge and FtF activity increase SOC. The model appears to hold; however, because there is no direct link between Listserv membership and SOC, the indirect effects of Listserv membership on SOC approach zero. Therefore, the parts of knowledge and FtF activity that are related to SOC are independent of those related to Listserv membership.

Why is there no effect of the Listserv on SOC? Reflecting upon the findings, we conclude that Listserv membership increases participants' amount of knowledge. Members of the Listserv have access to more information simply by being members of the Listserv. This is verified in the data by increased knowledge for the Listserv group compared to both its internal and external control groups. Thus, the Listserv increased how much people knew about what was going on in the department.

The data also support the conclusion that information is related to SOC. That is, as people know more about the department (i.e., they have

Table 6. Testing Hypothesis 5: Listserv increases SOC

Variable	B	SE B	β
Step 1			
Academic Year	.00	.07	-.02
Step 2			
Academic Year	-.01	.06	-.06
Pre-SOC	.67	.06	.55***
Step 3			
Academic Year	-.01	.14	.00
Pre-SOC	.67	.07	.55***
Psych List to No List	.00	.04	.02
Psych List to Sociology	.00	.05	.04

*Note: R^2 = .00 for Step 1. ΔR^2 = .30 for Step 2 ($p<.001$). ΔR^2 = .00 for Step 3 ($p=.70$). * $p<.05$, ** $p<.001$*

Table 7. Exploratory analysis regressing sense of community and Listserv membership on knowledge

Variable	B	SE B	β
Step 1			
Pre-Knowledge	.38	.07	.35***
Pre-SOC	.36	14	.17**
Step 2			
Pre-Knowledge	.30	.06	.27***
Pre-SOC	.00	.14	-.04
Post-SOC	.78	.11	.45***
Step 3			
Pre-Knowledge	.30	.06	.27***
Pre-SOC	-.00	.14	-.03
Post-SOC	.75	.11	.44***
Psych List to No List	.13	.07	.10$^{\tau}$
Psych List to Sociology	.17	.08	.12*

*Note: R^2 = .20 for Step 1 ($p<.001$). ΔR^2 = .14 for Step 2 ($p<.001$). ΔR^2 = .03 for Step 3 ($p<.001$). $^{\tau}p$=.06, * $p<.05$, ** $p<.01$, *** $p<.001$*

Table 8. Exploratory analysis regressing sense of community and Listserv on FtF activities

Variable	B	SE B	β
Step 1			
Pre-Activities	.77	.07	.61***
Pre-SOC	.00	.03	.00
Step 2			
Pre-Activities	.78	.07	.62***
Pre-SOC	.00	.03	.11$^{\tau}$
Post-SOC	-.01	.01	-.19***
Step 3			
Pre-Activities	.76	.06	.60***
Pre-SOC	.00	.02	.08
Post-SOC	-.01	.02	-.17**
Psych List to No List	.00	.02	.01
Psych List to Sociology	-.01	.02	-.23***

*Note: R^2 = .37 for Step 1 ($p<.001$). ΔR^2 = .02 for Step 2 ($p<.01$). ΔR^2 = .05 for Step 3 ($p<.001$). $^{\tau}<.10$, * $p<.05$, ** $p<.01$, *** $p<.001$*

more objective knowledge), they have a greater SOC. However, because Listserv membership (with more objective knowledge) did not have a direct effect on SOC, we questioned the direction of this relationship. Instead, we considered whether SOC could lead to a perception of more knowledge. That is, people with a strong SOC believe they know more about what is going on in the department apart from any objective increases in knowledge.

Therefore, we questioned whether the direction of the relationship between SOC and information in the original model is correct. We regressed the relationship of Listserv membership and SOC on knowledge. Table 7 presents the results of this exploratory analysis.

These analyses show that, indeed, Listserv participation and SOC are both independently related to the participants' knowledge about the department. SOC is strongly related (β=.44, $p<.001$) while Listserv membership is more moderately related; knowledge is higher for the Listserv members compared to psychology nonmembers of the Listserv (β=.10, p=.06) and the sociology students (β=.12, $p<.05$).

We also questioned the direction of the relationship between SOC and FtF activity. We regressed SOC along with Listserv membership on FtF activity. Table 8 presents the results of this analysis. While FtF activities in the first time period have the strongest relationship to FtF activities at the end of the study (β=.60, $p<.001$), SOC (β=-.17, $p<.001$) and Listserv membership for the psychology students as compared to the sociology students (β=-.23, $p<.01$) are also strongly related to FtF activity at the end of the study.

The two exploratory analyses challenge our research model. In particular, the data are better explained by using information and FtF activity as criteria instead of predictor variables. We propose, then, that the direction of the relationship between information with FtF activity and SOC was incorrect. Knowledge and FtF activity are not mediators of the relationship between Listserv membership and SOC. Instead, knowledge and FtF activity are outcomes of Listserv participation and SOC. Model 2 revises the original model to reflect the results.

Figure 2. Revised proposed model

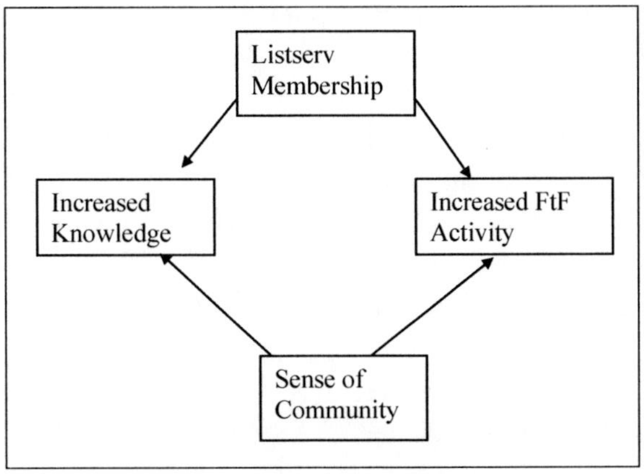

DISCUSSION

The purpose of this research was to examine the effects of Listserv implementation on the sponsoring organization's members. It was expected that the Listserv would increase participants' knowledge about as well as their FtF activity in the Listserv's sponsoring organization, in this case, an academic department. It was then expected that knowledge and FtF activity would increase the members' SOC, thus mediating the relationship between the Listserv and the members' SOC.

The results show that although the Listserv is related to participants' knowledge and FtF activity, and that subsequently knowledge and FtF activity are related to participants' SOC, there was no relationship between the Listserv and sense of community. Instead and ironically, the Listserv and SOC together independently affect knowledge and FtF activity. A new model is proposed that reverses the relationship between SOC and FtF activity and knowledge.

This model challenges much of the current research about the antecedents of SOC. Previously, research has argued that knowledge and FtF activity increase SOC (Burroughs & Eby, 1998; García et al., 1999; Royal & Rossi, 1999; Zaff & Devlin, 1998). However, much of this research has been conducted qualitatively or through cross-sectional survey research. Thus, it is possible that SOC researchers have confused the direction of this correlational relationship.

Although the Listserv did not increase SOC as anticipated, it did affect knowledge about and potentially FtF activity in the sponsoring organization, which are in and of themselves worthy outcomes. Thus, the results of this study concur with other research demonstrating the effects of participation in electronic media on increased information and FtF interaction. Therefore, increased knowledge and FtF activity may be important goals for organizations who want to keep their members informed and active. We argue that the Listserv potentially affected FtF activity because the psychology Listserv members had higher FtF activity than the sociology students but not more than the psychology nonmembers. It is possible that the psychology department simply became more active without relation to the Listserv. However, it is also possible that the Listserv's ability to advertise about upcoming events and activities increased participation among members as well as nonmembers as they discussed during regular student interactions.

These findings are especially encouraging because the Listserv, as an electronic collaboration technology, itself was quite simple. It was not interactive in that members did not hold discussions on it. Instead, the Listserv consisted of one-way communications about opportunities and events. A more interactive Listserv might elicit more of an affective response from its members (e.g., Koh & Kim, 2003; Rodgers & Chen, 2005; Weis et al., 2003). However, interactive Listservs are harder to develop and maintain due to the time requirements on their members. Additionally, the focus of the effects in a more interactive Listserv may be directed toward the Listserv and not the sponsoring organization (Liao, Troth, & Griffith, 2002).

That the Listserv increases FtF activity is particularly intriguing. An information-dispersing Listserv is a very passive medium for its members. That it was able to increase FtF interactivity for the department in which the Listserv was implemented is encouraging. It also points out where organizations may be able to increase the organizational attachment of their members. For instance, if the FtF interactions foster the development of relationships and exchange of support, which were not measured in this study, then the interactions could help develop SOC as predicted by the model. Thus, FtF interactions and SOC could be part of a reinforcing cycle in which the Listserv could draw in new members for FtF interactions. Although not examined in this study, increased information and FtF activity may affect other desirable organizational outcomes such as organizational commitment or involvement. Future research should consider what other desirable outcomes for dispersed organizational members may be affected by Listservs.

Finally, this research can be used as a lesson for organizations who wish to increase their members' affiliation through an information-dispersing Listserv only. It may not be the best use of their time, resources, and money to use a Listserv solely for increasing their members' affiliation. Even though it can increase their members' knowledge, it will not directly affect their affiliation with the organization. Additional or different media may be required.

Limitations and strengths. Participants in this study were not randomly assigned to be members of the Listserv. Even though there were no significant differences between the groups at the start of the study, participants were self-selected into Listserv membership. Clearly, the members of the Listserv are different from the nonmembers because they had the opportunity and chose to join the Listserv. An alternative explanation of the findings is that Listserv joiners are more likely to be more curious about and involved in the organization. Thus, there could be some personality trait that explains their changes in knowledge and FtF activity.

However, the members who joined this Listserv may, in fact, be similar to members of other Listservs in which members must choose to join. Most Listservs require that the members join them instead of being assigned to them. Additional research should address this issue by examining both randomly assigned and self-selected Listserv members.

Additionally, the student population may limit some of the study's generalizability. For example, students and traditional employees are likely to differ in their feelings of affiliation with their work groups, departments, and overarching organizations. For employees, work groups and organizations may be the primary focus of their affiliation outside the home, whereas these commuter students may have less developed affiliation because they do not spend as much time on campus as employees do at an organization. This research may be most generalizable to nontraditional employees such as telecommuters who rely more on collaboration technologies to stay informed or to membership groups such as voluntary social or professional organizations who do not interact with each other on a daily basis. As with all research, findings must be replicated in different populations to strengthen their conclusions.

A strength of this research, however, is that it controls for levels of the variables of interest before and after implementation of the Listserv. We are in a better position to assess the relationship between SOC, and information and FtF activity after the Listserv implementation because we have a better understanding of what occurred before the Listserv.

CONCLUSION

This research examined how implementing a Listserv affects sense of community for the sponsoring organization's members: It did not. Instead, it demonstrated that a Listserv can increase knowledge about and face-to-face activity within an organization; interactivity and knowledge do not directly increase feelings of community.

What then are the broader implications for this study? Managers, school administrators, and membership group coordinators can and should use their Listservs to share information and promote FtF interactions. This is a simple use of an easy collaboration technology that can take advantage of the active involvement of few to reward the passive involvement of many.

However, we should not expect that an information-distributing Listserv alone will promote feelings of affiliation for the sponsoring organization. Instead, at this time, research suggests that affiliation will increase during exchanges of information and particularly support (Clark, 2002; Schuster, 1998). The question remains as to whether these exchanges need to occur primarily face to face or if exchanging information and support through collaborative technologies will create feelings of affiliation (like those suggested by virtual communities) to the sponsoring organization. Managers and researchers who wish to increase affiliation for their members will need to pursue these questions further.

Electronic collaboration technologies vary drastically in how much members interact within them and how much effort is required for them to be successful. Information-dispersing Listservs have a place in the arsenal of communication technologies as a very simple tool to keep members informed about and, potentially, active in their organizations.

REFERENCES

Adams-Price, C. E., & Chandler, S. (2000). The star fleet ladies auxiliary: Evolution of an online women's mailing list. *CyberPsychology and Behavior, 3*(5), 811-816.

Baron, R. M., & Kenny, D. A. (1986). The moderator-mediator variable distinction in social psychological research: Conceptual, strategic and statistical considerations. *Journal of Personality and Social Psychology, 51*, 1173-1182.

Bess, K. D., Fisher, A. T., Sonn, C. C., & Bishop, B. J. (2002). Psychological sense of community: Theory, research, and application. In A. T. Fisher, C. C. Sonn, & B. J. Bishop (Eds.), *Psychological sense of community: Research, applications and implications.* New York: Kluwer Academic/Plenun Publishers.

Blanchard, A., & Markus, M. L. (2003). The experienced "sense" of a virtual community: Characteristics and processes. *The Data Base for Advances in Information Systems.*

Burroughs, S. M., & Eby, L. T. (1998). Psychological sense of community at work: A measurement system and explanatory framework. *Journal of Community Psychology, 26*, 509-532.

Chavis, D. M., & Pretty, G. H. (1999). Sense of community: Advances in measurement and application. *Journal of Community Psychology, 27*(6), 635-642.

Chipuer, H. M., & Pretty, G. H. (1999). A review of the Sense of Community Index: Current uses, factor structure, reliability and further development. *Journal of Community Psychology, 27*, 643-658.

Clark, S. C. (2002). Employees' sense of community, sense of control, and work/family conflict in Native American organizations. *Journal of Vocational Behavior, 61*, 92-108.

Connaughton, S. L., & Daly, J. A. (2004). Identification with leader: A comparison of perceptions of identification among geographically dispersed and co-located teams. *Corporate Communications, 9*(2), 89-103.

Foster, P. M. (2004). Psychological sense of community in groups on the Internet. *Behaviour Change, 21*(2), 141-146.

García, I., Giuliani, F., & Wiesenfeld, E. (1999). Community and sense of community: The case of an urban barrio in Caracas. *Journal of Community Psychology, 27*, 727-740.

Harris, J. (1999). The idea of community in the study of writing. In L. Ede (Ed.), *On writing research: The Braddock essays 1975-1998* (pp. 260-271). Boston: Bedford.

Huff, C. W., Sproull, L., & Kiesler, S. (1989). Electronic communication and organizational commitment: Tracing the relationship in a city government. *Journal of applied social psychology, 19*, 1371-1391.

Koh, J., & Kim, Y.-G. (2003). Sense of virtual community: A conceptual framework and empirical validation. *International Journal of Electronic Commerce, 8*(2), 75.

Liao Troth, M. A., & Griffith, T. L. (2002). Software, shareware and freeware: Multiplex commitment to an electronic social exchange system. *Journal of Organizational Behavior, 23*(5), 635-653.

Long, D. A., & Perkins, D. D. (2003). Confirmatory factory analysis of the Sense of Community Index and development of a brief SCI. *Journal of Community Psychology, 31*(3), 279-296.

McMillan, D. W., & Chavis, D. M. (1986). Sense of community: A definition and theory. *Journal of Community Psychology, 14*, 6-23.

Meier, A. (2000). Offering social support via the Internet: A case study of an online support group for social workers. *Journal of Technology in Human Services, 17*(2-3), 237-266.

Mesch, G. S., & Levanon, Y. (2004). Community networking and locally-based social ties in two suburban localities. *City & Community, 2*(4), 335-351.

Nah, S., Veenstra, A. S., & Shah, D. V. (2006). The Internet and anti-war activism: A case study of information, expression, and action. *Journal of Computer Mediated Communication, 12*(1).

Obst, P., & White, K. M. (2005). An exploration of the interplay between psychological sense of community, social identification and salience. *Journal of Community & Applied Social Psychology, 15*, 127-135.

Obst, P. L., Smith, S., & Zinkieqicz, L. (2002). An exploration of sense of community: Part 3. Dimensions and predictors of psychological sense of community in geographical communities. *Journal of Community Psychology, 30*(1), 119-133.

Obst, P. L., & White, K. M. (2004). Revisiting the Sense of Community Index: A confirmatory factor analysis. *Journal of Community Psychology, 32*(6), 691-705.

Roberts, L. D., Smith, L. M., & Pollock, C. M. (2002). MOOing till the cows come home: The sense of community in virtual environments. In C. C. Sonn (Ed.), *Psychological sense of community: Research, applications, implications*. New York: Kluwer Academic/Plenum.

Rodgers, S., & Chen, Q. (2005). Internet community group participation: Psychosocial benefits for women with breast cancer. *Journal of Computer Mediated Communication, 10*(4).

Royal, M. A., & Rossi, R. J. (1999). Predictors of within-school differences in teachers' sense of community. *Journal of Educational Research, 92*, 259-267.

Sarason, S. B. (1974). *The psychological sense of community: Prospects for a Community Psychology*. San Francisco: Jossey-Bass.

Schuster, E. (1998). A community bound by words: Reflections on a nursing home writing group. *Journal of Aging Studies, 12*(2), 137-148.

Sproull, L., & Kiesler, S. (1991). *Connections: New ways of working in the networked organization*. Cambridge, MA: MIT Press.

Tabachnik, B. G., & Fidell, L. S. (2001). *Using multivariate statistics* (4th ed.). Needham Heights, MA: Allyn & Bacon.

Walther, J. B. (1996). Computer mediated communication: Impersonal, interpersonal and hyperpersonal interaction. *Communication Research, 22*, 33-43.

Weis, R., Stamm, K., Smith, C., Nilan, M., Clark, F., Weis, J., et al. (2003). Communities of care and caring: The case of MSWatch.com®. *Journal of Health Psychology, 8*(1), 135-148.

Zaff, J., & Devlin, S. (1998). Sense of community in housing for the elderly. *Journal of Community Psychology, 26*, 381-398.

APPENDIX A

Are you a member of the following groups in the psychology/sociology* department? If so, please indicate your level of activity in these groups:

PSI CHI/PSSA*

Other (Please Indicate): _______________

How many semesters have you been supervised in an internship by a psychology/sociology* professor?

How many semesters have you participated in a research activity or project with a professor outside of class?

How many meetings or activities sponsored by the psychology faculty or staff have you attended?

How many meetings or activities sponsored by the psychology students have you attended?

Note: This part of the survey was tailored to each department and either contained psychology or sociology as well as the specific departmental clubs Psi Chi or PSSA.

Chapter V
Agile IT Outsourcing

Boris Roussev
University of the Virgin Islands, USA

ABSTRACT

Agile methods are lightweight, iterative software development frameworks used predominantly on small and mid-sized software development projects. This chapter introduces a project structure and management practices creating agile conditions for large software projects outsourced either offshore or onshore. Agility is achieved by slicing a large project into a number of small projects working in agile settings. Development is divided into research and development activities that are located on-site, and production activities located off-site. The proposed approach makes agile methods applicable to the stressed conditions of outsourcing without compromising the quality or pace of the software development effort. Creating an agile environment in an outsourcing project relies on maintaining a balance between the functions and sizes of on-site and off-site teams, on redefining the developers' roles, and on reorganizing the information flow between the different development activities to compensate for the lack of customers on-site, team colocation, and tacit project knowledge.

INTRODUCTION

We live in a digital world where any activity not requiring a physical presence can be outsourced to any place that is connected to the Internet. Even the term *physical presence* comes with a qualification. Information and communication technologies enable cooperation in a distributed mode. Technologies, such as groupware and videoconferencing, are increasingly becoming feasible for organizations to use in distributed projects.

Advances in ICT have been essential for loosening the spatial constraints on software development. The largely digital nature of software development allows changing its geography of provision. The combination of low labor costs, technological sophistication, satisfactory (but not outstanding) project management skills, and successful software establishment makes south

Asia a particularly attractive location for software production outsourcing (Dossani & Kenney, 2003). However, the decision to outsource software production is not a simple matter of upside (Natovich, 2003). Product and project managers have been looking for ways to mitigate risks in outsourcing projects to maximize reward (Craumer, 2002). Even though the total contract value of IT outsourcing transactions continues to increase (Hale, Souza, Lo, Adachi, & Babaie, 2005; Technology Partners International [TPI], 2005), many large organizations are bringing key IT projects back in house because outsourcing contracts have failed to meet expectations (Huber, 2005).

Agile methods (Agile Alliance, 2001; Beck, 1999; Schuh, 2005; Sridhar, Mahapatra, & Mangalaraj, 2005) are popular software development processes designed for use on small to mid-sized software projects. They are based on the notion that object-oriented software development is not a rigidly defined process, but an empirical one that may or may not be repeated with the same success under changed circumstances.

Agile methods are based on four critical values: simplicity, communication, feedback, and courage, informing a set of key practices (Pollice, 2004), which will be considered in more detail later. Boehm and Turner (2004, p. 27) define agile methods as "very light-weight processes that employ short iterative cycles; actively involve users to establish, prioritize, and verify requirements; and rely on tacit knowledge within a team as opposed to documentation." However, many of the agile practices seem to be incompatible with the context of outsourcing, especially when outsourcing offshore; for example, agile practices work better when there are customers on-site, team colocation, short life cycles, and the value of embracing change. Above all, agile methods are applicable mainly to small and mid-sized projects because of many of the characteristics of agile development. For example, smaller, colocated teams require fewer lines of communication and less coordination efforts, can use less formal development processes, can reach an agreement faster, and can adapt to change more rapidly (Schuh, 2005; Stephens & Rosenberg, 2003).

In the past, there have been several attempts to reproduce the conditions for agility in large projects. To the best of our knowledge, no such attempt has been made for outsourcing projects. Pollice (2001a, 2001b), Evans (2004), and Boehm and Turner (2004) propose to scale up agile methods by balancing agility and discipline. Pollice and Evans, for instance, look for common ground between agile and RUP (Jacobson, Booch, & Rumbaugh, 1999), while Boehm and Turner try to get the best of both agile and plan-driven (waterfall) worlds. In contrast, Kruchten (2004) proposes to scale down large projects to meet the agile "sweet spot" based on experience reported in Brownsword and Clements (1996) and Toth, Kruchten, and Paine (1993). The sweet spot, or in other words the ideal agile context, is characterized by a small team of developers sharing common values, team colocation, customers on-site, and a short life cycle.

In this work, we show how to reengineer large outsourcing projects to benefit from the sweet spot of agile methods while avoiding its "bitter spot." The proposed approach makes agile methods applicable to the stressed condition of both offshore and onshore outsourcing environments without compromising the quality of the software development effort. Creating a context congenial to agile methods hinges on balancing the functions and sizes of on-site and off-site teams, on redefining the developers' roles, and on reorganizing the information flow between the different development activities.

The rest of the chapter is structured as follows. Next it examines some critical issues in outsourcing software development activities. Then it presents the structure of the agile outsourcing project. The chapter then elaborates on the inception and architectural activities in agile outsourcing, which are crucial to slicing large projects into multiple, relatively independent agile projects. Next it discusses some of the challenges likely to be experienced by outsourcing suppliers when applying the proposed approach, and finally it concludes.

OUTSOURCING SOFTWARE DEVELOPMENT ACTIVITIES

What can be Outsourced?

When the software development process is considered in its totality, it appears to resist relocation because software production requires face-to-face interactivity with clients (and among codevelopers), for example, during user requirements elicitation and testing. The development workflow has to be parsed into activities requiring different levels of interactivity, and the less client-communication-intensive activities can be potentially outsourced.

The software complexity chain is frequently described as a series of processes, each of which is less complex than the earlier one (Akella & Dossani, 2001). The initial processes are less labor intensive but more complex than the later ones. The complexity of a process is roughly inverse proportional to its labor intensity. The pyramid model in Figure 1 gradates the processes in terms of labor, complexity, risk, and communication intensity.

Theoretically, outsourcing is possible for all the levels of the complexity pyramid, but there are important differences of how outsourcing for each level is likely to be implemented. The processes at or closer to the pyramid's apex, such as domain knowledge acquisition, architecture design, and technology determination, are objectively more difficult to outsource (King, 2005). Moving up the complexity pyramid entails establishing an intimate client-supplier rapport and acquiring knowledge about the client's core and critical activities.

IT client activities can be divided into core and noncore on the one hand and critical and noncritical on the other hand. Core activities are services that differentiate a firm from its competitors and they are key to its continued growth. Noncore activities are the firm's noncompetitive services, such as payroll and human resources. Within both core and noncore activities, there are activities crucial to their functioning, termed critical activities. The higher the pyramid level is, the more communication intensive the activities become, and the bigger the demand grows for domain, architectural, and design knowledge. Adopting agile methods can help outsourcing suppliers move up the value chain because agile methods address the issues enumerated above.

Figure 1. Activities and pyramid of labor

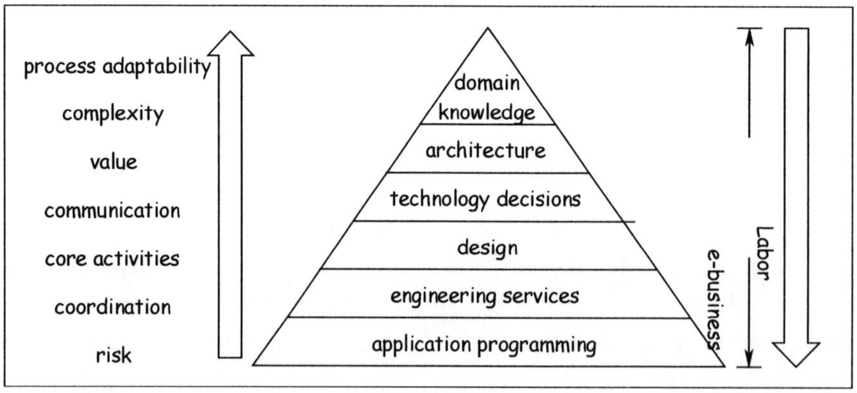

Interactivity

Interactivity always comes across as a stumbling block for outsourcing (King, 2005). Interactivity has two dimensions: interaction among codevelopers and interaction with clients. Requirements elicitation and acceptance testing are the activities requiring the most active involvement on the part of the client, which makes them impossible to outsource.

The greater the need of codevelopers to interact across a set of different activities of the software process, the higher the risk threshold of outsourcing a subset of the activities is. Outsourcing the entire set of activities might be considered as a way of retaining interactivity at the new location. However, if some activities cannot be outsourced because that would disrupt the interaction with the client, then outsourcing the others might need careful consideration.

Rethinking of Earlier Cost-Benefit Decisions

The lower cost of highly skilled personnel calls for the rethinking of established cost-benefit decisions. For example, the much lower cost of software engineers in India compared to the United States (Dossani & Kenney, 2003; Outsourcing Institute, 2004) makes it feasible to increase the density of software inspections (Gilb & Graham, 1993) or to dispel managers' doubts about the feasibility of pair programming (CeBASE, 2004). Other activities that are candidates for reconsideration are regression and integration testing, iteration planning, and metrics for tracking project progress. In all of the above-mentioned practices, lower labor costs can change the break-even point, and thus create new sources of revenue.

Risks in IT Outsourcing

A recent study based on the managerial assessment of overall risk in in-house IT development projects identifies six critical risk factors: (a) the lack of customer involvement, (b) requirements volatility, (c) ill-fitting software methods, (d) the lack of formal project management practices, (e) dissimilarity to previous projects, and (f) project complexity (Laplante & Neill, 2004; Tiwana & Keil, 2006). In addition, there are some unique risks at play for clients of outsourcing projects related to controlling vendor involvement (Choudhury & Sabherwal, 2003). These unique risks can be divided into intractable risks and unforeseen risks (Taylor, 2006). Intractable risks are those risks that resist mitigation actions despite managers' best efforts to address them. Unforeseen risks appear unlikely to affect project outcomes, even if a project involves active risk assessment, but they have significant impact in later development stages.

The most common intractable risk is overoptimistic schedules and budgets, or in other words, schedule and budget management. The most likely unforeseen risk is inflated client expectations, or trust. Understanding and managing overoptimistic schedules and budgets, and client trust are critical for both clients and vendors.

The best strategy to avoid underestimation of budgets and schedules is to do costly presale requirements analysis. However, even if carried out, the results of such analysis can be compromised by the presale team's desire to secure the project by placing a winning bid, which is characteristic of both small and large companies (Jiang, Klein, & Discenza, 2002). A more effective risk mitigation approach is to factor out the presale requirements specification activity into a stand-alone, chargeable consulting activity (Taylor, 2006).

In general, iterative and incremental software methods employ project scoping to assess the functionality that must be delivered, the resources that must be allocated to deliver this functionality, and the time to implement the software system (Kruchten, 2000). Typically, both in-house and outsourced software projects are overscoped. According to the Standish Group (2003), 53% of all projects cost 189% of their estimates. The project scope is the underpinning for project planning. The project plan creates the breakdown work structure, defines iterations, sets targets as major and minor milestones, allocates resources, and creates incentives.

Agile methods, in particular, do not rely on a detailed project plan. Their assumption is that as the project unfolds, some features will evolve while other features will be dropped altogether because they do not bring the expected value, the external environment has changed, or the customers' understanding of these features has evolved. In other words, agile methods do not overpromise unrealistic schedules and budgets.

Client trust and satisfying client expectations are extremely important to vendors as they depend on client referrals and references. One of the main factors for the success of agile methods is the improved communication line between clients and developers. Agile methods rely on a shared vocabulary and improved communication. This results in improved traceability between client needs and resultant software artifacts. In agile methods, the goal of the development team is to become part of the customer's discourse, or context. From a communications standpoint, agile software development can be viewed as a discursive communication model in which the vendor responds to client requests with increments—pieces of executable functionality produced in one iteration (Roussev & Rousseva, 2006). The customer tests each increment, after which the communication cycle repeats. In the testing subprocess, or the ensuing discourse between client and vendor, the customer may become aware of a compelling necessity to redefine needs. The process of constant reassessing of client needs decreases the extent of the client's disconfirmation of expectations, thus leading to improved customer satisfaction.

Problems Experienced by Outsourcing Suppliers

Levina and Ross' (2003) research on outsourcing vendors' value proposition indicates that cost, quality, and project management skills are the three most important attributes of value to clients. Success in IT development outsourcing projects is predicated on factors related to vendors' core competencies (Prahalad & Hamel, 1990), complementarities of the core competencies (Milgrom & Roberts, 1990), and the following capabilities of the development team: technical knowledge, domain expertise, and management skills, including client management (Akella & Dossani, 2001; Levina & Ross, 2003). Offshore developers face problems in acquiring advanced domain knowledge as they are geographically separated from their clients' business environments and more often than not have a different background and culture (Dossani & Kenney, 2003; Outsourcing Institute, 2004). The lack of advanced domain knowledge can lead to misunderstanding of client expectations and misinterpretation of client needs (Kruchten, 2000). The latter may be further compounded by the suppliers' business interaction, client coordination and communication style, and the shortage of project management expertise, including planning, scheduling, and progress monitoring (Akella & Dossani). The end result is inadequate quality.

Quality in outsourcing contexts should be considered on a much larger scale than merely in terms of defects, usability, maintainability, reliability, and performance (Akella & Dossani, 2001). Outsourcing clients define quality of work as the possession of advanced domain knowledge, adherence to schedules, high coordination between client and supplier, adaptability, and mature software processes, for example, SEI/CMM (Paulk, Curtis, Chrissis, & Weber, 1993) and ISO 9001 (International Organization for Standardization) certifications, coupled with transparency (i.e., observability) and outcome measurability (Kirsch, Sambamurthy, Ko, & Purvis, 2002). As a consequence, clients attribute project failure mostly to poor coordination, communication, and scheduling. In addition, clients expect suppliers to be very responsive and to push beyond the stated goals by engaging actively in identifying new problems and opportunities, and offering innovative solutions. Clients anticipate outsourcing suppliers to become part of their teams and to advise them on which components and activities to outsource and which ones to keep in house. The major impediment to satisfying these client expectations is the low value-added levels of activities (see Figure 1) on which most offshore software firms operate. Typically, offshore vendors undertake projects in the areas of applications development, testing, and maintenance, while the high-end work, such as developing the IT strategy, building the software architecture, designing the system, integrating the project with enterprise packages, and designing custom components, are all discharged by firms in developed countries (Akella & Dossani, 2001; R. Akella & R. Dossani, personal communication, 2004).

Another issue with outsourcing suppliers is employee turnover. The rapid growth of outsourcing suppliers at popular offshore locations has created a dynamic labor market with a big demand for lead developers for new projects. As a consequence, turnover levels can be extremely high, which could have an adverse impact on product quality as later hires may not have adequate qualifications or time to make the transition to the new development environment (technology, software process, and domain expertise; Mezak, 2005).

In an environment where communication lines with clients and end users suffer by time difference, cultural issues, language barriers, business practices, and lack of advanced domain knowledge, project management skills and coordination with clients become even more critical to project success. However, management at offshore locations is often characterized by culture-bound, hierarchical administrative styles instead of the more efficient and beneficial leadership through role modeling (Akella & Dossani, 2001).

To sum up, the major causes for outsourcing failure can be attributed to the lack of advanced domain knowledge, poor communication and coordination with clients, immature management and scheduling practices, insufficient experience in higher-value-added activities (e.g., architecture design), and last but not least, inadequate quality. There is also the danger of losing touch with the local and national environments.

The problems experienced by outsourcing suppliers can be objective or subjective. For instance, it is objectively difficult to communicate effectively with a remote client. The lack of expertise in project scheduling is, on the other hand, subjective because it is not universally true. More often than not, a lot more could be done to alleviate subjective problems rather than objective ones.

Based on the above analysis, we can divide the problems experienced by outsourcing suppliers into three categories: (a) communication problems, involving the relationship and coordination with remote clients, (b) problems due to less technical know-how, such as a lack of domain expertise, experience in building software architectures, design experience, and quality thereof, and (c) management problems, involving project controlling and scheduling. In order to deliver successful software products, outsourcing supplier firms need to address the communication, technical know-how, and management issues listed above.

THE AGILE OUTSOURCING PROJECT

In this section, we present an agile method of software development geared toward the context of outsourcing.

Core Agile Practices

Agile methods are iterative and incremental. In an iterative process, the activities that were performed earlier in the process are revisited later. Revisiting activities provides developers in areas such as requirements engineering and software architecture with the opportunity to correct mistakes they have made. The iterative model implies more interactions among the various management and technical groups. It is a means of carrying out exploratory studies early on in a project when the team develops the requirements and discovers a suitable architecture.

Some important agile practices are the following (Beck, 1999).

- **Embracing change:** Respond to change quickly to create more value for the customer.
- **Fast cycles:** Deliver small releases frequently, implement the highest priority functions first, speed up requirements elicitation, and integrate daily.
- **Simple design:** Strive for a lean design and restructure (refactor) the design to improve communication and flexibility or to remove duplication, while at the same time preserve the design's behavior.
- **Pair programming:** To quote from Beck (1999), "if code reviews are good, we'll review code all the time." With pair programming, two programmers collaborate side by side at one machine. This quality-control practice also helps disseminate tacit **knowledge among the team members.**
- **Test-driven development:** This is a quality-control technique where a developer first writes a test, and then writes the code to pass that test.
- **Tacit knowledge:** There is a preference for project knowledge in team members' heads rather than in documents.
- **Customers on-site:** Continuous access to a customer helps to resolve ambiguities, set priorities, establish scope and boundary conditions, and provide test scenarios.
- **Colocation:** Developers and on-site customers work together in a common room to enhance tacit project knowledge and deepen members' expertise.
- **Retrospection:** This is a postiteration review of the work done and work planned. This reflective activity facilitates learning and helps improve the estimates for the rest of the project.

The enumerated practices (or disciplines) can be divided into three broad categories: (a) communication, for example, in pair programming and sharing tacit knowledge, (b) management, for example, in planning, scrums (short, daily planning sessions in which the whole team takes part), short cycles, and frequent delivery, and (c) technical, for example, in test-driven design, simple design, and refactoring. These categories correspond to the three groups of problems commonly experienced by outsourcing suppliers, as discussed previously.

The fast cycle of agile methods gives business stakeholders multiple and timely feedback opportunities, which makes agile methods explorative (aiding in architecture discovery and requirements verification) and adaptive to change. There is a broad consensus that agile methods perform well on small and mid-sized projects in dynamic environments (Pollice, 2001a, 2001b), for example, in e-commerce. What is seen as problematic with agile methods is that they lack formal project management practices, such as formal and sufficiently detailed plans, schedules, budgets, and milestones. Tiwana and Keil's (2004) study strongly suggests that formal project management practices have the power to reduce project risk substantially.

The feasibility-impact grid in Figure 2 shows how we map the core agile practices to the outsourcing suppliers' concerns discussed earlier.

Agile practices incompatible with outsourcing are highlighted. A checkmark in a cell indicates that the agile discipline listed in this cell's row heading can alleviate the concern listed in this cell's column heading.

One concern, not listed in Figure 2, is employee turnover. Agility could break the impact of high turnover as several core practices, for example, preference for tacit knowledge, pair programming, and retrospection, make it possible for new team members to be brought quickly up to speed.

Structuring the Agile Outsourcing Project

The main question we address below is how to reproduce the conditions ideal for agility in an outsourcing project.

Even a quick look at the core agile practices above would reveal that some of them are incompatible while others are not entirely consistent with the context of outsourcing, for example, having customers on-site, colocation, short life cycles, and embracing change. Above all, agile methods can be applied only to small projects.

Kruchten (2004) proposes to scale down large projects to meet the agile sweet spot. The author describes the organization of a large project as an evolving structure, starting out as one, colocated team, which over time is transformed to a team of teams (with a tree-like structure). Kruchten also suggests organizing the iterative process into four typical RUP phases as shown in Figure 3 (Kruchten, 2000). Each phase shifts the focus of the development effort onto a different activity. A phase consists of one or more iterations, where iterations can be thought of as mini waterfalls.

The structure of the agile outsourcing project is based in Kruchten's approach. However, in an agile outsourcing project, the primary goal is not to slice a big project into multiple agile subprojects (which might be required anyway), but to outsource the project to one or more agile teams, which are colocated at a remote site and share common culture and educational background.

The structure of the development process is illustrated in Figure 4. The phases of the life cycle

Figure 2. Impact of agile practices on outsourcing suppliers' concerns

Practices\Concerns	Domain expertise	Architecture	Design	Scheduling	Communication	Coordination	Quality
Customer onsite	√				√		
Fast cycle		√	√		√		√
Embracing change		√		√			√
Test-driven development		√					√
Refactoring			√	√			
Simple design			√				√
Retrospection		√			√	√	√
Risk mitigation				√			
Tacit knowledge			√		√	√	
Co-location			√		√	√	
Planning game				√		√	
Scrums				√		√	
Pair-programming			√				√
Immediate customer feedback	√	√			√		√

Figure 3. Typical RUP phases of the iterative process

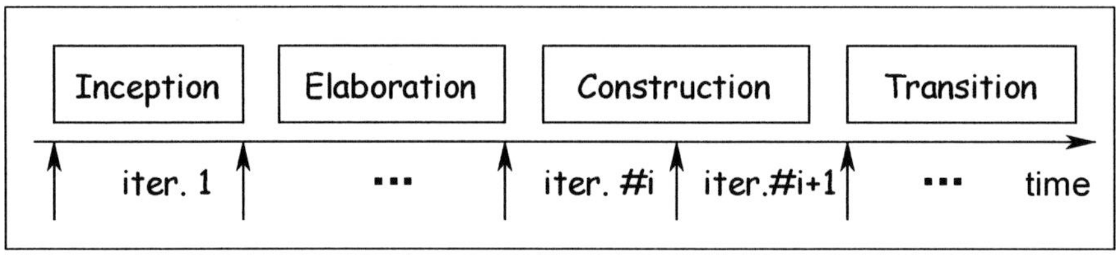

Figure 4. Team structure of agile outsourcing

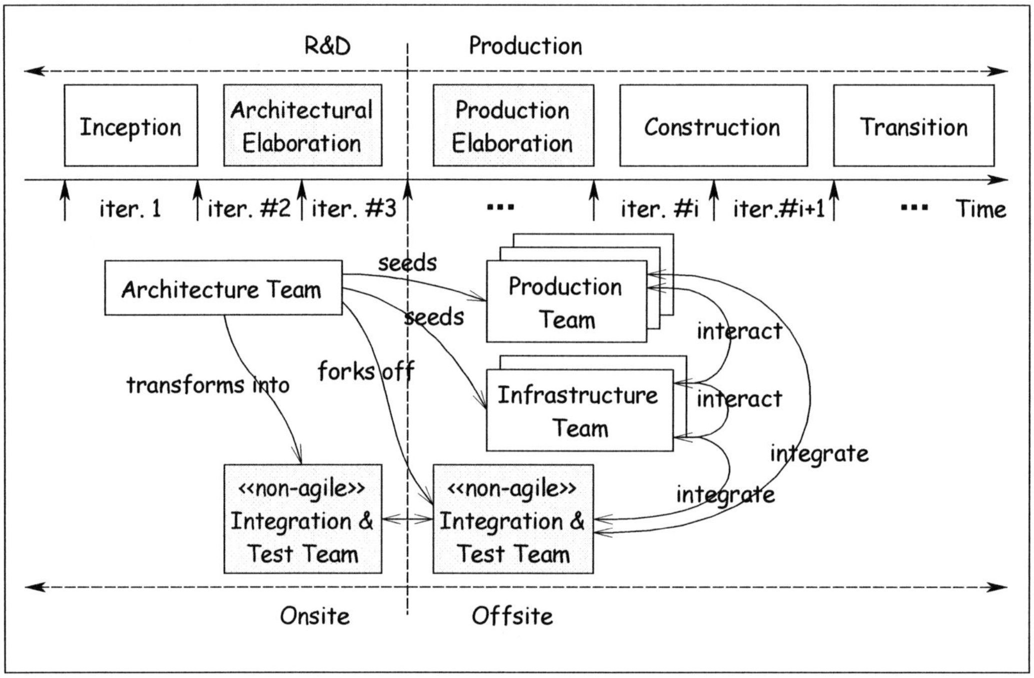

are separated between research and development (R&D) activities and production activities, as suggested in Royce (2002). R&D is carried out on-site, that is, close to the client, while production takes place off-site. Elaboration is split between R&D and production. The two new phases are called architectural elaboration and production elaboration.

A team comprised of requirements engineers and architects starts the development process. The team is a mix of developers from the client's organization and from the outsourcing supplier. Initially, this team focuses on the business case, vision document, and system requirements. The team works closely with the client in a typical agile setting. The team's objective, however, is not to deliver executable code (a must in agile methods), but to set up an architectural prototype, including prioritized system-level use cases.

When the team has enough clarity on key architectural choices, it can set about building and testing an architectural prototype. Toward the end of the architectural elaboration phase, when the architecture stabilizes, additional teams are created off-site. Each new team is seeded preferably with two members of the architecture team. For large projects, the early architectural prototype is used to slice the project into smaller, considerably independent agile projects. Each agile project owns part of the system architecture.

The seed developers transfer domain expertise from the client to the supplier's site. Some of them are employees and managers from the client's organization. The seed developers take on several roles. They lead the teams, serve as local architects, act as customer surrogates, communicate with the initial architecture team, and, if necessary, communicate directly with the client. This organization creates near-perfect conditions for agility in each off-site team.

Each agile off-site team works on a subsystem and develops its detailed subsystem use-case model, hence the second elaboration phase, named production elaboration in Figure 4.

For small projects, one off-site production team will suffice to develop the complete code. For large projects, however, several production teams might be necessary to do the job. In addition, one or more infrastructure teams must be set up to develop common elements such as middleware, services, and any reusable assets. The customers for the infrastructure teams are all other teams. Thus, no matter how many teams are spawned off-site, they all work in agile conditions even though predominantly with customer surrogates.

It is important to use a common software tool to reduce the risk of miscommunication, track and communicate project requirements, identify replicated modules, map test cases to requirements, and successfully integrate the outputs from the agile teams into builds.

The context diagrams in Figure 5 model the environments in which the different off-site teams operate. Note the dual nature of seed developers.

Figure 5. Context diagrams for off-site teams

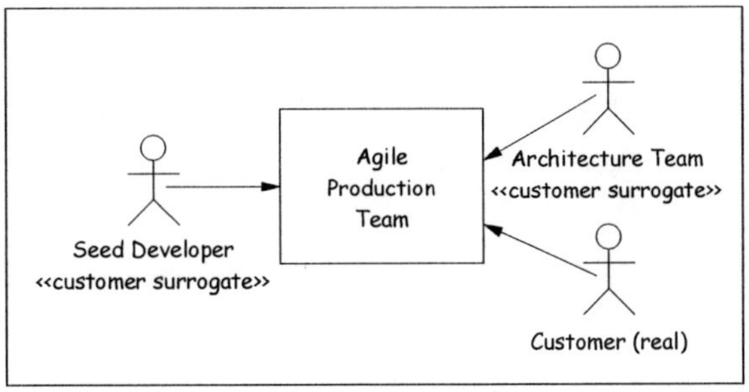

(a) Production team

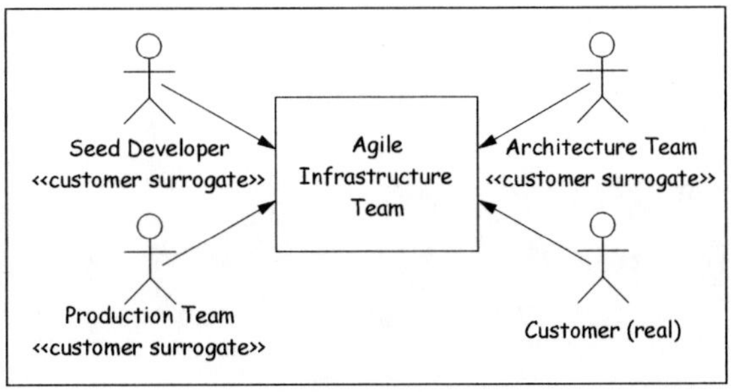

(b) Infrastructure team

On the one hand, a seed developer impersonates a customer, and on the other hand, he or she is part of the team.

The interactions of off-site teams with the real customer are supposed to be infrequent and mediated by the seed developers and the architecture team.

For large projects, somebody needs to put together the builds delivered by the production and infrastructure teams, and test the assembled system. This job is assigned to the integration and test team, or integration team for short. The testing engages the client so that the client's feedback can situate the project and steer the future effort.

The problem with the integration team is that it gets input from the off-site teams but receives feedback from the client. Normally, the input and the feedback are coming from the same place, that is, the customer. To account for this anomaly, we split the integration team into two teams: one located on-site and the other off-site (see Figure 4). Both integration teams derive directly from the initial architecture team. This guarantees that the developers in charge of integration and testing thoroughly understand the client's needs.

A problem with the proposed organization is that of fault propagation. For example, a defect committed by one of the production or infrastructure teams can propagate through the integration process before the client detects it. Owing to the late stage of its detection, isolating and removing such a defect is expensive.

We have provisioned for two floodgates preventing fault propagation: (a) the seed developers in each off-site team and (b) the test engineers in the integration team. Since they all come from the primary architecture team, it is very likely that they would be able to detect many defects, which are normally revealed with help from customers.

The only two teams operating in a nonagile environment, but still in an iterative and incremental mode, are the on-site and off-site integration teams.

The iterations (heartbeats) of both the agile and integration teams can be best illustrated with the

Figure 6. Nested iterations

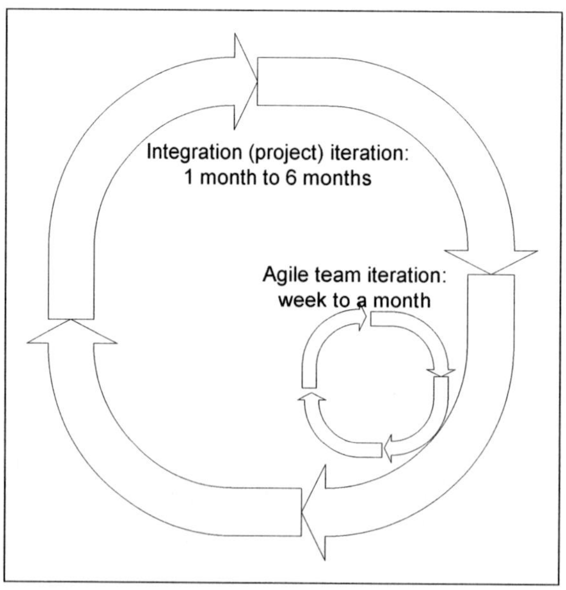

dual-beat structure shown in Figure 6. Striking a balance between the lengths of the agile and integration iterations is the underlying objective of the management team.

The management team, comprised of all local team managers, is led by the managers of the integration teams. We do not show the management team as a separate box in Figure 4 because management is thinly distributed across all teams.

Since all off-site teams reside in the same country, and most probably in the same city, there are no cultural differences to overcome, and communications among production, infrastructure, and integration teams are not as ineffective as they are with geographically distributed teams. The management team is in a favorable position because good communication is a prerequisite for effective coordination.

ACTIVITIES AND WORKFLOWS

In this section, we discuss the activities in an agile outsourcing project.

Inception Phase

During inception, the emphasis is on the user requirements. A user requirement is a specification of what the system must do. As user requirements are elicited, they are organized into use cases (Jacobson, 1987). Use-case descriptions can smoothly scale to large and small systems alike. They promote refinement and traceability from system-level usage goals down to low-level (subsystem, component, and instance) usage goals. Use cases are sufficiently flexible to be used in highly iterative and incremental development environments (Jacobson et al., 1999), as the one proposed in this work.

If a system consists of several subsystems, use-case analysis can be applied recursively to the subsystems as well. This defines clearly the requirements and responsibilities of each subsystem. Subsystem-level use cases are derived from the system-level use cases and the system architecture (the architectural decomposition of the system into subsystems).

Architectural Elaboration

The on-site team carries out the architectural elaboration. The goals of the architecture team are to partition the system into multiple semantic domains centered on different subject matters, to define the architectural decomposition of the system into subsystems, and to map system-level use cases to subsystem use cases.

Domain Modeling

A domain is a subject area with a shared vocabulary (Mellor, Kendall, Uhl, & Weise, 2004), for example, as in a user interface (UI) or payment transaction management. Each domain contains many classes organized around a single subject matter. Most domains require specialized expertise, for example, experience and knowledge in UI design or in payment transaction management. It makes sense to allocate the modeling of a domain to a developer with domain knowledge in that particular subject area.

Since a domain model captures precisely the conceptual entities of a single subject matter, it can be said that domain models are logical models. Logical models are in sharp contrast to subsystem models, which are pieces of the physical system. Typically, a physical subsystem is constructed from instances of several logical models. For example, a collaboration realizing a system-level use case would involve instances from a UI domain, a business logic domain, a transaction management domain, a persistent storage domain, and a security domain. Domain modeling leverages scarce domain knowledge normally limited to very few team members and shields the rest of the team from the domain implementation detail. For example, to make an object persistent, a developer needs only to mark its class or one of its attributes as persistent, and a persistent software entity is automatically generated at compile time using the knowledge locked in the persistent storage domain model. The result is a simplified development process, where only a few developers need detailed knowledge about domain technicalities.

Domains, unlike objects, are not elemental, but just like objects they are cohesive. The classes and components in a domain are tightly coupled and interdependent, and yet the domain is autonomous; that is, its classes and components are decoupled from entities lying outside the domain boundary. Once constructed, domain models have greater longevity than an application because they evolve independently of other domain models out of which the application is built; in other words, they become corporate assets and the biggest units of reuse.

In UML (unified modeling language), developers represent the containment hierarchy of a system through an aggregation of subsystems. A subsystem is defined as a subordinate system within a larger system. The subsystems define the large-scale physical architecture of the system.

Model Organization

For small systems, the subsystem structure can be organized by use cases. The system model is divided into subsystems of related use cases. This model organization is straightforward and allows tracing requirements easily to model elements. The downside of the use-case-based model organization is that it does not scale up well and encumbers reuse. Developers are forced to reinvent similar classes in collaborations realizing different use cases.

For large systems, we propose to derive the subsystem structure from the requirements model and the domain model.

A UML package is a container for modeling elements (Jacobson et al., 1999). It is an organizational unit defining a name space for the modeling elements it contains. The top-level packages of the domain-based system model are (a) a system use-cases package, (b) domains package, (c) infrastructure package, (d) subsystems package, and (e) builds package. The system use-cases package contains system-level use cases and their actors. The domain package has one subpackage for each domain. The infrastructure domain is a special type of domain. Infrastructure domains contain services and extension points for system communication and infrastructure, and they are dependent on the selected implementation technology, for example, RMI/J2EE or CORBA. The infrastructure package contains one subpackage for each infrastructure domain. Each infrastructure subpackage is assigned to an infrastructure team. Several production teams may use the classes and components realizing the services of an infrastructure subpackage. The subsystem package contains one package for each subsystem, that is, the classes and components of a subsystem. A large subsystem package can be recursively divided into subpackages until the size of the leaf packages becomes manageable. A subsystem package normally refers to the classes of several domain packages. Finally, the builds package, containing the system test model, is divided into subpackages, one per prototype, to allow for easy management of incremental builds.

The domain-based model organization scales up well to large projects for three main reasons. First, subsystem package decomposition can be applied recursively, resulting in subsystems realizing subsystem use cases. Second, classes from different domains may be reused in different settings and deployments of the designed system, and third, domain models are assets that can be reused across projects.

There are two problems with the domain-based model organization. Developers tend to blur the distinction between domain models and subsystem models, and there is an overhead associated with maintaining two separate types of models: domain models (logical) and subsystem models (physical).

Production Elaboration

All strategic decisions for the overall organization structure of the system have been made in the architectural elaboration phase. Production elaboration drills down inside the subsystems to specify the semantic objects whose collaborations deliver the system's structure and behavior. In production elaboration, the subsystem use cases are detailed and mapped to collaborations of components and objects using communication diagrams (Jacobson et al., 1999).

Next we show how packages are assigned to teams; for example, the architecture team is in charge of developing the system use cases and therefore owns the system use-cases package.

CHALLENGES TO AGILE OUTSOURCING PROJECTS

Managers of agile outsourcing projects ought to be aware of the following challenges to project success.

The architecture team is the key to project success. It is of paramount importance that this team be a good mix of requirements analysts and architects. Members of the architecture team elicit requirements, build the primary system architecture, and subsequently take the roles of surrogate customers, local managers, and integration and test engineers, and communicate with customers throughout the entire life cycle.

Interteam dependencies reduce teams' abilities (especially affected is the infrastructure team) to test code and deliver builds. Production teams should provide the infrastructure team(s) with stubs early on so that the work of the infrastructure team(s) proceeds smoothly.

With multiple teams, there is always the danger of replicating functionality across teams. The proximity and common culture of the off-site teams work against the chances of duplicating functionality. The participation of the members of the architecture team, and later of the integration teams, in design reviews and software inspections helps detect replication early.

CONCLUSION

In the late 1990s and early 2000s, agile methods took the imagination of software developers by storm. This fact clearly indicates that heavyweight methods have not been embraced wholeheartedly by developers and managers alike and are found either impractical or costly (or both) in many environments. In this chapter, we introduced a novel project structure creating agile conditions for large outsourcing software projects. We showed how to slice a large software project into multiple agile projects. We proposed to separate the development activities into R&D activities, carried out on-site

Table 1.

Packages	Team
System use-cases package	Architecture team
Domains package	Architecture team or reuse
Infrastructure domain packages	Infrastructure teams
Subsystem packages	Production teams
Builds package	Integration and test teams

and close to the client, and production activities, carried out off-site and possibly offshore. The on-site architecture team starts work on the project. In cooperation with the client, the architecture team develops the high-level use-case model of the system and completes an architectural prototype with the strategic decisions for the overall system structure. The agile off-site teams are seeded with members of the architecture team and start functioning toward the end of the architectural elaboration. The seed developers transfer domain expertise to the supplier's site and act as customer surrogates to help reproduce agile conditions off-site. Outsourcing entire sets of related production activities retains interactivity at the off-site location and reduces the project risk significantly.

REFERENCES

Agile Alliance. (2001). *Agile Alliance manifesto.* Retrieved from http://www.aanpo.org

Akella, R., & Dossani, R. (2001). *IT outsourcing and the software value chain: The Indian supplier during the downturn.* Proceedings of the IT Conference, Bangalore, India.

Beck, K. (1999). *Extreme programming explained: Embrace change.* Boston: Addison-Wesley.

Boehm, B., & Turner, T. (2004). *Balancing agility with discipline.* Boston: Addison-Wesley.

Brownsword, L., & Clements, P. (1996). *A case study in successful product line development* (Tech. Rep. No. CMU/SEI-96-TR-035). Software Engineering Institute.

CeBASE. (2004). *eWorkshop on software inspections and pair programming.* Retrieved from http://www.cebase.org

Choudhury, V., & Sabherwal, R. (2003). Portfolios of control in outsourced software development projects. *Information Systems Research, 14*(3), 291-314.

Craumer, M. (2002). How to think strategically about outsourcing. *Harvard Management Update, 7*(5).

Dossani, R., & Kenney, M. (2003). *Went for cost, stayed for quality? Moving the back office to India.* Stanford University, Asia-Pacific Research Center. Retrieved from http://APARC.stanford.edu

Evans, G. (2005). Agile RUP: Taming the Rational™ unified process. In L. Liu & B. Roussev (Eds.), *Management of object-oriented software development.* Hershey, PA: Idea Group.

Gilb, T., & Graham, G. (1993). *Software inspection.* Boston: Addison-Wesley.

Hale, K., Souza, R., Lo, T., Adachi, Y., & Babaie, E. (2005). *Forecast: IT services, worldwide, 2005-2009.* Gartner. Retrieved June 2005 from http://www.gartner.com

Huber, N. (2005, May 3). Outsourcing "backlash" highlights need for IT leaders to sharpen management skills. *Computer Weekly*, pp. 24-27.

Jacobson, I. (1987). Object-oriented development in an industrial environment. *ACM SIGPLAN Notices, 22*(12), 183-191.

Jacobson, I., Booch, G., & Rumbaugh, J. (1999). *The unified software development process.* Boston: Addison-Wesley.

Jiang, J., Klein, G., & Discenza, R. (2002). Pre-project partnering impact on information systems projects, project team, and project manager. *European Journal of Information Systems, 11*(2), 86-97.

King, W. R. (2005). Outsourcing becomes more complex. *Information Systems Management, 22*(2), 89-90.

Kirsch, L. J., Sambamurthy, V., Ko, D., & Purvis, R. L. (2002). Controlling information systems development projects: The view from the client. *Management Science, 48*(4), 484-498.

Kruchten, P. (2000). *The rational unified process: An introduction.* Boston: Addison-Wesley.

Kruchten, P. (2004). Scaling down large projects to meet the agile "sweet spot." *The Rational Edge.* Retrieved from http://www.therationaledge.com

Laplante, P. A., & Neill, C. J. (2005). "The demise of the waterfall model is imminent" and other urban myths. *ACM Queue*, pp. 10-15.

Levina, N., & Ross, J. W. (2003). From the vendor's perspective: Exploring the value proposition in information technology outsourcing. *MIS Quarterly, 27*(3), 361-364.

Mellor, S. J., Kendall, S., Uhl, A., & Weise, D. (2004). *MDA distilled.* Boston: Addison-Wesley.

Mezak, S. (2005). *The seven deadly dangers of outsourced software development.* Retrieved April 2005 from http://www.accelerance.com

Natovich, J. (2003). Vendor related risks in IT development: A chronology of outsourced project failure. *Technology Analysis & Strategic Managements, 15*(4), 409-419.

Outsourcing Institute. (2005). Retrieved April 2005 from http://www.outsourcing.com

Paulk, M. C., Curtis, B., Chrissis, M. B., & Weber, C. V. (1993). Capability maturity model, version 1.1. *IEEE Software, 10*(4), 18-27.

Pollice, G. (2001a). RUP and XP: Part I. Finding common ground. *The Rational Edge.* Retrieved from http://www.therationaledge.com

Pollice, G. (2001b). RUP and XP: Part II. Valuing differences. *The Rational Edge.* Retrieved from http://www.therationaledge.com

Pollice, G. (2004). RUP and extreme programming: Complementing processes. In L. Liu & B. Roussev (Eds.), *Management of object-oriented software development.* Hershey, PA: Idea Group Publishing.

Roussev, B., & Akella, R. (n.d.). Agile outsourcing projects: Structure and management. *International Journal of e-Collaboration, 2*(4), 37-52.

Roussev, B., & Rousseva, Y. (2006). *Unreliable interlocutors in information systems development.* Proceedings of San Diego International Systems Conference, San Diego, CA.

Royce, W. (2002). The case for results-based software management. *The Rational Edge.* Retrieved from http://www.therationaledge.com

Russel, B., & Chattargee, S. (2003). Relationship quality: The undervalued dimension of software quality. *Communications of the ACM, 46*(8), 85-89.

Schuh, P. (2005). *Integrating agile development in the real world.* Hingham, MA: Charles River Media, Inc.

Sridhar, N., Mahapatra, R., & Mangalaraj, G. (2005). Challenges of migrating to agile methodologies. *Communications of the ACM, 48*(5).

Standish Group. (2003). *CHAOS research report.* Retrieved from http://www.standishgroup.com

Stephens, M., & Rosenberg, D. (2003). *Extreme programming refactored: The case against XP.* San Francisco: Apress.

Taylor, H. (2006). Critical risks in outsourced projects: The intractable and the unforeseen. *Communications of the ACM, 49*(11).

Technology Partners International (TPI). (2005). *Technology Partners International Inc. quarterly index.* Retrieved from http://www.tpi.net

Tiwana, A., & Keil, M. (2004). The one-minute risk assessment tool. *Communications of the ACM, 47*(11), 73-78.

Tiwana, A., & Keil, M. (2006). Functionality risk in software development. *IEEE Transactions on Engineering Management, 53*(3), 412-425.

Toth, K., Kruchten, P., & Paine, T. (1993). *Modernizing air traffic control through modern software methods.* Proceedings of the 38th Annual Air Traffic Control Association Conference, Nashville, TN.

Section II
Distributed Project Management

Chapter VI
Using Computer-Mediated Groups to Improve University Processes:
An Action Research Study in New Zealand

Ned Kock
Texas A&M International University, USA

James Corner
University of Waikato, New Zealand

ABSTRACT

We describe in this chapter an action research study of a computer-mediated business process redesign (BPR) group in a New Zealand university. The BPR group used an integrated BPR framework, which comprises a group process methodology, called MetaProi, and an asynchronous groupware tool. BPR group members were from two different departments and successfully redesigned two course-related processes. The study reveals some possible effects of computer mediation on groups that are particularly relevant for managers of distributed BPR projects, namely, lower demand for leadership skills, much lower overall running cost, and much lower degree of interaction. No impact on group effectiveness was observed. The study also indicates that computer mediation lowers barriers to and, in turn, fosters interdepartmental communication, which creates a suitable context for the occurrence of other BPR groups involving different departments. On the other hand, the study indicates that those groups lead to more threats to management, an effect that may lead to lack of support from managers for future BPR groups. Finally, the study suggests that strategic BPR groups, as opposed to those dealing with local operational issues, can better benefit from computer mediation when this is combined with face-to-face and other types of vocal interaction.

INTRODUCTON

Business process redesign (BPR) has been the basis of several world-class organizational development approaches (French & Bell, 1990; Kock, 2003, 2005, 2006). Common characteristics of these approaches are their high impact on shaping management thinking and their controversial history of success and failure. Also, whether we consider the approaches to increasing domestic and international competitiveness of organizations in the economic Japanese revolution (Deming, 1986; Walton, 1989) or the business process reengineering movement (Hammer, 1990; Kock, 1999, 2005), there seems to be a clear focus on business processes.

There is a large growing body of normative frameworks to redesign business processes (Guha, Kettinger, & Teng, 1993; Harrington, 1991; Kock, 1995, 2006; Kock & Murphy, 2001; Tapscott & Carston, 1993). Those frameworks suggest that the groups that typically carry out BPR share some common characteristics (Kock, 2001a). The groups are typically small, having from 3 to 12 members (Soles, 1994); follow a group process or methodology (King, 1990); and have defined roles (Hammer & Champy, 1993).

The literature on groupware support for groups indicates the potential of groupware technology to augment the efficiency and effectiveness of BPR groups (Brothers et al., 1992; Chidambaram & Kautz, 1993; Clement, 1994; Kock, 2001a, 2005; Nunamaker, Dennis, Valacich, Vogel, & George, 1991; Pietro, 1992; Sheffield & Gallupe, 1993; Sproull & Kiesler, 1991; Wilson, 1991). This potential accrues from some effects observed in computer-mediated groups in the past, such as the following.

- Better support for group activities, such as making communication faster and cheaper, reducing paper flow, recording group discussion data in a more efficient way, and increasing cross-departmental communication.
- Positive effects on individual behavior, such as reducing participation stress, and making individuals communicate more openly.
- Positive effects on group behavior, such as fostering a more balanced distribution of individual contributions, separating ideas from individuals, reducing repetition of old ideas, and increasing commitment toward group decisions.

The limited amount of empirical research on computer-mediated BPR groups so far, however, contrasts with the potential of groupware technology to support these groups (Kock, 2005; Kock & Davison, 2003). There have been some representative examples of empirical research on groupware-supported BPR or BPR-like groups, such as the study by Pietro (1992) of quality improvement groups, the study by Dennis, Daniels, Hayes, and Nunamaker (1993) of one business process reengineering group, the study by Dennis, Hayes, and Daniels (1994) on business process modeling groups, and the more recent study by Dennis, Carte, and Kelly (2003) of business process reengineering groups. These studies, however, have focused largely on synchronous groupware tools, predominantly group decision support systems (GDSS). Very few studies addressed the impact of asynchronous groupware-supported BPR groups, most notably those conducted by Kock (2001a, 2005), and Kock and Davison.

In this chapter we try to contribute to filling the research gap above with an action research study of an asynchronous computer-mediated BPR group conducted at SCHOOL (pseudonym), a school of studies at the University of Waikato in New Zealand. This BPR group is an initial step toward the implementation of a campus-wide computer-supported BPR project, and had the first author of this chapter as its facilitator.

The chapter is set up as follows. Initially, a discussion of the BPR framework used is conducted with emphasis on its two main elements: a BPR group methodology, MetaProi, and a simple asynchronous computer mediation tool. A description of the research study follows, using the action research cycle (Jonsonn, 1991; Kock, 2007; Peters & Robinson, 1984) proposed by Susman and Evered (1978) as a structural framework. This action research cycle comprises five stages: diagnosing, action planning, action taking, evaluating, and specifying learning. Diagnosing is concerned with the identification and specification of an opportunity for improvement in the client organization. During action planning, the researcher and client organization consider alternative courses of action to attain the improvement identified, and agree upon the adoption of one of those courses of action. Action taking refers to the realization of the chosen course of action. Evaluating comprises the assessment of the outcomes of the selected course of action. Learning specification happens when the information compiled in the previous stage is structured in the form of a set of research findings, which are further refined into implications for future practice.

THE BPR FRAMEWORK USED

Two main components formed the BPR framework used in this study: MetaProi, a group methodology (Kock, 2005), and the asynchronous computer mediation tool. MetaProi includes a metaprocess for BPR group discussions. This metaprocess was devised based on published BPR methodologies (Harrington, 1991; Kock, 1995), normative studies (Caron, Jarvenpaa, & Stoddard, 1994; Dingle, 1994; Guha et al., 1993; Hall, Rosenthal, & Wade, 1993; Venkatraman, 1994; Wastell, White, & Kawalek, 1994), and a previous study of the effects of asynchronous groupware on BPR groups (Kock & McQueen, 1995). MetaProi combines its BPR group metaprocess with activity guidelines and a simple graphical modeling tool. Three main stages form the core of MetaProi: definition, analysis, and redesign. A more detailed description of MetaProi's activities, along with respective guidelines and the modeling tool, is provided by Kock (1999, 2005).

The computer tool used in this research embodied three main common asynchronous communication support functions, and was implemented by the researcher using Novell Groupwise (trademark by Novell Corporation). First, it allowed BPR group members to send and receive electronic messages to and from any other SCHOOL staff. Second, the computer tool allowed BPR group members to attach electronic documents to the electronic messages. These documents comprised a wide range of application formats, such as spreadsheets, flowcharts, presentations, and graphs. Third, the tool incorporated a distribution list. Using this function, BPR group member messages were automatically distributed to the other members of the BPR group, irrespective of their physical location in the organization. Additionally, the computer tool stored all messages associated with the BPR group discussion in a separate folder for further analysis.

Diagnosing

Interviews with a course coordinator at SCHOOL revealed several problems related to the teaching of a newly created course, which involved two distinct departments: an academic department and a computer support department. The academic department housed the lecturers of the course. The computer support department was responsible for the quality of equipment and software used in the course's laboratory "practicals," a term used to refer to laboratory sessions addressing specific application features. Most of the problems were related to software application breakdowns during the course practicals. The academic department blamed the computer support department for those breakdowns, alleging negligence. The computer support department, in turn, blamed the academic department for the problems, alleging poor design of the practicals and use of unauthorized software packages.

The importance of the course was identified during interviews with lecturers and the quality manager of SCHOOL. These interviews unveiled the fact that the course was among the most successful courses of SCHOOL, in terms of student enrollment, and the most successful among the elective courses of the academic department. This happened in spite of the fact that the course had been offered for the first time in the previous semester. Additionally, interviews with students suggested that problems related to the course had been having a negative impact on students' perceptions of SCHOOL as a whole, not only of the departments involved.

Some of the staff members in the two departments were contacted. They were introduced to the BPR framework and asked for their perceptions of its application in the redesign of the course. Some of them showed interest in the idea. One of the possible reasons for this, which was not evident at the outset, was that there was very little communication between the two departments at that stage. One lecturer suggested that it might be easier to start a computer-mediated group discussion since it was not "so formal as a face-to-face meeting." Moreover, some of the staff members in the computer support department were involved in equipment and software support activities, which required their full-time presence at the laboratories. As this department was understaffed at that stage, having some of its staff tied up in face-to-face meetings was seen as likely to negatively affect service quality even more.

Action Planning

The person who most closely felt the problems related to the course, the course laboratory coordinator, decided to set up and lead a BPR group involving members of both departments. Rather than training all group members, it was decided that only the leader would be briefly trained in the use of MetaProi. It was decided that the leader, with the help of the facilitator, would set the context for the group discussion by means of electronic messages sent to the group. Each of these messages would mention the stage in which the BPR group was, describe what was expected to be achieved in that stage, situate the stage within the overall BPR group discussion, and ask members to reply with their comments. Those comments would then be summarized and used as a starting point for the next stage until the BPR group agreed on the changes to be implemented.

The leader invited four group members, personally or by phone, to participate in the group. One member received this invitation in the body of the first message sent by the group leader to the group. One group member asked to be included in the group during its first stage after the group was started. To avoid misunderstandings, it was decided that the process and goal of the BPR group would be outlined in the first message sent to the group.

Action Taking

The BPR group started its discussion based on a message sent by the leader to the group explaining the group's main goals and stages. That message also included a list of problems observed in the course during the previous semester and a list of three possible causes for those problems. Each of the causes defined a business process. Two of these business processes were entirely contained in each of the departments while the other comprised communication activities between the two departments.

The activities that were chosen by the group to be redesigned involved some of the business processes suggested in the first message, but were not entirely contained in any of them. The leader and facilitator split those activities into two business processes, which were then modeled using flowcharts. Those models were posted to the group attached to a message. The message contained information extracted from a student survey with opinions about the course in the previous semester. It also included a list of possible changes in the processes.

A list of changes was summarized as a result of the discussion based on the information described above. This summary included the changes and those responsible for their implementation. The summary implicitly stated that those changes would be implemented within approximately 1 month, before the start of the next semester. A final list was agreed on by the BPR group after a discussion over the feasibility of some of the changes proposed. That list comprised changes to be implemented by both departments.

Evaluating

The BPR group discussion lasted 33 days. The group involved seven members: four from the academic department and three from the computer support department. Interviews indicated that part of the discussion had taken place orally, mainly by means of brief face-to-face and phone conversations between members of the same department.

Emerging patterns of asynchronous computer mediation effects on BPR groups have been noticed in the evaluating stage. These patterns led to the identification of variables related to two units of analysis: the organization and the BPR group. The variable identified in the organization was interdepartmental communication. The variables identified in the BPR group were the degree of interaction, demand for leadership skills, cost, effectiveness, and individual influence. A description of the effects related to those variables is provided next.

Interdepartmental Communication

The group leader, as a course laboratory coordinator, had been consistently complaining about laboratory problems to the staff of the computer support department prior to the start of the BPR group. This contributed to a widening of the communication gap between him and the staff of that department. Complaints were mainly about the quality of the equipment and software used in the course laboratory practicals. These complaints had been typically met with a defensive reaction from the computer support department. After the BPR group took place, however, the course laboratory coordinator stated that his relationship with the members of the computer support department had improved considerably. He mentioned the following about one of the members of the computer support department: "[H]e had been avoiding greeting or talking to me...probably because of my complaints about lab problems....After this group, though, our relationship improved considerably."

The above improvement was in the course laboratory coordinator's relationship with line staff of the computer support department. A senior manager of that department, on the other hand, seemed to feel uncomfortable with the group discussion. Regarding this feeling, the BPR group leader mentioned that the senior manager "seemed to feel uncomfortable with the fact that we were discussing his department's problems so openly..." This had been observed by the BPR group leader during brief conversations at the cafeteria, and it seemed to find support in defensive messages sent by the senior manager to all members of SCHOOL. Those messages addressed problems that had been raised during the BPR group discussion.

The senior manager denied, however, in an interview with the researcher, any feeling of discomfort caused by the BPR group discussion. Nevertheless, he added that "if the subject is particularly sensitive, I tend not to trust [computer mediation]." When asked whether the BPR group discussion had addressed particularly sensitive issues, the senior manager said, "I could have felt that the problem was because one of [the computer support department staff] was causing the system to run poorly, was incompetent...if I had sent a message about it and the message was distributed to the group, I would have felt annoyed." The main reason for this, according to him, was the fact that messages could be saved, printed out, and later used for "other purposes." For example, an admission by the computer support department that it was causing some of the problems discussed by the BPR group was seen as something that could be used later on as an excuse for top-down intervention and dismissals.

Degree of Group Interaction

A compilation of the group discussion suggested a drastic reduction in the interaction between members in comparison with face-to-face meetings. The level of interaction was defined by the ratio between the number of contributions to the group discussion and the number of members. Seven members exchanged 21 messages, an average of three messages per group member. Interviews with group members indicated that, in their opinion, the level of interaction would have been considerably higher had the discussion been done orally. Two group members declared that they had been more selective in their participation than they would have been in a face-to-face meeting, in some cases giving it low priority over other routine activities.

Also, two of the group members did not provide any electronic input to the group. Interviews with those members, however, revealed that they had read the messages and had been involved in related oral discussions with other members. Those discussions, as mentioned before, took place between members of the same department only.

Some group members discussed group-related issues with staff members outside the group, which indicates that the group discussion involved more members than the ones included in the system's distribution list (although indirectly).

Demand for Leadership Skills

All messages from group members, except some of the messages from the group leader, were related to previous messages. Moreover, most of the messages were replies to messages from the group leader. The messages from the group leader clearly set the context and progress of the discussion despite the fact that the group leader was the most junior person in the group.

The group leader also admitted having severely limited oral communication skills. He declared, based on this, that "leading a face-to-face meeting would be considerably more stressful for me." The leader perceived his limited oral communication skills and his junior position in the organization as likely to considerably hinder him from leading a face-to-face BPR group addressing similar issues.

Cost of Group Interaction

Five group members were invited by the leader to participate in the group through face-to-face or phone contact prior to the start of the group discussion. One member received this invitation in the body of the first message sent by the group leader to the group. One group member asked to be included in the group during its first stage after the group was started. The group leader saw this as a much less costly setup procedure, from a time perspective, than the one usually employed in face-to-face meetings. The latter often involves negotiating an agreed time for the meeting with each group member, and finding and booking an appropriate venue.

The BPR group completed its activities with very little individual cost for the group members, with much of the burden concentrated on the group leader and facilitator. However, the leader and the facilitator were of the opinion that a face-to-face BPR group discussing the same issues would have required slightly more time from them. Both spent about 9 hours on the group discussion. The lengths of messages exchanged by the group, as well as interviews with the group members, indicated that the average time spent by the other group members was 1 hour or less during the entire group discussion.

Effectiveness of the Group

Four months after the BPR group was completed, all redesign proposals had been implemented. The redesign proposals involved five activities to be carried out: three by the members of the academic department and two by the members of the computer support department. Interviews with the group leader and some of the group members suggested that the results achieved by the group were perceived as slightly better than the ones likely to have been achieved without computer mediation.

Also, two lecturers carried out a survey at the end of the course semester with students who had attended the course. That survey indicated a drastic improvement in the perceived quality of the course compared with the same survey performed at the end of the previous semester when the course had been offered for the first time. Several points of the survey had been directly or indirectly addressed by the BPR group.

Individual Influence on Other Members

One of the lecturers of the academic department and a senior manager of the computer support department had previously aired concerns about whether the BPR group should have continued its discussion. Each of them aired their concerns at different times early in the group discussion.

The senior manager stated in a message to the group, "I find [the computer support department] to be pretty good and getting better..." He also hinted to the group leader (orally) that the issues raised by the BPR group could well be dealt with within his department. Nevertheless, the BPR group continued its discussion. That was the last contribution of the senior manager to the group discussion.

The lecturer, on the other hand, said both orally to the BPR group leader and in an electronic message sent only to the members of the academic department, "I don't think we should involve [the computer support department] in the discussion of those issues..." Most of the issues referred to by the lecturer, however, were clearly related to communication problems between the two departments. Again, this happened early in the group discussion and did not prevent the group from proceeding with its BPR work. The group's proceeding against the lecturer's wishes had no effect on the lecturer's further participation in the group.

Specifying Learning

In the specifying-learning stage, explanations were developed by the researcher to account for the effects observed in the evaluating stage. The explanations are summarized in the next subsections. As suggested by Yin (1994), the explanations stipulated a set of causal links between the variables identified in the evaluating stage. Moreover, the explanations led to the identification of new variables, which were essential to the establishment of those causal links.

Interdepartmental Communication

The improvement in the personal relationship between the laboratory coordinator and the line staff of the computer support department might have been influenced by the impersonal characteristic of computer-mediated communication (Chidambaram & Kautz, 1993; Dallavalle, Esposito, & Lang, 1992; Kock, 1999, 2005), which can make it easier to start and conduct BPR groups involving staff from conflicting departments. BPR groups can help establish new communication channels between departments and, in turn, increase the amount of interdepartmental communication. This can foster the occurrence of new BPR groups.

It is important to stress, however, that previous studies suggest that the impersonal characteristic of computer-mediated communication can also lead to more conflict. People seem to express themselves more frankly and less politely when interacting electronically (Alonzo & Aiken, 2004; Easterbrook, Beck, Goodlet, Plowman, Sharples, & Wood, 1993; Sproull & Kiesler, 1991).

The BPR group discussion caused some discomfort for the computer support department's senior manager. This can be explained based on the notion that the establishment of new communication channels between his department and others led the senior manager to feel that his control over departmental staff had been reduced. Previous studies show that computer-mediated communication can make managers feel as though they are losing control over their subordinates, particularly in hierarchical organizations (Grudin, 1994; Sproull & Kiesler, 1991).

The above explanation, however, was not supported by the interview with the senior manager. A competing and perhaps more acceptable explanation is that asynchronous computer-mediated discussions generally pose more threats to participants than face-to-face meetings. Group members may see electronic messages as records that can be used by others against them (i.e., the group members) in the future. In some organizations, this threat increases when different departments are represented in the same BPR group. At SCHOOL, the departments involved were expected to be defensive due to the previous mutual criticisms. This may have led the senior manager to feel threatened by the sudden exposure brought about by participation in the BPR group, as indicated by his defensive messages to SCHOOL staff. This feeling would likely prevent him from supporting future BPR groups involving his departmental staff.

The effect of higher perceived threats to BPR group participants, particularly managers who typically are the ones blamed for departmental problems, conflicts with the positive effect of lower barriers to interdepartmental communication caused by computer mediation in BPR groups.

It is difficult to establish whether the increase in departmental communication is likely to offset the negative reaction from managers. That reaction could be weak in organizations where departmental boundaries are not so well defined, such as "adhocracies" (Toffler, 1970) and "spider web" organizations (Quin & Paquette, 1990). On the other hand, that negative reaction could be sufficiently strong enough to considerably limit the occurrence of new BPR groups in organizations with a high number of separate departments with well-defined boundaries, such as bureaucracies (Madsen, 1989). We could not find enough evidence to suggest that either structure is predominant in universities in general.

Degree of Group Interaction

The seemingly drastic reduction in the interaction between BPR group members can be explained in four main ways. First, it might have been caused simply by lack of interest from group members in the BPR discussion. Second, it might have been induced by the ideas being better thought out before they were typed out and sent to the group as messages. This could have decreased the need for requests for clarification and subsequent replies. Third, it might have been related to the discussion structure suggested by MetaProi. That structure seems to push the discussion toward a sequence of three to four main messages sent by the leader to the group, and related replies from group members. This structure may limit the free flow of ideas and thus the number of individual contributions. Fourth, as the discussion is computer-mediated and asynchronous, group members cannot be forced to provide contributions. Whether they feed in their inputs or not is at their discretion, which may tend to make their participation in the several stages of the discussion more selective than it would have been in a face-to-face meeting.

The first explanation was ruled out because the problems with the course had become serious enough to warrant top-down intervention at SCHOOL. As discussed before, those problems were affecting SCHOOL's image, not only those of the departments involved. This was seen as a strong appeal to engage in the group discussion as some of the staff involved believed that the situation could be improved locally, which was seen as a much more desirable alternative than a top-down intervention. The second explanation finds confirmation in the behavior displayed by the laboratory coordinator. He tried to be as precise and clear as possible because he knew that others could misinterpret his messages. This led him to provide better-thought-out input than he would normally have provided in a face-to-face meeting. The third explanation is also supported by the fact that all replies from group members were closely related to the main messages from the group leader. Finally, the fourth explanation was corroborated by the fact that some of the group members declared being more selective in providing input to the group discussion than they would have been in a face-to-face meeting.

Demand for Leadership Skills

There was arguably a low demand for leadership skills from the group leader to effectively lead the group discussion. This can be explained by the fact that the computer mediation conceals individual characteristics, such as rank in the organizational hierarchy (Chidambaram & Kautz, 1993; Dallavalle et al., 1992; Kock, 1999, 2005). Had those characteristics not been filtered by the system, the group leader could have felt uncomfortable in a leadership role. He would have been constantly reminded, by the other members' very presence, that he was probably not the best suited to lead the group.

A complementary explanation is the low stress on members, including the leader, to provide their input to a computer-mediated group discussion. This more relaxed characteristic of computer-mediated communication has been identified in previous studies as an important factor influencing the quality and length of a member's contributions (Pietro, 1992; Walther, 1995). This explanation is consistent with the leader's view about the increased stress of leading a face-to-face BPR group meeting compared with a computer-mediated one.

Cost of Group Interaction

There seemed to be a considerable reduction in the overall cost of running the BPR group as a result of the computer mediation. While there seemed to be a slight reduction in the demand of time from the leader in comparison with a verbal BPR meeting, the biggest gains were experienced by the other group members. Considering that a face-to-face discussion would have lasted 10 hours (1 hour more than the total time spent by the leader and the facilitator), the relative economy of time accruing to each of the other group members was about 90% since the average time spent by a member was estimated at 1 hour.

The reduction in cost can be explained by looking at the three main types of direct costs in face-to-face meetings. First, there are the setup (or gathering) costs, which are related to bringing people physically together for the meetings. These costs seem to increase with the physical distance between group members. Second, there are disruption costs since in face-to-face meetings members are required to interrupt their routine activities to be in the meetings. Third, there are the participation costs as typically each of the individual members cannot engage in parallel activities during face-to-face meetings. These three types of cost are obviously reduced by asynchronous computer mediation, often in a combined way. In the BPR group studied, there was a slight reduction in setup costs (the two departments were located on the same campus), a reduction in the disruption costs (especially in the computer support department), and a drastic reduction in the participation costs (except for the leader and facilitator).

Effectiveness of the Group

It is difficult to establish a strict causal relationship between the BPR group outcome (i.e., the process redesign proposals) and the improvement in the quality of the course. The latter was indicated by the surveys but cannot be traced back only to the redesign proposals. Some other factors, such as the laboratory coordinator's experience acquired during the previous semester, might have had a direct impact on the quality perceptions in the survey.

The BPR group, on the other hand, targeted several of the issues in the survey. Also, the perceptions of the group leader and some of the participants were that the quality of the BPR group outcomes was slightly better than it would likely have been without computer mediation. These results suggest that the drastic reduction in the degree of interaction between group members had very little, if any, impact on the effectiveness of the BPR group.

Small group theory offers an explanation for the above finding based on the type of tasks the group performs. BPR group members perform primarily additive tasks, whereby they provide individual input on information that was structured before. MetaProi defines the structure of that information. Only the leader and the facilitator perform conjunctive tasks, such as generating basic lists for discussion and modeling processes.

Previous studies suggest that group effectiveness increases with higher interaction in additive tasks and decreases in conjunctive tasks (Shaw, 1981). Therefore, given the key role played by the leader and facilitator, a low degree of interaction is unlikely to have a strong negative impact on the quality of the BPR group outcome, especially if the leader and facilitator have a great deal of information about the problems being addressed by the group. Conversely, it is reasonable to assume that poor facilitation and negligent leadership may have a strong negative impact on that variable.

A complementary explanation for the apparent lack of impact of a much lower degree of interaction on BPR group effectiveness is that the individual contributions tend to be more carefully prepared in computer-mediated than in face-to-face discussions, a phenomenon that Kock (2001b) refers to as "compensatory adaptation" (for a more recent study, see Kock, Lynn, Dow, & Akgün, 2006). Also, there are indications that merely forcing members to participate in group discussions may decrease group effectiveness (Zaltman, 1994). This often occurs in face-to-face meetings and may cause certain types of undesirable group behavior, such as groupthink (Janis, 1972).

Individual Influence on Other Members

There is some evidence, from what happened when the senior manager and the lecturer aired their concerns about the group discussion, that the individual influence of certain members is reduced in computer-mediated BPR groups. This phenomenon can be explained by the impersonal characteristic of computer mediation. Individual characteristics that foster control over groups, such as organizational status and verbal skills, are filtered by the system. This makes it much more difficult for individual members to control the group discussion or impose their views on the group.

The explanation above, however, seems to contradict the perception by the leader that he had led the discussion more easily than in a face-to-face situation. This calls for a complementary explanation as the group leader also had a low control of the group. The fact that the group followed a smooth path can be explained by the fact that the group's goal and process had been presented to the group members in the first message by the leader. The fact that the group members proceeded with the group discussion suggests that they had agreed with the group goal and process. Had that agreement not have been reached by the group, the discussion would probably have stalled given the low individual influence effect of computer mediation.

ORGANIZATIONAL IMPLICATIONS

The set of explanations above suggests several causal links between variables related to the BPR-group unit of analysis. Three main dependent variables have emerged from the explanation building process: demand for leadership skills, overall cost, and degree of interaction.

Two of the main dependent variables identified at the group level can foster the occurrence of new BPR groups. These variables are the demand for leadership skills and overall cost. A lower demand for leadership skills is likely to allow for a decentralization of improvement initiatives, which may foster the occurrence of local BPR groups. These BPR groups, however, are unlikely to target core processes and strategic issues. The breadth of knowledge and the authority to initiate a discussion over such issues is rarely found in lower managerial levels. Therefore, a lower demand for leadership skills is probably going to foster the occurrence of new BPR groups dealing with issues at the operational, rather than strategic, level in universities. While the drastic reduction of cost can affect all types of BPR groups, from the operational to strategic, those groups involving the most expensive staff members (which is typical in strategic BPR groups) are likely to attain the most significant savings in participation costs.

Finally, the lower degree of interaction in computer-mediated BPR groups can have a negative impact on the discussion of issues where there must be a high exchange of information between group members. This is likely to happen as BPR groups move toward business processes that cut across several different departments, and whose redesign involves risks. These characteristics tend to be found more in strategic than operational BPR. This suggests that, while strategic BPR groups may benefit from computer mediation, this is likely to be achieved as a result of the combination of this type of interaction with face-to-face and other types of oral interaction (for example, telephone conversation).

CONCLUSION

This chapter reports on an action research study of the effects of asynchronous computer mediation on one BPR group. That group was comprised of staff members from two collaborating departments of a school of studies at a New Zealand university. The group identified, analyzed, and redesigned two interrelated business processes related to the teaching of a practical laboratory course. The BPR group took 33 days to come up with a redesign proposal, which was fully implemented. Four months later, a survey with students indicated that a drastic improvement in the perceived quality of the course had been achieved. The survey covered several points that had been directly and indirectly addressed by the group.

Regarding the BPR-group unit of analysis, the study identified three main dependent variables that had been affected by the asynchronous computer mediation. The variables were the demand for leadership skills, overall cost, and degree of interaction. The study suggested that computer mediation caused a decrease in the demand for leadership skills, a drastic decrease in the overall cost, and a drastic decrease in the degree of interaction in the BPR group. No clear impact on the group's effectiveness was observed.

The study also indicated that asynchronous computer mediation lowers barriers to interdepartmental communication, which in turn fosters more departmental communication. This favors the occurrence of new BPR groups involving different departments. On the other hand, the study indicates that those groups tend to be perceived as bringing about more threats to management, which may induce negative reactions from managers and thus hinder the occurrence of new BPR groups.

Finally, the study suggests that, while strategic BPR groups, as opposed to those dealing with local operational issues, may benefit from computer mediation, this is likely to be achieved as a result of the combination of this type of interaction with face-to-face interaction (e.g., a combination of face-to-face meetings and computer-mediated discussions) and other types of oral one-to-one interaction, such as telephone conversation.

ACKNOWLEDGMENT

We would like to thank Dr. Robert McQueen for his helpful comments on an earlier version of this chapter. Thanks are also due to the BPR group members and other SCHOOL staff who were interviewed for this study. An earlier version of this chapter was previously published in Volume 14 of the journal *Campus-Wide Information Systems*; we thank the journal's publisher, MCB Press, for permission to use that article as a basis for this chapter.

REFERENCES

Alonzo, M., & Aiken, M. (2004). Flaming in electronic communication. *Decision Support Systems, 36*(3), 205-214.

Brothers, L., Hollan, J., Nielsen, J., Stornetta, S., Abney, S., Furnas, G., et al. (1992). Supporting informal communication via ephemeral interest groups. *Proceedings of CSCW'92 Conference*, 84-90.

Caron, J. R., Jarvenpaa, S. L., & Stoddard, D. B. (1994). Business reengineering at CIGNA Corporation: Experiences and lessons learned from the first five years. *MIS Quarterly, 18*(3), 233-250.

Chidambaram, L., & Kautz, J. A. (1993). Defining common ground: Managing diversity through electronic meeting systems. *Proceedings of the 14th International Conference on Information Systems*, 1-11.

Clement, A. (1994). Computing at work: Empowering action by low-level users. *Communications of ACM, 37*(1), 53-63.

Dallavalle, T., Esposito, A., & Lang, S. (1992). Groupware: One experience. *Proceedings of Groupware'92 Conference*, 470-477.

Deming, W. E. (1986). *Out of the crisis.* Cambridge, MA: MIT, Center for Advanced Engineering Study.

Dennis, A. R., Carte, T. A., & Kelly, G. G. (2003). Breaking the rules: Success and failure in groupware-supported business process reengineering. *Decision Support Systems, 36*(1), 31-47.

Dennis, A. R., Daniels, R. M., Jr., Hayes, G., & Nunamaker, J. F., Jr. (1993). Methodology-driven use of automated support in business process reengineering. *Journal of Management Information Systems, 10*(3), 117-138.

Dennis, A. R., Hayes, G. S., & Daniels, R. M., Jr. (1994). Re-engineering business process modeling. *27th Hawaii International Conference on System Sciences*, 244-253.

Dingle, M. E. (1994). *Business process reengineering: A New Zealand perspective.* Palmerston North, New Zealand: Massey University, Department of Executive Education.

Easterbrook, S. M., Beck, E. E., Goodlet, J. S., Plowman, L., Sharples, M., & Wood, C. C. (1993). A survey of empirical studies of conflict. In S. Easterbrook (Ed.), *CSCW: Cooperation or conflict?* (pp. 1-68). New York: Springer-Verlag.

Elden, M., & Chisholm, R. F. (1993). Emerging varieties of action research. *Human Relations, 46*(2), 121-141.

French, W. L., & Bell, Jr. (1990). *Organization development.* Englewood Cliffs, NJ: Prentice Hall.

Grudin, J. (1994). Eight challenges for developers. *Communications of the ACM, 37*(1), 93-105.

Guha, S., Kettinger, W. J., & Teng, J. T. (1993). Business process reengineering: Building a comprehensive methodology. *Information Systems Management*, 13-22.

Hall, G., Rosenthal, J., & Wade, J. (1993). How to make reengineering really work. *Harvard Business Review*, 119-131.

Hammer, M. (1990). Reengineering work: Don't automate, obliterate. *Harvard Business Review, 68*(4), 104-114.

Hammer, M., & Champy, J. (1993). *Reengineering the corporation.* New York: Harper Business.

Harrington, H. J. (1991). *Business process improvement.* New York: McGraw-Hill.

Janis, I. L. (1972). *Victims of groupthink: A psychological study of foreign policy decisions and fiascos.* New York: Houghton Mifflin.

Jonsonn, S. (1991). Action research. In H. Nissen, H. K. Klein, & R. Hirschheim (Eds.), *Information systems research: Contemporary approaches and emergent traditions* (pp. 371-396). New York: North-Holland.

King, N. (1990). Innovation at work: The research literature. In M. A. West & J. L. Farr (Eds.), *Innovation and creativity at work* (pp. 15-59). New York: John Wiley & Sons.

Kock, N. (1995). *Process reengineering, PROI: A practical methodology.* Sao Paulo, Brazil: Editora Vozes.

Kock, N. (1999). *Process improvement and organizational learning: The role of collaboration technologies.* Hershey, PA: Idea Group Publishing.

Kock, N. (2001a). Asynchronous and distributed process improvement: The role of collaborative technologies. *Information Systems Journal, 11*(2), 87-110.

Kock, N. (2001b). Compensatory adaptation to a lean medium: An action research investigation of electronic communication in process improvement groups. *IEEE Transactions on Professional Communication, 44*(4), 267-285.

Kock, N. (2003). Communication-focused business process redesign: Assessing a communication flow optimization model through an action research study at a defense contractor. *IEEE Transactions on Professional Communication, 46*(1), 35-54.

Kock, N. (2005). *Business process improvement through e-collaboration: Knowledge sharing through the use of virtual groups.* Hershey, PA: Idea Group Publishing.

Kock, N. (2006). *Systems analysis and design fundamentals: A business process redesign approach.* Thousand Oaks, CA: Sage Publications.

Kock, N. (Ed.). (2007). *Information systems action research: An applied view of emerging concepts and methods.* New York: Springer.

Kock, N., & Davison, R. (2003). Can lean media support knowledge sharing? Investigating a hidden advantage of process improvement. *IEEE Transactions on Engineering Management, 50*(2), 151-163.

Kock, N., Lynn, G. S., Dow, K. E., & Akgün, A. E. (2006). Team adaptation to electronic communication media: Evidence of compensatory adaptation in new product development teams. *European Journal of Information Systems, 15*(3), 331-341.

Kock, N., & McQueen, R. J. (1995). Integrating groupware technology into a business process improvement framework. *Information Technology & People, 8*(4), 19-34.

Kock, N., & Murphy, F. (2001). *Redesigning acquisition processes: A new methodology based on the flow of knowledge and information.* Fort Belvoir, VA: Defense Acquisition University Press.

Madsen, C. (1992). Approaching group communication by means of an office building metaphor. In D. Marca & G. Bock (Eds.), *Groupware: Software for computer-supported cooperative work* (pp. 316-328). Los Alamitos, CA: IEEE Computer Society Press.

Nunamaker, J. F., Dennis, A. R., Valacich, J. S., Vogel, D. R., & George J. F. (1991). Electronic meeting systems to support group work. *Communications of the ACM, 34*(7), 40-61.

Peters, M., & Robinson, V. (1984). The origins and status of action research. *The Journal of Applied Behavioral Science, 20*(2), 113-124.

Pietro, C. (1992). Groupware meetings that work. *Proceedings of Groupware'92 Conference*, 50-58.

Quin, J. B., & Paquette, P. C. (1990). Technology in services: Creating organizational revolutions. *Sloan Management Review*, 67-77.

Shaw, M. (1981). *Group dynamics: The psychology of small group behavior.* New York: McGraw-Hill.

Sheffield, J., & Gallupe, B. (1993). Using electronic meeting technology to support economic development in New Zealand: Short term results. *Journal of Management Information Systems, 10*(3), 97-116.

Soles, S. (1994). Work reengineering and workflows: Comparative methods. In E. White & L. Fischer (Eds.), *The workflow paradigm* (pp. 70-104). Alameda, CA: Future Strategies.

Sproull, L., & Kiesler, S. (1991). Computers, networks and work. *Scientific American*, 84-91.

Susman, G. I., & Evered, R. D. (1978). An assessment of the scientific merits of action research. *Administrative Science Quarterly, 23*, 582-603.

Tapscott, D., & Carston, A. (1993). *Paradigm shift.* New York: McGraw-Hill.

Toffler, A. (1970). *Future shock.* New York: Bantam Books.

University of Waikato. (1994). *Human Research Ethics Committee: Procedures.* Hamilton, New Zealand: University of Waikato, Human Research Ethics Committee.

Venkatraman, N. (1994). IT-enabled business transformation: From automation to business scope redefinition. *Sloan Management Review, 35*(2), 73-87.

Walther, J. B. (1995). Relational aspects of computer-mediated communication: Experimental observations over time. *Organization Science, 6*(2), 186-203.

Walton, M. (1989). *The Deming management method.* London: Mercury.

Wastell, D. G., White, P., & Kawalek, P. (1994). A methodology for business process redesign: Experiences and issues. *Journal of Strategic Information Systems, 3*(1), 23-40.

Wilson, P. (1991). *Computer supported cooperative work.* Oxford, UK: Intellect.

Yin, R. K. (1994). *Case study research.* Newbury Park, CA: Sage.

Zaltman, G. (1994). Knowledge disavowal in organizations. In R. H. Kilmann, K. W. Thomas, D. P. Slevin, R. Nath, & S. L. Jerrel (Eds.), *Producing useful knowledge for organizations* (pp. 173-187). San Francisco: Jossey-Bass.

Chapter VII
Applying Pattern Theory in the Effective Management of Virtual Projects

Ilze Zigurs
University of Nebraska at Omaha, USA

Deepak Khazanchi
University of Nebraska at Omaha, USA

ABSTRACT

The management of virtual projects is fundamentally different from that of traditional projects. Furthermore, the research in this area comes from different reference disciplines and perspectives, and a unified view or theory of best practices does not yet exist. We use the theoretical frame of patterns to propose a unified view. We focus on three concepts as the underlying theoretical elements for identifying patterns of effectiveness in virtual project management: (a) coordination, (b) communication, and (c) control. As a first step in the identification of specific patterns, we conducted a series of virtual focus groups with participants from industry who had real experience with virtual projects. The brainstorming data from the focus groups were analyzed to develop an initial set of patterns. Based on this first step, we also present a structured process for the discovery and continuing validation of patterns of effectiveness in virtual projects, and discuss the issues involved in applying the process.

INTRODUCTION

Project management (PM) is a challenging activity in the best of circumstances, and it has become even more so in the virtual world. The increasingly popular use of virtual teams for dispersed projects has resulted in new challenges for both research and practice. We use the term *virtual projects* to refer to any project in which team members are at least partly geographically dispersed and rely on information and communication technologies to accomplish their work. The project team may be dispersed on other dimensions as well, for example, culturally or organizationally, but geographic dispersion is a minimal condition. The challenge in virtual projects is to go beyond a simple transfer of knowledge from traditional environments by developing a theoretically sound set of practices that are relevant for the virtual domain.

We use the theoretical frame of patterns to address this challenge in a novel way. Pattern theory was introduced in architecture (Alexander, 1965; Alexander, Ishikawa, Silverstein, Jacobson, Fiksdahl-King, & Angel, 1977) and was later applied to software design (Gamma, Helm, Johnson, & Vlissides, 1994) as a way of developing accepted solutions for specific problems in a defined context. We propose that patterns of effective management for virtual projects can be identified, applied, and validated. We focus on three concepts as the underlying theoretical elements for identifying such patterns, namely, communication, coordination, and control. Different types of projects can be expected to have different patterns for successful project management. The key research question for the study is the following: What patterns of communication, coordination, and control can be identified for the successful management of virtual projects? The answer to this question is important because it advances theory in a significant research domain while also providing practical advice to managers on a question of real importance.

Based on the theoretical foundation just described, we conducted an empirical study in order to identify patterns. Brainstorming comments and questionnaire data from a series of virtual focus groups provided the data for textual analysis. Themes in the text were identified and related to the theoretical model. This analysis was used to extract patterns of effective virtual project management. The next section provides the theoretical development of patterns and the definition and background of key concepts. The method is then described, followed by the data analysis and results. The discussion section highlights the key findings and elaborates on additional issues related to validation and a process for the continuing discovery of patterns. The chapter concludes with a summary of the contributions as well as implications for research and practice.

THEORETICAL FOUNDATION

The management of virtual projects is a complex phenomenon, and the relevant theory and concepts that govern that phenomenon come from different domains. We begin with a definition of key concepts in order to set the boundaries for the specific study that is described here. First, projects are defined and characterized in terms of a parsimonious typology. Second, virtuality is defined, and the role and nature of technology are developed. Third, key factors for managing virtual projects are presented. Fourth, the concept of patterns is defined. Each of these separate pieces is built on existing literature and presented in the context of our overarching theoretical frame.

Typology of Projects

Projects are the lifeblood of organizational activity. A project can be defined as a "temporary endeavor undertaken to create a unique product or service" (Project Management Institute [PMI] Standards Committee, 1996, p. 4). Projects vary on many dimensions, including purpose, size, time span, urgency, scope, and complexity, and these dimensions are often overlapping. For example, are scope and complexity two independent characteristics of projects, or do they interact, or does one lead to or contribute to the other? These are not mere semantic arguments since a coherent characterization of projects is the first step to understanding and managing them.

A number of different typologies of projects exist based on dimensions such as cultural differences (Carmel & Agarwal, 2001), uncertainty vs. scope (Shenhar, 1998), type of coordination structure (Gassmann & Von Zedtwitz, 2003), and organizational characteristics (Evaristo & Munkvold, 2002). Three consistent themes can be observed in much of this literature, and we use these three themes as dimensions that characterize projects for the current study. First is complexity, which we define as the issues that have to be managed for successful completion of a project. Specifically, complexity is affected by team attributes such as size, culture, language, gender composition, personal characteristics, complementarity of resources, and the nature of project knowledge (Gassmann & Von Zedtwitz; Grant, 1996; Powell, Piccoli, & Ives, 2004; Royce, 1998). The second dimension is project scope, which we define as the boundaries of a project, including its duration and level of innovation (Gassmann & Von Zedtwitz). The third dimension is project risk, which encompasses unanticipated events that may affect successful completion. Risk may be programmatic, technical, quality related, logistical, or deployment related (Christensen & Thayer, 2001; IEEE, 2004).

If each dimension is characterized as low, medium, and high, the resulting typology of projects would have 27 different types. Our interest is in a more parsimonious examination so that we can ascertain key differences among the types that are at either end of the continuum vs. somewhere in the middle. Thus, we define three types of projects based on extreme and mixed values of each of the three dimensions. Table 1 shows the project typology for this research.

Virtuality and Technology

The term *virtual* is generally defined in degrees or extent of virtuality rather than as a binary condition (Fiol & O'Connor, 2005). The greater the dispersion on various dimensions, the greater the virtuality of the entity, whether it is a team, a project, or an organization. Dimensions of dispersion include such factors as geography, time, function, organizational affiliation, culture, continuity of the relationship, or technology used for communication (Dubé & Paré, 2004; Espinosa, Cummings, Wilson, & Pearce, 2003; Katzy, Evaristo, & Zigurs, 2000; Watson-Manheim, Chudoba, & Crowston, 2002). Consistent with these generally accepted views, we define virtuality as the extent to which project members are dispersed on geographical and other dimensions and rely on information and communication technologies to carry out project goals.

Virtuality is not possible without information and communication technologies, and the

Table 1. Project typology

Project Type/ Dimension	Com-plexity	Scope	Risk	Example
Lean	Low	Narrow	Low	In-house software development project with multiple segments, within one organization though across multiple locations, having clarity of goals and resource allocation, and relatively established teams
Hybrid	Mixed levels of complexity, scope, and risk			Significant enhancement of customer relationship management application using systematic development approach, but with global heterogeneity in outside partnerships
Extreme	High	Wide	High	Multinational implementation of global supply chain application involving multiple units, varied cultural orientations, conflicting goals, different personalities, and varied resource infrastructures

nature and capabilities of those technologies vary widely. Media richness theory defines technology in terms of fixed characteristics (Daft & Lengel, 1986). Channel expansion theory suggests that perceptions and the use of a channel can evolve over time based on such characteristics as team members' knowledge of one another and the task context (Carlson & Zmud, 1999). Adaptive structuration theory views technology as being malleable through group interaction (DeSanctis & Poole, 1994).

It seems that both fixed and emergent characteristics should be accommodated in any definition of technology. In addition, the main functions of communication, process structure, and task support need to be provided (McGrath & Hollingshead, 1994; Nunamaker, Dennis, Valacich, Vogel, & George, 1991; Zigurs & Buckland, 1998). Hence, we define technology for virtual projects as consisting of an integrated and flexible set of tools for communicating among project members, structuring the process, and supporting task analysis and performance.

Factors for the Management of Virtual Projects

Although there are many ways to classify key issues for the management of virtual projects, we propose that three major issues capture the essence of the majority of those different views, namely, communication, coordination, and control. We use these three concepts because they are intuitive and they have been consistently used in previous research and in practice (e.g., Goodbody, 2005). At the same time, we recognize the difficulty inherent in using these constructs since they are closely tied to each other. For example, it is well understood that control is a mechanism for mitigating coordination and communication challenges in virtual project teams. We are aware, as were previous researchers, that communication in all its forms is essential to achieving effective control and coordination. This apparent interaction suggests a level of confounding. However, this does not in any way take away from the fundamental differences between these concepts and their relevance separately and together to the management of either virtual or traditional projects.

In the following paragraphs, we briefly discuss each of these concepts in turn, along with a justification for their importance and a summary of what is known from prior research. Given that our focus is on the identification of patterns, we keep the review at a summary level, sufficient to develop a broad understanding of the concepts that are relevant to this study.

We define communication as the process by which people convey meaning to one another via some medium through which they exchange messages and information in order to carry out project activities. Communication is fundamental to virtual projects, and a large body of research has accumulated from the study of virtual teams in a variety of contexts. Virtual team members can find it difficult to deal with different interaction styles and preferences (Sarker & Sahay, 2001), and they sometimes make rapid and negative attributions based on infrequent communication and perceptions of unresponsiveness (Cramton, 2001). Periodic face-to-face (FTF) meetings help to overcome communication problems by serving as reinforcement points for the confidence and trust that are required to work remotely (Maznevski & Chudoba, 2000; Shani, Sena, & Stebbins, 2000). Cultural differences can exacerbate communication problems due to differences in such things as preferences for interaction and debate (Massey, Hung, Montoya-Weiss, & Ramesh, 2001), expectations of compatibility (Rutkowski, Vogel, Bemelmans, & Van Genuchten, 2002), and frames of reference (Van Ryssen & Godar, 2000). Appropriate communication is also needed to develop and sustain trust (Jarvenpaa & Leidner, 1999; Pinsonneault & Caya, 2005) and to set and reinforce norms that support attention and commitment from team members (Cramton, 2001; Watson-Manheim & Belanger, 2002). Overall, the existing research reinforces the importance of communication and the explicit attention that must be paid to communication issues throughout the life of the virtual team (Martins, Gilson, & Maynard, 2004; Schubert, Leimstoll, & Romano, 2003). This is even more important in the context of research that has found the amount of communication declines as teams move higher on the virtualness continuum (Martins et al.).

Coordination is the second major issue for the management of virtual projects. We define coordination as the mechanisms through which people and technological resources are combined to carry out specified activities in order to accomplish stated goals (Crowston, 1991; Grant, 1996). Coordination is a wide-ranging concept that requires action related to the task, team member roles, member relations, time, norms or values, language and culture, and even media (Zigurs, Evaristo, & Katzy, 2001). Zalesny, Salas, and Prince (1995) suggest that there are four components of coordination: identifying goals (i.e., the objectives of joint activities), mapping goals to activities (what specific activities are needed), mapping activities to actors (who does what), and managing interdependencies (sequencing and synchronizing activities). Contextual and organizational factors such as training, trust, and team cohesion can affect coordination (Chinowsky & Rojas, 2003). Dependencies within teams need to be managed (Malone & Crowston, 1994), and appropriate structures must be put into place (Gassmann & Von Zedtwitz, 2003). Cultural differences can also negatively impact virtual team coordination (Maznevski & Chudoba, 2000). Finally, reward systems can affect coordination (Burke, Aytes, & Chidambaram, 2001). Achieving and sustaining coordination occurs over time rather than as a single event (Turvey, 1990); as such, it requires the expertise and experience of team members (Zalesny et al.). In sum, coordination presents significant challenges to virtual teams, not the least of which is that it occurs through communication and thus includes interaction effects.

The third and final issue is control. We define control as the process of monitoring and measuring project activities so as to anticipate and manage variances from project plans and organizational goals (Hendersen & Lee, 1992; Kirsch, 1996; Project Management Institute, 2004). In this definition, monitoring may require the development of a variety of mechanisms for assessing behavioral actions and project outcomes so as to take corrective actions as needed. For example, in the virtual project context, the challenge of control could relate to establishing standards for assessing team member performance, communication of progress, establishment of norms for team member interaction, structuring of teams, or use of collaborative technology. It should also be noted that our definition of control does not preclude the use of a portfolio of control modes as suggested by Kirsch. Finally, the challenges associated with controlling projects are closely tied with coordination and communication issues. For example, temporal distance exacerbates coordination and control problems directly or indirectly through its negative effects on communication (Carmel & Agarwal, 2001). Thus, some organizations move toward reducing collaboration complexity by giving up control and transferring ownership to foreign entities, or by taking the full project ownership to the domestic entity (Carmel & Agarwal). Other challenges that impact control in virtual projects include reinforcing project objectives (Chinowsky & Rojas, 2003), monitoring and measuring issues, the collaborative infrastructure (Evaristo & Munkvold, 2002), the client's knowledge of the systems development process (Kirsch et al., 2002), and group leadership (Homsky, 2003).

This necessarily brief review highlights the many factors that can have an impact in virtual project environments, showing the complexity involved in seeking generalizations. Our approach in dealing with this complexity was to start with the typology and identify the three key dimensions for management. We recognize that our choice of communication, coordination, and control as the key dimensions of our typology could be argued or made differently. However, the prevalence of these three dimensions in the literature of project management, including PMI (2004), supports their importance.

We argue that the managerial practices of communication, coordination, and control combine with virtuality effects to imply different types of technology needs. We also expect differences across the three project types. Table 2 shows the implications for technology needs for each project type in terms of managerial concerns and virtuality effects. The next step is to use pattern theory to look for significant practices that could make a difference.

Pattern Theory

Pattern theory is a key starting point for our research and a natural perspective for understanding, at a somewhat abstract level, effective practices for virtual project management. Patterns help to make sense of complex behavior by looking for the regularities in such behavior. Pattern theory arose in architecture and the work of Alexander (1965) and Alexander et al. (1977), who developed patterns for common architectural design problems, for example, the "bathing room" or "bed cluster." To quote Alexander (1965),

When we build something good, when we build a system that works well, we must ask what is it about this that makes it good? Why is it good?

Table 2. Project types and implications for technology needs

Project Type	Dominant Mana-gerial Concern	Virtuality Effects	Technology Needs
Extreme (High complexity, broad scope, high risk, e.g., mission-critical virtual project)	Commun-ication	Relatively difficult to build shared context while virtual project team is dispersed and physical contact is rare Differences in culture, team experience, language, gender, personalities, resources, infrastructure, and historical knowledge exacerbate the difficulties of communication.	High communication: Rich context communication tools, lateral communication channels, e.g., videoconferencing, Web-based information-management portal
Hybrid (Varying levels of complexity, scope, and risk, e.g., Y2K [Year 2000] bug)	Coordi-nation	Units/partners that do not share mutual project knowledge might underperform due to miscommunicated needs. Differences in culture, team experience, language, gender, personalities, resources, infrastructure, and historical knowledge are moderated by the existence of rapport among some project members.	High process structure: Virtual collaboration systems and knowledge management tools, lateral and vertical communication channels
Lean (Low complexity, narrow scope, low risk, e.g., maintenance virtual project)	Control	Virtual project team already has an established shared context; however, absence of physical interactions might hinder successful project completion Heterogeneity on various dimensions (culture, team experience, language, gender, personalities, resources, infrastructure, and knowledge management) is not a critical factor because it is managed by prior experience within and across team members and through sharing of historical repository of project experiences	High information processing: Project management/workflow tools, CASE tools, software configuration management tools, vertical communication channels

What are its essential qualities that will allow us to build something completely different but which is good in the same way.

Patterns are analogous to recurring themes, familiar processes, rules of thumb, or standard procedures. Patterns provide holistic "abstractions of experiences" that are profound in some way and can be implemented to solve problems in a specific context. To some extent, patterns are a means of communicating insights about a problem domain to others. Patterns do not have to be distinct from each other; in fact, if they are linked in some way, then that allows us to develop a pattern language.

Formally, a pattern is defined as a three-part rule that expresses a relationship among a specific context, a problem, and a solution (Alexander et al., 1977). Alexander's work was carried over into software engineering and popularized in object-oriented design by the "Gang of Four" (Gamma et al., 1994). There are many ways to document specific patterns, but common practice is to include the key elements of the context, problem, and solution. An example of a pattern from object-oriented development is shown below (adapted from Gamma et al.) in the same format that we later use to describe our derived virtual project patterns.

Singleton

- **Context:** There are certain actions that need to be coordinated by a single object across the entire application, for example, the print spooler and file manager.
- **Problem:** How do you provide a single instance of a class that is easily accessible?
- **Solution:** Ensure that only one instance of a class is created, and provide a global point of access to it. (Note that at this point, the actual code for implementing the Singleton pattern could also be provided.)

Some work has been done on patterns in the context of collaborative work. Schuemmer (2003) proposed a structure for sociotechnical patterns that could be used to support collaboration. Fernandez, Holmer, Rubart, and Schuemmer (2002) specified patterns for designing groupware tools with the goal of developing a common vocabulary. Völter (2002) presented patterns specifically for project management, naming them "anti-patterns" because they represented the antithesis of common knowledge for basic project management techniques. To our knowledge, no one has applied pattern theory in a systematic way to the management of virtual projects.

Essentially, we are arguing that the major components of the framework—project type, technology, and virtuality—all affect the managerial dimensions of communication, coordination, and control, which in turn affect project outcomes. Considerable existing research addresses these major components, but it is not our intent to reargue that research here. Instead, we view pattern theory as providing a new way to bring these components together. Thus, and as stated earlier, our research question asks what patterns of communication, coordination, and control can be identified for the successful management of virtual projects.

RESEARCH METHOD

We devised a study that represents a first step in examining our research question. We used a grounded theory approach because such an approach is particularly useful to explore complex and dynamic phenomena in organizational settings (Glaser & Strauss, 1971). Participants in the study were businesspeople who had experience with being members of a virtual team. They interacted asynchronously in a series of virtual focus groups, brainstorming ideas on the factors that contributed to the success or failure of virtual projects. A presession questionnaire and the brainstorming ideas from the focus groups served as the data from which we inductively derived our patterns.[1]

Twenty-nine individuals from five different firms committed to participate in the virtual focus groups, with 14 people completing all phases of the study. Each focus group was an asynchronous brainstorming session conducted via Web-based groupware, with a separate session conducted for each company. All five firms were global companies: Two were software and service providers, one was a technology manufacturer, one a services company, and one a research and engineering firm. Each participant was asked to respond about a project in which he or she had participated within the last 12 months, thus each participant was responding about a different project. The project descriptions ranged from Web site development to systems integration, to the development of customer support applications.

The virtual focus groups were conducted using WebIQ™ (http://www.webiq.net), a Web-based meeting support application that includes capabilities for building an agenda, conducting electronic brainstorming, and administering questionnaires. Each participant was given an individual log-in and password. After logging in, participants filled out a questionnaire that asked about a specific virtual project in which they had participated within the last year. Responses to brainstorming questions were instructed to be about that same project. Participants then had a 72-hour window in which to brainstorm ideas about the following two questions.

1. What specific management and team member practices contributed to the effectiveness of your project?
2. What specific management and team member practices contributed to the ineffectiveness of your project?

The instructions asked participants to think broadly to include individual behaviors, processes, technologies, and tools as they applied to both of the questions.

The questionnaire data provided the basis for classifying each project with respect to the typology. The following characteristics of the project were derived.

- Project complexity (average of eight questions related to complexity)
- Project scope (average of three questions related to scope)
- Project risk (average of six questions related to risk)

Project type was calculated as the mean of project complexity, scope, and risk. There was a natural break point between the top four projects (highest complexity, scope, and risk) and the bottom three projects (lowest complexity, scope, and risk). The four projects with the highest scores were identified as extreme projects; those with the lowest scores were identified as lean projects, and the remaining ones were identified as hybrid projects. The overall mean for extreme projects ranged from 3.94 to 4.08, while the overall mean for lean projects ranged from 2.00 to 2.68 (on a five-point scale). Hybrid project means ranged from 3.21 to 3.75, thus each break point between the three different project types was half a point between lean and hybrid and nearly a quarter of a point between hybrid and extreme. A total of 14 unique projects were reported: a different project by each participant who completed all the phases of the study. Two examples from each type of project show the diversity: (a) Lean projects may include the development of a Web site, or installing and testing a new version of an application, (b) hybrid projects include customer support and the enhancement of an application, and (c) extreme projects include large dual-shore development projects, or adding a new product line.

Project success and virtuality were also calculated from questionnaire items. Based on prior research, overall virtuality was assessed by asking participants to rate the number of organizations or firms represented by project team members and their temporal dispersion. Similarly, project success was measured by assessing the extent to which participants perceived that the project was completed on schedule and within budget, met goals and specifications, and was on the whole successful. These variables were used to evaluate the impact of virtuality and success across the proposed virtual project typology.

We developed a coding scheme to analyze the brainstorming text (see Table 3). Each complete idea from the two brainstorming questions was coded for references to communication, coordination, or control. Each idea could have more than one reference to a theme, as well as a reference to multiple themes. Each complete idea was also coded for any reference to a technology, and a plus or minus sign was used to show whether the technology was being referred to as having a positive or negative impact. The authors worked together initially to identify the codes, resolving disagreements through discussion. Remaining data were coded independently and then reviewed, and discrepancies were resolved through discussion, with few disagreements.

ANALYSIS AND RESULTS

This section provides descriptive information from the questionnaire data, followed by the key patterns that relate to coordination, control, and communication as derived from the virtual focus group data.

Use of Technology

In the presession survey, participants were asked to rate the extent to which they used specific technologies to work with team members on the project. Our data confirm that e-mail is still the most-often-used technology for communication among virtual team members regardless of whether the project type is lean, hybrid, or extreme (see Table 4). The next most-used technology was various forms of the telephone, including conference calling and voice mail. The frequency rankings of e-mail and telephone use were true across all project types. Next in importance were face-to-face meetings, including an especially interesting result. The data show that team members make the most use of face-to-face communication in extreme and hybrid projects and very minimal use in lean projects. This is probably because lean projects are clear and have little complexity and scope, and could potentially be dealt with collaboratively via e-mail alone.

Table 3. Coding scheme with examples

Text of Brainstorming Idea	Theme	Subcategory (if applicable)	Technology (if applicable)	+/-
Utilized the phone for discussion and error diagnosis/ resolution	Communication	Meaning	Telephone	+
Good error logging capabilities of tools	Coordination		Distributed PM tools	+
Daily checkpoint meetings amongst the developers and the architecture folks were crucial and added a lot value	Control			

Table 4. Use of technologies during virtual projects

Technology	Mean for Lean Projects	Mean for Hybrid Projects	Mean for Extreme Projects	Overall Mean	Overall Std. Dev.
E-mail	4.65	5.00	5.00	4.93	0.27
Telephone	3.00	4.71	4.50	4.29	0.91
Conference calling	2.67	4.57	4.50	4.14	0.95
Voice mail	2.33	4.00	4.00	3.64	0.84
Face-to-face meetings	1.33	3.14	3.25	2.79	1.12
Tools for group work, distributed PM tools, EMS, IM, shared whiteboard, and others	Ratings ranged from 1.00 to 2.00				

Scale: 1 = Never; 5 = Almost Always

Interestingly, there is also a clear break in the frequency with which participants used the more traditional e-mail and voice tools vs. the tools for group work that have been developed more recently. Such tools as simultaneous document editing and shared whiteboard were rarely used. It is also worth noting that distributed project management and electronic meeting systems were used very little across all project types. The means of usage are highest in hybrid projects for both of these tools, but even so, the means are still low. These two tools in particular support structure for group processes, but they require a greater learning curve and continuing reinforcement.

Patterns for Effective Management of Virtual Projects

We argued that three theoretical elements should help to define patterns of project management, namely, communication, coordination, and control. Furthermore, technology was expected to constrain and enable how each element would be handled and the balance or pattern among elements. Thus, a potential design pattern for virtual project management would include descriptions of processes, best practices, factors, tools, and/or techniques that impinge upon coordination, communication, and control.

In this section, we detail some of the patterns that were identified for lean, hybrid, and extreme projects. Each pattern is based on the brainstorming data from the virtual focus groups. That is, for each type of project (lean, hybrid, extreme), we examined all the comments that were coded for each dimension of management practice (communication, coordination, control) and developed a pattern based on that set of comments. Multiple patterns could be generated from one set of comments.

Each pattern is described in terms of (a) the pattern's name, a descriptive word or phrase that captures its essence, (b) the context, a description of the situation to which the pattern applies, (c) the problem, a question that captures the essence of the problem that the pattern addresses, and (d) the solution, a prescription for dealing with the problem. The patterns presented here are a subset of all the patterns discovered during this study, showing examples of the key concerns in each type of project. The goal here is to provide the most critical patterns for a virtual context: those that could potentially be used as a check or safeguard against ineffective project management practices. Appropriate attention and management of communication, control, and coordination via the application of these patterns may offer help in improving or assuring the effectiveness of virtual

Table 5. Patterns by project type

Lean	Hybrid	Extreme
• ChangeControlCoordination • CommTime • FlexWorkTime • Gatekeeping • ManagerialProjectControl • SharedResources • TaskCoordination • TeamProjectControl • VersionControl	• ConflictResolution • HumanExpertise • MeetingDesign • ProjectLeadership • RelationshipCoordination • RoleCoordination • SharedResources	• CoordinateHumanResource • ManageCommitment • ManageKnowledge • ManageTeamTraining • ManageVirtuality • SharedUnderstanding • Standardize
Common across all project types: CommunicationCheck, FaceTimeCheck, and ScopeCreepCheck		

project management practices. Table 5 shows the names of all patterns identified in our study; a subset is illustrated in the subsequent sections.[2]

Common Patterns

Communication is not only critical for all types of projects, but it impacts effective coordination and control as well. The importance of communication is reflected in the two example patterns presented below, both of which are common to all three types of projects. The patterns relate to communication among team members either generally via various media and/or by using face-to-face meetings. Participants working in lean projects were particularly concerned about communication. Since lean projects are neither complex nor large in scope, the study's participants handled them mostly via a virtual mode and predominantly used e-mail and regularly scheduled telephone conferencing for communicating with stakeholders. However, it was evident from participant comments that these patterns were also applicable to hybrid and extreme projects.

CommunicationCheck

- **Context:** Team members do not have a shared understanding of project issues and solutions.
- **Problem:** How do you ensure effective communication among team members?
- **Solution:** Schedule periodic conferences using technologies that emphasize communication, for example, telephone and telephone conferencing, e-mail, and videoconferencing.

FaceTimeCheck

- **Context:** Team members neither agree nor have a shared understanding of project issues, solutions, work processes, and documentation requirements.
- **Problem:** How do you ensure effective communication and build trust among team members?
- **Solution:** Schedule periodic face-to-face conferences by flying some team members, possibly by rotation, to different locations. Though costly, even occasional participation in FTF meetings over the lifetime of a project is very effective. FTF meetings can engender increased trust and engagement among team members and also help clarify various facets of the project and resolve issues and conflicts.

Patterns in Lean Projects

Patterns identified in lean projects related to resource sharing, work schedule flexibility, and task, managerial, and team control issues. Issues included the sharing of information across virtual stakeholders, management of rework, change control and coordination, and management of scope creep. Participants in our study were particularly concerned about the negative impact of rework requests that cropped up without warning, primarily due to the absence of good communication and established coordination among stakeholders, project managers, and virtual team members. We provide examples of two patterns for lean projects that relate to control and coordination issues that are often encountered in projects of this type: ManagerialProjectControl and TaskCoordination.

ManagerialProjectControl

- **Context:** The progress of the project is impeded due to inadequate information sharing between the project team members and the manager responsible for interacting with the client.
- **Problem:** How do you monitor project changes based on interactions with your client?
- **Solution:** Schedule periodic (weekly or daily, as needed) project review meetings for all or some team members with the project manager or manager interacting with your client(s). These meetings are used both to update the manager regarding the project status and to cull new information obtained from the project client(s) that may have a direct impact on project tasks.

TaskCoordination

- **Context:** There is a complete disconnect between team members as new members are added.
- **Problem:** How do you ensure task coordination as new members are added to a team?
- **Solution:** Coordinate task assignment to new team members by clearly communicating revised roles and responsibilities along with timelines and tasks to all the team members. Ensure that everyone understands the assignment of new members in the team and convey this to all stakeholders and team members.

Patterns in Hybrid Projects

Patterns in hybrid projects are related to such issues as meeting design, shared resources and infrastructure capabilities, team member role and relationship coordination, human expertise, conflict resolution, and project leadership. The following two patterns provide examples of critical issues in hybrid projects. The MeetingDesign pattern addresses the creation of an effective meeting environment for a variety of stakeholders in different situations. The RoleCoordination pattern addresses the issue of ensuring clarity in role responsibilities, another area where virtual projects are likely to suffer given the distributed nature of at least some project team members.

Meeting Design

- **Context:** During meetings conducted via conference calls, your team gets bogged down in details that do not necessarily apply to many team members. This becomes increasingly complicated with a large number of project stakeholders.
- **Problem:** How do you develop a meeting environment that stimulates effective communication among all team members?
- **Solution:** Schedule periodic conferences using a variety of technologies that emphasize communication (e.g., telephone and telephone conferencing, e-mail, and videoconferencing). Design meetings

based on the following guidelines: (a) Use the participation of all stakeholders when the goal is to inform and develop a shared understanding of broad project goals and issues, (b) use selective participation of relevant stakeholders to deal with specific issues and challenges, and (c) keep meeting agendas short. Remember that people have short attention spans, particularly when you cannot see them. Anything more than an hour is probably better suited to a focused small meeting; consider having more meetings rather than long ones. Also consider which format would work best at meetings for the issues at hand.

RoleCoordination

- **Context:** Your team members are unclear about their roles and responsibilities in the project. This is causing misunderstandings about project goals and resulting in a delayed project.
- **Problem:** How do you provide team members with a clear understanding of their individual roles and responsibilities in the project?
- **Solution:** Cleary define team members' roles and responsibilities and work processes at the outset. If new members are added, clearly communicate revised roles and responsibilities to all team members. Ensure that they all understand their assignments and provide them with the tools to deliver. Communicate roles, responsibilities, and work processes to all stakeholders and team members. If feasible, consider rotating members through different roles. Use technologies with a high process structure (such as virtual collaboration systems and knowledge management tools) to share information on the team's work processes and the roles and responsibilities of team members. Include team members in designing work processes and delineating roles and responsibilities. This will increase team ownership.

Patterns in Extreme Projects

Extreme projects are likely to need all of the patterns that were identified for hybrid projects. In addition, we identified patterns for extreme projects that related to the coordination between remote and local sites, management of virtuality, management commitment, standardization of processes and documentation, knowledge management, building a shared understanding of project requirements and processes, and appropriate and consistent training of all team members. We present two patterns for extreme projects that were particularly interesting and need specific attention from managers.

First, the ManageVirtuality pattern is particularly significant for extreme projects, which by definition involve a combination of high complexity, scope, risk, and varying levels of virtuality. Overcoming geographic and time-zone differences is not just critical for global teams, but can also matter within a single country. For example, one of the participants from the United States said that project notifications from the Pacific time zone would reach the central time zone later in the day, leaving less time for addressing issues and/or requiring team members to work outside of normal hours.

ManageVirtuality

- **Context:** Your team is having difficulty with time-zone differences at both the national and global levels. This problem occurs especially during crunch time or crisis situations when communication is not prompt. As a result, the problem resolution process is delayed.
- **Problem:** How do you overcome time-zone and geographic differences and effectively engage all team members?

- **Solution:** Overcome or eliminate distance barriers due to time-zone and geography by providing activities that require intensive interaction and coordination (e.g., project initiation) through temporarily collocating team members. Require periodic site visits and travel by team members to different sites. Designate team member liaisons as focal points of coordination who spend some time in the home office location to become acculturated and informed about technical issues; liaisons can then transfer knowledge to local sites for day-to-day coordination. Assign team members in one geographic region (e.g., North and South America) to tasks requiring telephone or video-based interactions because they share time zones and thus can more easily schedule conferences.

The second example for extreme projects, the ManageKnowledge pattern, focuses attention on the importance of knowledge management and information sharing among stakeholders and the organization as a whole. One participant pointed to the effectiveness of a knowledge portal in the project as follows: "[W]e have a homegrown tool—knowledge portal which has features for collaboration, e-learning and knowledge management which was found really useful. It was a challenge to implement it but once the team started using it everyone saw the power and usefulness of the same."

ManageKnowledge

- **Context:** Team members are unable to share intelligence and best practices, or to simultaneously edit master documents. In some instances, team members are not following established processes.
- **Problem:** How do you mobilize and share knowledge across the team and your organization?
- **Solution:** Start with input from project team members across the organization and build a repository of best practices, templates, learning tools, workflow standards, and examples of processes within standard methodologies. Make sure all members of the team have access to and can contribute to the knowledge portal.

DISCUSSION

One issue that was reinforced throughout the projects in this study is the importance of communication. Regardless of the type of project, communication was mentioned time and time again as a fundamental necessity. Both prior research and our study reinforce the idea that communication is important in and of itself as well as through its relationship to coordination and control. All of these teams relied heavily on e-mail and voice media, which emphasize communication. Thus, the communication dimension of technology had the greatest priority, more so than process structure or information processing.

Our study also reinforced the importance of periodic face-to-face communication for virtual teams, a finding that is consistent with prior research (Maznevski & Chudoba, 2000). Participants emphasized the advantages of regular face-to-face meetings for some or all team members to help resolve issues and monitor progress. Periodic collocation of team members can help to establish ground rules and common understanding, which in turn facilitates communication and coordination when team members return to their distant home locations. This practice allows team members to build a social network, as well as stimulate the development of team identity, cohesion, and commitment that help to sustain members during dispersed periods (Davidson & Tay, 2003). Another interesting result related to technology was the relatively low use of distributed project management tools. Indeed, there was generally low use for all of the technologies that we would categorize as providing support for process structure or information processing. Clearly, there is much room for improvement in providing better tools and training for virtual teams in these areas.

In order to develop appropriate training, apply the right tools and use patterns effectively; one thing that managers must first be able to do is to assess the nature of the project and its context. This study provides an initial set of such assessment tools. We have described specific measures to identify the nature of a virtual project in terms of complexity, scope, risk, and virtuality. In addition, a technology inventory can be taken to identify the extent and type of technology support available. Managers can be provided with a simple project dashboard type of tool that elicits information on each of the aspects of the project and displays calculations on the measures (Khazanchi & Zigurs, 2007). The assessment can help to build a common vocabulary for the project team that will help to identify the key issues around which patterns will be identified.

The discovery and validation of patterns should be an ongoing process that is part of a continual learning approach for project managers and members. We recommend a five-step process: (a) Recognize and abstract an attribute or feature of virtual project management that affects the key issues of communication, coordination, or control, (b) define the recurring problem or system of conflicting forces that the feature solves, (c) define the context in which the feature is appropriate, (d) name and describe the pattern, and (e) continue to validate and refine the pattern. This five-step approach is described in more detail elsewhere (Khazanchi & Zigurs, 2007), but we have provided an example of one round of that process in this study. The brainstorming tools that were used in this project are a good example of tools that can support the important first step of this process.

CONCLUSION

This study makes several contributions. First, we developed a project typology based on theoretically founded characteristics of projects. Second, we applied pattern theory to the practice of virtual project management in terms of coordination, control, and communication. Third, we collected data that identified a starting set of patterns for the effective management of lean, hybrid, and extreme virtual projects. Fourth, we developed a structured approach that is usable by other researchers.

The contributions to theory come from the development of concepts, the typology, and the pattern approach. We have elaborated the concepts of virtuality, communication, control, and coordination as they relate to virtual project management and developed a new typology and descriptions of extreme, lean, and hybrid projects. In addition, if at its core, pattern theory is the basis for developing a solution to a problem in a certain context; our study has identified an initial set of patterns in virtual project management across various types of projects. Indeed, there is still much to learn from the original concept of patterns. For example, in his speeches and other writings, Alexander (1996) has argued that patterns should also have a "moral component" and "create coherence, morphological coherence in the things which are made with it," and "be generative, [that is], allow people to create coherence, morally sound objects, and encourage and enable this process." We have not dealt with these aspects directly in our study, but they have the potential to contribute to a higher level of understanding. Also, patterns might be viewed as providing a bridge between virtual and traditional contexts. Earlier, we argued that virtuality is a continuum. Viewed that way, patterns can expand one's capability to operate effectively along the entire continuum, that is, to connect to what we already know from traditional practices and bridge the gap to virtual environments.

The contributions to research methodology arise from the measures and coding scheme. We have developed and implemented measures for the various concepts that drive the typology of virtual projects. We have also developed a coding scheme to analyze brainstorming text from virtual focus group sessions. In fact, we believe that our study is a relatively unique example of conducting research using asynchronous virtual focus groups with globally dispersed participants. This required the development of a complete set of protocols for managing, organizing, and conducting the data collection elements of the research.

Contributions to practice are in the patterns themselves. The patterns identified in the study can be utilized as critical checks and/or design principles for use in managing virtual projects. Managers of virtual projects can follow either a deductive or inductive process. A deductive approach would start with identifying the type of project (using the measures developed), then searching the patterns for that project type and applying the prescribed solution for each relevant problem. An inductive approach would start with a search of the pattern library, looking for any patterns that apply. If the problems fall primarily into one project type, then a manager can reasonably infer that this is the type of the project.

As with any study of this kind, several limitations apply. Firms and participants were selected from a convenience sample based on contacts developed by the authors. We recognize the limitations inherent in such a sample, but given that the study is only a first step and that the firms represented a good cross section of global industries, we believe that the data provide a good starting point. The patterns were derived from a limited data set. Even though we had participants from five different companies and a sufficient diversity of project types for our categorization, the generalizability of our conclusions is limited. In addition, even though the coding scheme was based on theory, there may be other relevant dimensions besides the ones we identified and used in the development of patterns. The next steps for research follow naturally from the limitations of the study. Additional settings need to be examined and the concepts tested with additional data.

Some of our patterns are clearly related to each other in some way; this came about from the natural process of applying the grounded theory approach to our data. Inducing from the data that was provided during discussion in our virtual focus groups, we found that some patterns turn out to have similarities and apparent relationships. In fact, this is as it should be within the realm of pattern theory since it allows us to develop a pattern language as more patterns are discovered and a coherent structure of patterns in virtual project management evolves. Our study resulted in the identification of patterns for managing virtual projects; we do not lay claim to the uniqueness of all of these patterns. In fact, we are quite sure that many elements of these patterns already exist in some form in traditional PM or in other fields of endeavor. Further exploration of these patterns may result in the coalescing of some and/or expansion or redefinition of others.

We started with the goal of going beyond a cookbook approach to the management of virtual projects. The theoretical frame of patterns helps us make a large step toward that goal. One of the most intriguing aspects of the pattern approach is the idea that specific patterns themselves should fit together into a coherent pattern language that provides a higher level of understanding. A specific pattern is not a prescription, although a pattern could be used to create a prescription in a specific context. Patterns are generic and more akin to the idea of universal laws than to prescriptions. Consistent with Alexander's ideas, patterns should help to create new processes and help organizations to continue to change and adapt. Patterns as a set represent an abstraction of something important: They provide a language for communication and action, and they embody a value set. It is these aspects of patterns that can provide that higher level of understanding.

The results from this study provide immediate and practical guidance for managers of virtual projects, as well as providing a strong foundation for the necessary next steps in research. Those steps entail studying how people use patterns in a systematic way, how people adapt them, and how pattern use reveals what really matters across a variety of contexts. The pattern perspective provides an essential cross-level view—a view of specific practices for success as embodied in individual patterns, as well as the systemic view that comes from their combination in a pattern language.

ACKNOWLEDGMENT

The research was partially supported by a grant from the Project Management Institute.

REFERENCES

Alexander, C. (1965). *Notes on the synthesis of form.* Cambridge, MA: Harvard University Press.

Alexander, C. (1996, October). The origins of pattern theory, the future of the theory, and the generation of a living world. *ACM Conference on Object-Oriented Programs, Systems, Languages and Applications (OOPSLA).* Retrieved from http://www.patternlanguage.com/archive/ieee/ieeetext.htm

Alexander, C., Ishikawa, S., Silverstein, M., Jacobson, M., Fiksdahl-King, I., & Angel, S. (1977). *A pattern language: Towns, buildings, construction.* Oxford University Press.

Burke, K., Aytes, K., & Chidambaram, L. (2001). Media effects on the development of cohesion and process satisfaction in computer-supported workgroups: An analysis of results from two longitudinal studies. *Information Technology & People, 14*(2), 122-141.

Carlson, J. R., & Zmud, R. W. (1999). Channel expansion theory and the experiential nature of media richness perceptions. *Academy of Management Journal, 42*(2), 153-170.

Carmel, E., & Agarwal, R. (2001). Tactical approaches for alleviating distance in global software development. *IEEE Software*, 22-29.

Chinowsky, P. S., & Rojas, E. M. (2003). Virtual teams: Guide to successful implementation. *Journal of Management in Engineering, 19*(3), 98-106.

Christensen, M. J., & Thayer, R. H. (2001). *The project manager's guide to software engineering's best practices.* Los Alamitos, CA: IEEE Computer Society.

Cramton, C. D. (2001). The mutual knowledge problem and its consequences for dispersed collaboration. *Organization Science, 12*(3), 346-371.

Crowston, K. (1991). A coordination theory approach to organizational process design. *Organization Science, 8*(2), 157-175.

Daft, R. L., & Lengel, R. H. (1986). Organizational information requirements, media richness and structural design. *Management Science, 32*(5), 554-571.

Davidson, E. J., & Tay, A. S. M. (2003). Studying teamwork in global IT support. *Proceedings of the 36th Hawaii International Conference on Systems Sciences.*

DeSanctis, G., & Poole, M. S. (1994). Capturing the complexity in advanced technology use: Adaptive structuration theory. *Organization Science, 5*(2), 121-147.

Dubé, L., & Paré, G. (2004). The multi-faceted nature of virtual teams. In D. J. Pauleen (Ed.),

Virtual teams: Projects, protocols, and processes (pp. 1-39). Hershey, PA: Idea Group Publishing.

Espinosa, J. A., Cummings, J. N., Wilson, J. M., & Pearce, B. M. (2003). Team boundary issues across multiple global firms. *Journal of Management Information Systems, 19*(4), 157-190.

Evaristo, R., & Munkvold, B. E. (2002). Collaborative infrastructure formation in virtual projects. *Journal of Global Information Technology Management, 5*(2), 29-47.

Fernandez, A., Holmer, T., Rubart, J., & Schuemmer, T. (2002). Three groupware patterns from the activity awareness family. *EuroPLoP 2002: Seventh European Conference on Pattern Languages of Programs.*

Fiol, C. M., & O'Connor, E. J. (2005). Identification in face-to-face, hybrid, and pure virtual teams: Untangling the contradictions. *Organization Science, 16*(1), 19-32.

Gamma, E., Helm, R., Johnson, R., & Vlissides, J. (1994). *Design patterns: Elements of reusable object-oriented software.* Boston: Addison-Wesley.

Gassmann, O., & Von Zedtwitz, M. (2003). Trends and determinants of managing virtual R and D teams. *R & D Management, 33*(3), 243-262.

Glaser, B. G., & Strauss, A. L. (1971). *The discovery of grounded theory: Strategies for qualitative research.* Chicago: Aldine Publishing Company. (Original work published 1967)

Goodbody, J. (2005). Critical success factors for global virtual teams. *Strategic Communication Management, 9*(2), 18-21.

Grant, R. M. (1996). Toward a knowledge-based theory of the firm. *Strategic Management Journal, 17,* 109-122.

Henderson, J. C., & Lee, S. (1992). Managing IS design teams: A control theories perspective. *Management Science, 38*(6), 757-777.

Homsky, O. (2003). More patterns for group leadership. *EuroPLoP 2003: Seventh European Conference on Pattern Languages of Programs.*

IEEE. (2004). *IEEE Standard 1540-2004. IEEE standard for software life cycle processes: Risk management.* IEEE Press.

Jarvenpaa, S., & Leidner, D. (1999). Communication and trust in global virtual teams. *Organization Science, 10*(6), 791-815.

Katzy, B., Evaristo, R., & Zigurs, I. (2000). Knowledge management in virtual projects: A research agenda. *Proceedings of the 36th Hawaii International Conference on Systems Sciences.*

Khazanchi, D., & Zigurs, I. (2005). *Patterns of effective management of virtual projects: An exploratory study.* Newtown Square, PA: Project Management Institute.

Khazanchi, D., & Zigurs, I. (2007). An assessment framework for discovering and using patterns in virtual project management. *Proceedings of the 36th Hawaii International Conference on Systems Sciences.*

Kirsch, L. J. (1996). The management of complex tasks in organizations: Controlling the systems development process. *Organization Science, 7*(1), 1-21.

Malone, T., & Crowston, K. (1994). The interdisciplinary study of coordination. *ACM Computing Surveys, 26*(1), 87-119.

Martins, L. L., Gilson, L. L., & Maynard, M. T. (2004). Virtual teams: What do we know and where do we go from here? *Journal of Management, 30*(6), 805-835.

Massey, A. P., Hung, Y. C., Montoya-Weiss, M., & Ramesh, V. (2001). When culture and style aren't about clothes: Perceptions of task-technology "fit" in global virtual teams. *Proceedings of the 2001 International ACM SIGGROUP Conference on Supporting Group Work*, 207-213.

Maznevski, M. L., & Chudoba, K. M. (2000). Bridging space over time: Global virtual team dynamics and effectiveness. *Organization Science, 11*(5), 473-492.

McGrath, J. E., & Hollingshead, A. B. (1994). *Groups interacting with technology: Ideas, evidence, issues, and an agenda.* Thousand Oaks, CA: Sage Publications.

Nunamaker, J. F., Dennis, A. R., Valacich, J. S., Vogel, D. R., & George, J. F. (1991). Electronic meeting systems to support group work. *Communications of the ACM, 34*(7), 40-61.

Pinsonneault, A., & Caya, O. (2005). Virtual teams: What we know, what we don't know. *International Journal of e-Collaboration, 1*(3), 1-16.

Powell, A., Piccoli, G., & Ives, B. (2004). Virtual teams: A review of current literature and directions for future research. *The Data Base for Advances in Information Systems, 35*(1), 6-36.

Project Management Institute (PMI). (2004). *A guide to the project management body of knowledge* (3rd ed.). Newtown Square, PA.

Project Management Institute (PMI) Standards Committee. (1996). *A guide to the project management body of knowledge.* Newtown Square, PA: Project Management Institute.

Royce, W. (1998). *Software project management: A unified framework.* Reading, MA: Addison-Wesley, Inc.

Rutkowski, A.-F., Vogel, D., Bemelmans, T. M. A., & Van Genuchten, M. (2002). Group support systems and virtual collaboration: The HKNet Project. *Group Decision and Negotiation, 11*(2), 101-125.

Sarker, S., & Sahay, S. (2001). Information systems development by US-Norwegian virtual teams: Implications of time and space. *Proceedings of the 36th Hawaii International Conference on Systems Sciences.*

Schubert, P., Leimstoll, U., & Romano, N. C. (2003). Internet groupware systems for project management: Experiences from a longitudinal study. *Proceedings of the 16th Bled eCommerce Conference: eTransformation*, 611-631.

Schuemmer, T. (2003). Evolving a groupware pattern language. *ECSCW2003: Fifth International Workshop on Collaborative Editing.*

Shani, A. B., Sena, J. A., & Stebbins, M. W. (2000). Knowledge work teams and groupware technology: Learning from Seagate's experience. *Journal of Knowledge Management, 4*(2), 111-124.

Shenhar, A. J. (1998). From theory to practice: Toward a typology of project management styles. *IEEE Transactions on Engineering Management, 45*(1), 33-48.

Turvey, M. T. (1990). Coordination. *American Psychologist, 45*(8), 938-953.

Van Ryssen, S., & Godar, S. H. (2000). Going international without going international. *Journal of International Management, 6*, 49-60.

Völter, M. (2002). Hope, belief and wizardry: Three different perspectives on project management. *EuroPLoP 2002: Seventh European Conference on Pattern Languages of Programs.*

Watson-Manheim, M. B., & Belanger, F. (2002). Exploring communication-based work processes in virtual work environments. In *Proceedings of the 36th Hawaii International Conference on Systems Sciences.*

Watson-Manheim, M. B., Chudoba, K. M., & Crowston, K. (2002). Discontinuities and continuities: A new way to understand virtual work. *Information Technology & People, 15*(3), 191-209.

Zalesny, M. D., Salas, E., & Prince, C. (1995). Conceptual and measurement issues in coordination: Implications for team behavior and performance. *Research in Personnel and Human Resources Management, 13*, 81-115.

Zigurs, I., & Buckland, B. (1998). A theory of task/technology fit and group support systems effectiveness. *MIS Quarterly, 22*(3), 313-334.

Zigurs, I., Evaristo, R., & Katzy, B. (2001). *Collaborative technologies for virtual project management.* Washington, DC: Academy of Management.

ENDNOTES

1. See Khazanchi and Zigurs (2005) for a complete copy of the questionnaire.
2. For complete details on the full set of patterns, see Khazanchi and Zigurs (2005).

Chapter VIII
The Launch of Web-Based Collaborative Decision Support at NASA

Irma Becerra-Fernandez
Florida International University, USA

Matha Del Alto
NASA Ames Research Center, USA

Helen Stewart
NASA Ames Research Center, USA

ABSTRACT

Today, organizations rely on decision makers to make mission-critical decisions that are based on input from multiple domains. The ideal decision maker has a profound understanding of specific domains coupled with the experience that allows him or her to act quickly and decisively on the information. Daily, decision makers face problems and failures that are too difficult for any individual person to solve; therefore, teams are now required who share their knowledge in spontaneous collaborations. Since requisite expertise may not all reside in the same organization, nor be geographically colocated, virtual networked teams are needed. This chapter presents a case study describing the development and use of Postdoc, the first Web-based collaborative and knowledge management platform deployed at NASA.

INTRODUCTION

Knowledge-intensive organizations rely on decision makers to make mission-critical decisions based on input from multiple domains (Nonaka & Takeuchi, 1995). The ideal decision maker has a profound understanding of specific domains that influence the decision-making process coupled with the experience that allows quick and decisive action based on such information (Becerra-Fernandez, Gonzalez, & Sabherwal, 2004; Davenport & Prusak, 1998). The ideal decision maker is usually someone who has lengthy experience and implicit knowledge gained from years of observation (Leonard & Swap, 2004, 2005; Senge, 1990).

While the profile of today's ideal decision maker does not mark a significant departure from past practices, the following four underlying trends are raising the stakes in the decision-making scenario (Becerra-Fernandez et al., 2004).

1. **Increasing complexity:** The complexity of the underlying domains (internal, external, competitive, process, technology, etc.) is increasing.
2. **Accelerating volatility:** The pace of change (volatility) within each domain is increasing.
3. **Speed of responsiveness:** The time required to take action based upon subtle changes within and across domains is decreasing.
4. **Less experience:** Individuals with decision-making authority potentially have less tenure with the organization than ever before due to such factors as high employee turnover rates.

Today's technological environment is complex and changes at an ever-increasing pace. Many problems and failures are too difficult for any individual person or organization to solve. Teams are now required to share their knowledge in spontaneous collaborations. Since requisite expertise may not reside in the same organization, nor be geographically colocated, virtual networked teams are needed. Collaborative decision support technologies enable knowledge sharing and provide access to explicit organizational knowledge, so it is easy to learn from previous experiences. The use of adequate collaboration technology platforms results in the minimization of costly mistakes while reducing time to market in research and development projects (Majchrzak, Cooper, & Neece, 2004). Collaboration tools also help the organization make better decisions by capturing the knowledge from groups of experts and providing the means to mine this knowledge and experience (Malhotra & Majchrzak, 2005; Malone, Crowston, Lee, & Pentland, 1999).

In this chapter, we describe the characteristics of decision making in knowledge-intensive organizations (Becerra-Fernandez et al., 2004). This chapter is based largely on the case study published by Becerra-Fernandez, Del Alto, & Stewart (2006). Given the fact that increasingly complex decisions require the collaboration of individuals who many times are dispersed geographically and across organizations, Web-based collaboration technology platforms can effectively support decision making at such organizations. The balance of the chapter is organized as follows. The second section provides a description of one of the best-known knowledge-intensive organizations, the National Aeronautics and Space Administration (NASA). Given the characteristics of decision making at NASA, it provides for an excellent environment to study how this organization has been able to successfully coordinate complex projects through the use of Postdoc, a Web-based collaboration system. The third section describes the design, development, and implementation of Postdoc. The fourth section describes the use of Postdoc to manage complex projects such as Remote Agent, and the fifth section demonstrates the value of this application as a platform for collaboration in complex decision-making environments. Finally, the last section presents conclusions and lessons that could prove valuable to organizations considering the implementation of such systems, as well as a vision for the future of Web-based collaboration systems in general.

HISTORY OF COLLABORATIVE DECISION MAKING AT NASA

A recent NASA workforce study (*NASA Workforce Analysis Report*, 2003) reveals that the average number of years of service for all occupation groups at NASA has been increasing since 1995. The NASA workforce has in fact been aging since most recent science and engineering graduates are taking jobs outside of government, partly due to the lure of dot-coms and private industry in general, coupled with years of government downsizing and hiring freezes. Most of NASA's employees today are between the ages of 40 and 60 years, and less than 5% of NASA's scientists and engineers are younger than 30 years old. The aforementioned trends provide a backdrop for the decision-making scenario at NASA.

1. **Increasing complexity:** NASA is no exception to increasing complexity as it is being forced to diminish support for personnel devoted to safety assurance due to congressional pressures to downsize while fulfilling increasingly complex safety requirements associated with the launch of the space shuttle. A recent article reports that NASA's safety office manpower at its headquarters in 2003 is at 55% of its peak level in 1994 (Smith, 2003).
2. **Accelerating volatility:** This is certainly the case at NASA, where the rapid changes in technology continue to impact the design of new space vehicles. The volatility in technology development counteracts NASA's safety requirements, resulting in the preference for proven leading technologies over "bleeding," albeit unproven, technologies.
3. **Speed of responsiveness:** During the launch of the Space Shuttle Columbia, one of the aircraft wings' leading edges was hit by falling foam debris, which previously covered the liquid fuel booster.[1] Safety experts who evaluated this particular incident had to make a quick decision as to the impact of this event, which in the case of the Columbia, unfortunately resulted in an incorrect assessment. The required speed of responsiveness contributes an additional dimension to the complexity of making an accurate assessment.
4. **Less experience:** At NASA, there is growing concern that the scarcity of younger scientists and engineers provides few opportunities for mentorship and knowledge transfer to the organization's future decision makers.

NASA is a knowledge-intensive organization, dedicated to the mission of human space flight, space science, Earth observation, and aeronautics research. NASA, like many U.S. government agencies, is a heavily matrixed organization with many entities involved in its mission: NASA field centers, NASA programs carried out at these centers, and industrial and academic contractors. Effective collaboration among these entities is critical for NASA to effectively carry out its mission.

For example, the pioneering research to develop Remote Agent,[2] the innovative software that operated the Deep Space 1 (DS-1) spacecraft and its futuristic ion engine, involved three teams of artificial intelligence experts from Carnegie Mellon University in Pittsburgh, the Jet Propulsion Lab (JPL) in Pasadena, California, and the NASA Ames Research Center (ARC) in Silicon Valley, which jointly developed the system. Together, the scientists developed the intelligent software that operated the DS-1 spacecraft more than 60 million miles away from Earth.

When Remote Agent was designed, it required innovations in highly risky intelligent technologies for systems execution, fault tolerance and recovery, and autonomous planning systems. Due to budget constraints, the Remote Agent design team could not work at the same location. Given that e-mail would not provide an adequate infrastructure, the Postdoc Web-based collaborative document management system was developed to support the need for distributed collaboration. The first version of Postdoc supported the collaboration and project management needs of the 25-researcher team. In the words of computer scientist Kanna Rajan, who participated in the Remote Agent research project:

Postdoc enabled the team to develop a common language that we used to share our design ideas and start talking about them. We created a token dictionary that enabled the defined team to establish clear semantics that were used to exchange comments among the team members.

Although the evidence of Postdoc's success presented in this case is limited to usage statistics and anecdotal comments from users, this case is valuable because it demonstrates that virtual teams can effectively improve their collaboration know-how when doing development work (vs. routine work) if the information technology provides the necessary support for contextualizaton (Malhotra & Majchrzak, 2005). In fact, Postdoc's success can be explained by the finding that information technology support for contextualization may be of particular benefit to distributed teams that are structurally diverse, primarily virtual, and composed of members performing what they perceive as nonroutine tasks (Malhotra & Majchrzak, 2004). In other words, Postdoc enabled virtual teams to share not only their content, but also their context.

THE DESIGN, DEVELOPMENT, AND IMPLEMENTATION OF POSTDOC AT NASA

The need for collaboration tools at NASA dates back to 1995, when the New Millennium Program was tasked with dramatically improving the process by which spacecraft are designed, built, and operated. The vision to develop a collaborative system resulted from the need to support a geographically dispersed team. The New Millennium Program had impressive goals that had to be met using constrained resources due to then NASA administrator Daniel Goldin's[3] mandate for "faster, better, cheaper, smarter" interplanetary missions. Already, the program was beginning to experience budget cuts and desperately required the use of collaborative tools to ensure meeting mission objectives and schedules. The initial attempt at building such a system, the New Millennium Documentation System, demonstrated shortcomings, specifically (a) a lack of portability, (b) lack of integration flexibility with other systems within NASA, and (c) the inability for users to manage and control access to directories and files.

A team comprised of employees at NASA ARC and JPL, as well as partners from Stanford University's Center for Design Research and private industry, was tasked with the development of a new collaborative system. At that time, most NASA teams involved in software development across the agency were only using the file transfer protocol (FTP) for transferring, sharing, and integrating software code, which did not provide good visual representation or programmatic organization. Also at that time, there were no other commercial off-the-shelf software applications that used the Web for collaborative document sharing and archiving.[4] The team quickly recognized that an adequate collaborative structure would require the development of an application that (a) used the emerging Web infrastructure for document uploading, archiving, visualization, and integration, (b) could easily be implemented agency-wide with access controls and authentication capabilities, (c) had a more portable application source, and (d) had features that allowed users full control of their information anytime and anywhere. It was decided to build these functionalities into Postdoc. Postdoc, which required a development effort of 5 person-years of software coding and testing, is a live system that is currently being used to support teams throughout NASA in their collaborations with other NASA facilities, private industry, and academia.

The first step in designing technology that supports collaborative decision making is to understand the role of collaborative environments in science and engineering. This requires the users to partake in the proactive design and development of the system that will support this new way of doing research (Kumar & van Dissel, 1996; Majchrzak et al., 2004). In order to identify this initial set of user requirements, the user-interface work group worked directly with the New Millennium Program users to understand their needs for information sharing across the mission. The user-interface work group was integrated with the DS-1 mission software development group, with whom they first identified the requirements for effectively collaborating across NASA centers, but later continued work with other mission development teams such as the spacecraft instrumentation team and the communication team. From this study, it was defined that the system needed to provide a specific set of capabilities, which are described in Table 1.

Postdoc was designed to meet these foundation requirements for knowledge management and collaboration by enabling data capture, analysis, and fusion capabilities. Furthermore, the software architecture also allows other virtual collaborative interfaces to be easily integrated

Figure 1. Postdoc architecture

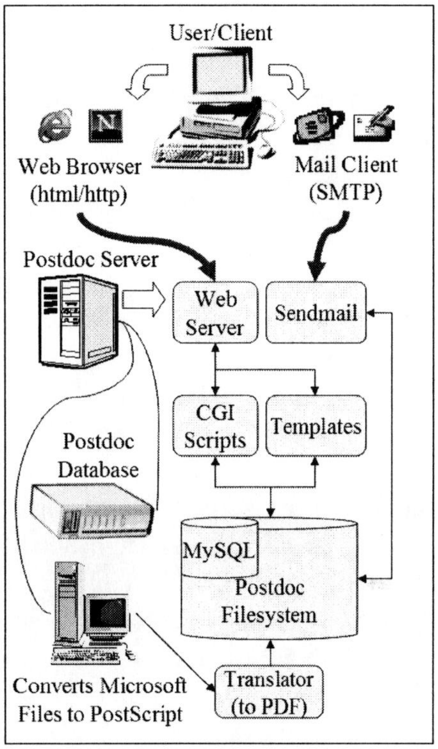

Table 1. Postdoc required capabilities

Name	Description
Access Control	Allow users to define and control access to their information. Inter-organizational users should be able to invite, and grant access to potential team members to collaborate as part of a geographically dispersed group. Postdoc would allow users to define the types of access to the information and manage their own information spaces.
Revision Control	Provide a platform for collaboration and a virtual document file system that supports versioning, document checkout, and revision control.
Audit Trail	Provide an audit trail for interactions with documents that the users can review. When users share documents, its important to know who has read or modified the documents and when they did so.
File-type Transparency	Be equipped with auto-conversion of file-type to portable document format (PDF), and on-demand document conversion to a broader option of document types (i.e., postscript, HTML, text, and digital assistant file types). Image files would need to be converted across different formats (e.g. JPEG, GIF, etc.).
Related Utilities	Have utilities such as fax capabilities, a fully featured email list manager, and an integrated online help.
System Administration	Provide administration features for system administrators monitoring the translations of the documents, maintaining statistics of events, and maintaining the software system.

into the structure. Figure 1 represents Postdoc's software architecture.

Postdoc is a multiuser, Web-based collaborative document management system that is used primarily for document storage and retrieval, including word processing documents, spreadsheets, slides, illustrations, images, video, audio, software archives, and others. The majority of Postdoc was written in the language Perl; the Web pages for the user interface are dynamically generated using ePerl. The HTML (hypertext markup language) generated conforms to Versions 3.2 and 4.0. Postdoc runs on a Unix operating system with a Web server (Apache), a relational database engine (MySQL), a full-text search engine (SWISH++), a document format conversion software (either Adobe Acrobat® Distiller® or Aladdin Ghostscript), and a facsimile software (HylaFax); most of these components are based on open-source standards. The conversion of Microsoft® Office, Adobe® Illustrator®, and other Macintosh or Windows documents to PostScript (and ultimately to PDF, Portable Document Format) uses a dedicated Windows system with custom software written in Python to perform the document conversions. Postdoc also allows converting image files among a choice of different formats, and it's this file conversion capability that allows Postdoc to stand out among many other similar systems. The Windows system communicates with the Unix system using Samba, an open-source environment that enables Linux and Unix servers to interoperate with Microsoft Windows clients.

In terms of the system's architecture, the Postdoc software uses a template-driven design, with HTML presentation code intermingled with macros and Perl instructions to provide interactivity and customizability with minimum effort. Preprocessor macros allow for the inclusion of common Web elements such as navigation. Libraries of common Perl code elements also reduce the maintenance overhead. Configuration of the software for unique installations is accomplished by redefining configuration parameters in a centralized file. Postdoc's dynamic data are stored both in the file system (for documents and e-mail messages) and in the MySQL relational database. Currently the system runs at NASA ARC on a Sun Solaris 2.6 server. As the system is written completely in Perl code, porting it to other modern Unix or Linux operating systems is a matter of installing the various off-the-shelf support components and ensuring that Perl is properly installed. The client interaction performance of Postdoc has maintained a high level of usability without the need for more complex Web application server architectures. Off-the-shelf components, such as Adobe Acrobat for PDF conversions, can be integrated into the system to take advantage of technology improvements. System reliability and availability of the Postdoc environment is key in supporting the adoption of the software nationally and internationally. The simplicity of the Apache Web server architecture and CGI interface, coupled with a variety of system monitors, helps to keep the system continuously available for months at a time without interruptions or outages.

POSTDOC IS LAUNCHED

With the initial development of Postdoc, NASA's first efforts toward the implementation of an innovative collaborative computing platform were initiated. The Postdoc team at ARC's Computational Science Division pioneered Web-based collaboration at NASA. Since its launch in 1997, Postdoc has proven to be an effective tool for organizing, storing, and retrieving data and information of all types. This multiuser, Web-based application has dramatically improved the communication of technical, managerial, administrative, and scientific information between members of widely dispersed teams, nationally and internationally. One of the first collaboration tools of its kind, Postdoc allowed NASA scientists and engineers to share mission-critical information in real time, across different organizations, geographies, and computing platforms, regardless of format. One of the unique features of Postdoc is that it can support the needs for secure interorganizational collaboration since the system provides built-in security and may reside outside of the organization's firewall. Given NASA's requirements for knowledge sharing across organizations, which include the agency and collaborating universities and private companies, this specific capability is included among its unique features.

The development of Postdoc focused on effectively supporting knowledge creation by directly enabling collaboration via the sharing of information resources among users. Users are allowed to add structure to those resources that are consistent with their own work practices. Soon after the software was launched, it was viewed as an enabler for sharing meaningful processes, information, and documents between the many geographically distributed programs and enterprise activities at NASA. Bob Kanefsky, staff scientist and lead of the Postdoc development team, recalls the following:

The prototype system that was used during the Remote Agent development itself was...developed rapidly, in the heat of the Remote Agent prototyping effort, and was further refined in parallel with the flight software development for the DS-1 mission. The original model was for the system to be a distributed editing system that would allow multiple authors to create living documents by adding paragraphs of text, links, images, and so on. It quickly became clear, though, that users preferred to use conventional editing tools such as Microsoft Office to create documents, and saw value in Postdoc as a means to pass these documents around for review and editing. We therefore switched to the model that it should act much like a distributed file system: "section headers" became "folders," "paragraphs" became "sticky notes," and documents became the most common type of item to add to a folder.

We experimented with specialized types of folder contents. We briefly used an Action Item object for tracking DS-1 workflow and deadlines, but it proved not to be the most convenient way to do this. We made more extensive use of a DS-1-specific "Message Dictionary" that automatically cross-linked DS-1 documentation by inter-process message names. For example, the EXECUTE_TURN message was documented by the Attitude Control, Remote Agent Executive, and Remote Agent Planner subsystem documentation, and also appeared in logs of test runs. Hyperlinks were automatically added to these documents, wherever the term EXECUTE_TURN appeared, to cross-link them whenever a new document was posted.

The most popular capabilities of Postdoc, though, were the ability to create and manage "folders," post documents to them, and automatically translate PC/Mac documents to Portable Document Format for use on Unix. Those capabilities, along with new ones [enabled] Postdoc to remain in continuous use into the 21st century.

The DS-1 Remote Agent team quickly became one of Postdoc's most active groups of users, using the system as a repository of source code, design parameters, and software models for the DS-1 Remote Agent. Successful early adoptions of Postdoc by the DS-1 developers demonstrated its ability to support aerospace designers to shorten and enhance the efficiency of the management and design-cycle processes. One of the program milestones was to demonstrate an increase by a factor of 10 times the number of geographically distributed participants using Web-based collaboration in support of science and engineering design, development, and management activities. User participation in Postdoc increased extensively (refer to Table 2) due to its ease of use and wide availability. The teams using Postdoc experienced significant reduction in cross-enterprise project management costs, both by enhancing the communication among various distributed programs, and by enabling project members to share the required information resources. The success of the DS-1 Remote Agent captured the whole New Millennium Program's attention, and Postdoc would become the infrastructure of choice to support this program of six missions and six technology teams. It is estimated that Postdoc resulted in $1 million dollars in savings due to streamlined documentation and reduced travel costs for software development and integration.

Another group of Postdoc's early users was the Combined Federal Campaign (CFC), an annual fundraising drive conducted by federal employees each fall. The CFC pools together millions of dollars from private contributions by federal and military employees to benefit thousands of nonprofit charities. At NASA ARC, approximately 50 federal employee volunteers, working in over 50 divisions, carry out the CFC. Prior to the deployment of Postdoc, the required information flow for decision making between the volunteers was primarily supported by e-mail. The debut of Postdoc at NASA ARC enabled government volunteers to share required documentation via this tool. Deepak Kulkani, a computer scientist in the Collaborative Assistant Systems group, describes the experience:

Bob Lopez and I decided to use Postdoc for sharing documents that were revised continuously over time. This had tremendous benefits in making sure that we had access to the latest information while running the campaign. It saved us significant amount of time.

Table 2. Usage statistics for Postdoc (Summary by Month)

Summary by Month										
Month	Daily Avg				Monthly Totals					
	Hits	Files	Pages	Visits	Sites	KBytes	Visits	Pages	Files	Hits
Sep 2004	4623	2716	917	177	3820	9499477	5143	26607	78789	134084
Aug 2004	4411	2714	764	168	4031	10009662	5235	23712	84143	136742
Jul 2004	4450	2617	803	164	3864	12662925	5086	24893	81156	137965
Jun 2004	6189	3861	837	176	8689	10898199	5297	25115	115835	185698
May 2004	4625	2888	828	176	4155	7584753	5463	25677	89545	143397
Apr 2004	5274	3308	970	207	4296	11154085	6221	29116	99266	158227
Mar 2004	6380	3586	1536	199	4171	10784534	6177	47646	111183	197804
Feb 2004	6252	3864	808	159	3057	12523316	4630	23450	112056	181317
Jan 2004	6177	3816	798	163	3259	18779325	5053	24747	118299	191491
Dec 2003	5042	2681	594	143	2836	11606001	4447	18435	83141	156324
Nov 2003	6666	4549	1056	179	3060	13189635	5384	31696	136482	199998
Oct 2003	11265	4840	1522	200	3787	9502325	6220	47200	150068	349220
Totals						138194237	64356	348294	1259963	2172267

Postdoc's success as an infrastructure to support collaboration continued to grow within the agency. The Postdoc team quickly recognized that user support would be key to its continued growth, therefore it set up an appropriate human- as well as computer-based infrastructure to ensure that users would have access to an effective and efficient user support system.

Once Postdoc demonstrated its usefulness as a collaborative information-sharing application for cross-agency management and design, there was one remaining issue: Who would take over the operations of this innovative system, considered to be one of the first agency-wide systems? The developers of Postdoc would need to assume this role for the agency. Postdoc had started out as a research project used to define how a Web-based collaboration system could be used to manage research and development projects and programs that need to continually share information with partners across the agency, as well across other organizations. However, gaining acceptance as the Web-based collaboration system of choice meant that Postdoc still had to be put through a last test: the capability maturity model (CMM) process. Since NASA requires software assurance of mission applications, the ARC operations branch performed an operational capability review to ensure the software was ready for operation, and Postdoc successfully achieved CMM status. Postdoc was also certified and accredited in meeting the federal IT security requirements using the National Institute of Standards and Technology's guidance in securing the confidentiality, integrity, and availability of information maintained in the system. Postdoc provided users with guidance on the type of information the system was accredited for, providing prompts within the system to remind users of their information-handling responsibilities in the use of the application.

THE IMPACT OF POSTDOC IN NASA

Postdoc provided a foundation of many key lessons learned for some of NASA's newer technologies that came in to enhance NASA's sharing of information. Perhaps the biggest barrier to the pervasive adoption and utilization of collaboration technologies is changing the way people work (Groth, 1999), even with respect to the notion of time (Saunders, Van Slyke, & Vogel, 2004). Although some barriers may exist to Web-based collaboration, primarily because of its availability and simplicity to use, as well as its flexibility, elegance, and sophistication in design, the use of Postdoc grew organically across the different NASA centers to more than 6,000 users by 2004. Elegance in design came from utilizing diverse resources in an integrated way to provide functionality. Postdoc was built on top of existing sophisticated server applications and capitalized on all the development efforts that have preceded its code, and its sophistication derived from using system resources in a combined fashion. Resource components were recruited from the system environment when they were required to accomplish specific tasks. Postdoc's simple and intuitive graphic interface was the result of creative ingenuity. Finally, Postdoc achieved simplicity of operation while driving powerful collaborations.

One of the most important lessons learned from this implementation is that users prefer simplicity and ease of use in the application, in particular when uploading and downloading documents for sharing on any platform and in any format. In addition, having a help desk with a live person on the line provided users with the necessary insight for adequate implementation of the collaborative functionality. This combination created the necessary user trust in the application as a platform for sharing data securely with colleagues in a team. Finally, the users as the primary customers contributed to the design requirements of system upgrades, and therefore felt ownership of the application via their ability to contribute to its design.

Collaboration technologies develop a perspective on how each person's work relies on the work of others. Users across NASA centers used Postdoc to conduct document sharing and decision support of all types, such as the following:

1. Programmatic functions including the review of proposals responding to NASA research announcements, program reporting and tracking, and scientific reporting
2. Calls for papers involving large conferences (i.e., American Association for Artificial Intelligence, International Joint Conference on Artificial Intelligence) and many other professional society activities
3. Program initiatives across several centers, supporting collaboration on technology and mission plans and proposals, and enabling the distribution of information throughout the centers

Some of the uses of Postdoc across NASA include the following:

1. **Scientific research:** The Astrobiology Institute. Postdoc supported around 500 users who use the system for sharing publications, data, and scientific results, and for conferencing support. The NASA Astrobiology Virtual Institute Experiment has used the system extensively. Program IT manager William Likens stated, "*An Institute with geographically dispersed staff can use [this] technology to operate effectively as if it was located within a single campus.*"
2. **Business process support:** Ames ISO (International Organization for Standardization) Certification Directorate. Ames used the software as a documentation system for storage of ISO documentation, correction action requests, and revisions.
3. **Field experiments:** The Marsokhod project, for example, used the system to share results in real time with geographically dispersed teams. The Marsokhod, a joint U.S. and Russian effort, is a tele-operated robotic vehicle. In the early part of 1993, a prototype of the rover was brought to NASA ARC from Russia to be interfaced with a virtual environment control system. The rover has since made several missions around the world at locations that exhibit Mars-like terrain, including the Kamchatka Peninsula in Siberia in 1993, the Mojave Desert in the United States in 1994, and in the Kilauea volcano in the United States in 1995. Postdoc provided a platform for automated capture of the data collected by the rover, and supported the creation of higher level data products. Postdoc allowed Marsokhod mission scientists to easily share and collaborate on scientific results internationally.
4. **Scientific proposal review:** Several programs, including the Earth Sciences Enterprise Advanced Information System Technology program, the Space Sciences Enterprise Cross Enterprise Technology Development program, and the Aviation Safety and Security Program at Langley Research Center used the system to review the proposals received as a response to the NASA research announcements. These programs continued to report that the software allowed them to communicate smoothly across platforms, software versions, and the country. George B. Finelli, director of the Aviation Safety and Security Program, made the following comment.

Postdoc has been invaluable to successfully managing the NASA Aviation Safety and Security Program. The system allows us to quickly and easily share information among the Program management team, to report accomplishments, and to prepare for technical meetings and reviews. Postdoc has also enabled the Program to completely demonstrate official record keeping and tracking as required by NASA procedures and guidelines. Because the application accepts virtually all file types and is so simple to use—anytime anywhere—we use it on a daily basis. Our independent reviewers even use it to assess the Program's financial and schedule archives, and it has become a "service" that NASA provides to groups like the Commercial Aviation Safety Team that includes many Government and industry organizations. No formal training required!

5. **Cooperation and collaboration across programs:** Postdoc fostered cooperation and synergy within NASA's Information Technology Base Program and the new Intelligent Synthesis Environments Initiative, which was aimed toward improving NASA's engineering, science culture, and creative process. According to Yuri Gawdiak, Level I program manager for Engineering for Complex Systems (ECS) at NASA Headquarters, Postdoc has been a key collaboration tool for coordination of the Aviation Safety Program design, Non Advocate Review (NAR), Internal Audit Review (IAR), initial implementation and general programmatic operations. The system is used extensively by all the NASA flight centers as well as for our national and international customer base for global aviation safety issues. It has been a great tool in facilitating teleconferences, workshops, and conferences. The Postdoc repository, with its sophisticated user interface, has become an indispensable corporate knowledge database for the entire Aviation Safety Program.

NASA's use of the Postdoc software sparked an awareness in the agency about how large programs can create virtual work spaces to be shared among geographically distributed teams.

It was estimated that during the period of 1995 to 2004, the use of Postdoc as a collaboration infrastructure resulted in savings to the agency of over $4 million a year. Many NASA programs across the NASA organization estimated savings of at least $100,000 and up to $200,000 in travel expenditures due to the use of Web-based collaboration. These savings do not consider the intangible efficiency gains achieved through this Web-based collaboration platform, such as eliminating the hardships associated with attaching large documents to e-mails, as well as increased document security and integrity. The agency continued to save as the user base on the current system continues to grow primarily due to the many benefits that Postdoc provides. Postdoc's use reached across the agency in supporting approximately 30 NASA programs. This even included partnerships across the federal government including programs at the Department of Defense, National Institutes of Standards and Technology, Naval Research Laboratories, and the National Imagery and Mapping Agency. Today Postdoc has been successfully migrated into NX, a new knowledge-management-based technology that leveraged Postdoc's foundation of lessons learned and user requirements. NX is the result of a partnership between NASA Ames Research Center's Computational Sciences Division (the same division that developed Postdoc) and Xerox Corporation. In this partnership, Ames Research is transferring the research and development in collaboration technologies to Xerox, who in turn is implementing this research into Xerox's DocuShare product to create the NX product specifically for NASA. NX incorporates critical collaboration functionality that was not available in the DocuShare product. The NX product will then be available as a commercial off-the-shelf product for use within NASA and other organizations with similar needs.

VISIONS FOR THE FUTURE OF WEB-BASED DECISION SUPPORT

Several projects under way at NASA seek to develop automated interfaces for data collection and analysis, Web-based collaborative and knowledge capture systems, and intelligent systems for autonomous reasoning. Postdoc demonstrated the ability to serve as the foundation to build NASA's collaborative information management capability by strengthening functionalities for knowledge sharing, and serving as the infrastructure for integration of other virtual collaborative interfaces. Postdoc allowed users to tailor their documentation systems to their processes and work habits, within their time constraints. Providing NASA engineers and scientists with a platform for Web-based collaborative decision support is of high importance. The future of this technology, yet to be determined, is defined by how we apply our own humanity. People are expanding their awareness of space and time in unprecedented ways. The farther our machines and minds explore leading-edge Internet content and tools for communication, collaboration, and visualization, the more we must look within ourselves.

Today the requirements from technologies providing information and knowledge sharing across federal agencies have increased, due in part to the increased information security threat that the Internet poses today. The new Federal Information Security Management Act of 2002 specifies that any technology development for the federal sector must adhere to strict design requirements that focus on the protection of information confidentiality, integrity, and availability. Thus, the federal government requires specific security criteria for technologies used for the sharing of information across agencies and their partners. Most these requirements are found in the National Institute of Standards 800 series.

Improvements in cost-to-performance ratios of IT have caused the cost of digitizing information to approach zero, and the cost of coordinating across individuals, organizational subunits, and organizations to approach zero as well (Grover & Segars, 1996). We envision the future of Web-based collaboration to include technologies that are spawned off by collaboration technologies similar to Postdoc in the exploration and building of virtual worlds on the Internet (Malone et al., 1999). For example, integrating three-dimensional (3-D) avatars[5] and intelligent agents within the system could provide researchers the opportunity to have their body doubles interacting and collaborating in virtual communities inside 3-D collaborative virtual worlds. Considerable progress is expected in the way in which the agents will evolve (i.e., change, develop, and act). Such evolutionary agents may be dramatically different in their abilities to (a) build theories and create a world of their own, (b) assume any virtual identity they wish, (c) possess free will, and (d) develop a moral code and value system of their own. Thus, the future of Web-based collaboration will be dramatically different due to the inevitable, unpredictable over any long period of time, and quantum changes in IT.

REFERENCES

Becerra-Fernandez, I., Del Alto, M., & Stewart, H. (2006). Postdoc: A model for Web-based collaborative decision support at NASA. *International Journal of e-Collaboration, 2*(3), 49-63.

Becerra-Fernandez, I., Gonzalez, A., & Sabherwal, R. (2004). *Knowledge management: Challenges, solutions, and technologies.* Upper Saddle River, NJ: Prentice Hall.

Davenport, T., & Prusak, L. (1998). *Working knowledge: How organizations manage what they know.* Boston: Harvard Business School Press.

Groth, L. (1999). *Future organizational design.* West Sussex, UK: John Wiley & Sons.

Grover, V., & Segars, A. H. (1996). IT: The next 1100102 years. *Database, 27*(4), 45-57.

Kumar, K., & van Dissel, H. G. (1996). Sustainable collaboration: Managing conflict and cooperation in inter-organizational systems. *MIS Quarterly, 20*(3), 279-290.

Leonard, D., & Swap, W. (2004). Deep smarts. *Harvard Business Review, 82*(9), 88-97.

Leonard, D., & Swap, W. (2005). *Deep smarts: How to cultivate and transfer enduring business wisdom.* Boston: Harvard Business School.

Majchrzak, A., Cooper, L., & Neece, O. (2004). Knowledge reuse for Innovation. *Management Science, 50*(2), 174-188.

Malhotra, A., & Majchrzak, A. (2004). Enabling knowledge creation in far-flung teams: Best practices for IT support and knowledge sharing. *Journal of Knowledge Management, 8*(4), 75-86.

Malhotra, A., & Majchrzak, A. (2005). Virtual workspace technologies. *MIT Sloan Management Review, 46*(2), 11-14.

Malone, T. W, Crowston, K., Lee, J., & Pentland, B. (1999). Tools for inventing organizations: Towards a handbook of organizational processes. *Management Science, 45*(11), 65-78.

NASA workforce analysis report. (2003). Retrieved August 14, 2003, from http://nasapeople.nasa.gov/workforce/

Nonaka, I., & Takeuchi, H. (1995). *The knowledge creating company.* New York: Oxford University Press.

Saunders, C. S., Van Slyke, C., & Vogel, D. (2004). My time or yours? Managing time visions in global virtual teams. *Academy of Management Executive, 18*(1), 19-31.

Senge, P. M. (1990). *The fifth discipline: The art and practice of the learning organization.* New York: Currency Doubleday.

Smith, J. (2003, July 13). *Mistakes of NASA touted up.* Washington Post.

ENDNOTES

1. Space Shuttle Columbia broke during reentry into the atmosphere on February 1, 2003, due to damage at lift-off to its protective tiles.
2. Remote Agent was a complex software system for controlling and monitoring autonomous spacecraft.
3. Dan Goldin was the NASA administrator during the period of 1992 until 2001.
4. The first Web-based document management systems would not appear on the market until late 1996, with the offerings of Livelink by Open Text (http://www.opentext.com) and DocuShare 2.0 by Xerox (http://www.xerox.com).
5. An avatar is a graphical icon that represents a real person in a cyberspace system. When you enter the system, you can choose from a number of fanciful avatars. Sophisticated 3-D avatars even change shape depending on what they are doing (e.g., walking, sitting, etc.; http://www.webopedia.com).

Chapter IX
Minimizing the Challenges of Risk Management in Distributed IT Projects:
The Importance of the Alignment of Strategic, Tactical, and Operational Levels

Rafael Prikladnicki
Pontifícia Universidade Católica do Rio Grande do Sul – PUCRS, Brazil

J. Roberto Evaristo
Knowledge Management Program Office, USA

Jorge L.N. Audy
Pontifícia Universidade Católica do Rio Grande do Sul – PUCRS, Brazil

Marcelo H. Yamaguti
Pontifícia Universidade Católica do Rio Grande do Sul – PUCRS, Brazil

ABSTRACT

Distributed IT projects exhibit certain features that make them fundamentally different from traditional colocated projects, not only involving additional steps and decisions, but also impacting the risk management process. The goal of this chapter is to discuss these impacts and to suggest the development of an integrated risk management process taking into account site dispersion, time-zone difference, and cultural boundaries not only at the operational but also at the tactical and strategic level. We also report results of an exploratory case study conducted in a software development center (a Brazilian subsidiary of a U.S. corporation) in support of such a model, and conclude with a discussion of theoretical and practical implications of our work.

INTRODUCTION

Project failure, particularly in information systems development, is unfortunately a very common occurrence (Schmidt, Lyytinen, Keil, & Cule, 2001). Many of these failures are well documented. Key reasons include the lack of top-management commitment to the project, lack of client responsibility, unstable corporate environment, failure to manage end-user expectations, failure to identify all stakeholders, lack of change management, poor risk management and control, unclear or misunderstood scope, staffing volatility, poor team relationships, and artificial deadlines. Sophisticated risk management techniques have been developed to address these problems (Kumar, 2002).

However, a new level of difficulty looms: Such risks are magnified when IS projects are distributed (Erickson & Evaristo, 2005). IS projects tend to be performed in a distributed fashion in offshore outsourcing arrangements, for instance. This is becoming increasingly more common for several reasons: the search for lower costs, higher quality, and better access to skilled resources (Herbsleb & Moitra, 2001). In fact, economic forces are relentlessly turning national markets into global markets; software development is becoming a multisite, multicultural, globally distributed undertaking (Boehm, 2006; Morstead & Blount, 2003). This phenomenon is impacting not only marketing and distribution, but also the way IS products are conceived, designed, constructed, tested, and delivered to customers (Karolak, 1998; Sengupta, Chandra, & Sinha, 2006). For these reasons, distributed software development (DSD) has attracted a large amount of research over the last few years (Carmel, 1999; Evaristo, Scudder, & Desouza, 2004; Herbsleb & Moitra, 2001; Kiel, 2003; Lanubile, Damian, & Oppenheimer, 2003; Prikladnicki, Audy, & Evaristo, 2003; Robinson & Kalakota, 2004). In this context, risk management is critical. According to Karolak, risk management in distributed IT projects should happen at the operational as well as at the strategic and tactical levels. In the strategic and tactical levels, the role of risk management is to help in the decision whether to distribute the development of an IT project across several locations and, once the decision is made, to help in identifying the risks in projects that will be developed by a particular subsidiary. In the operational level, the risk management process relates to the software development process, and it is performed by the project manager.

This study examines risk management problems that organizations face when going global in software development. In particular, there is a strong lack of alignment between the risk management approaches or decisions made at the top and how that is shared and implemented across tactical and operational levels. We therefore focus on the following research question: How can we integrate the risk management processes across strategic, tactical, and operational levels in distributed IT projects?

In order to analyze this problem, we first develop a model of the integration of risk management approaches across different organizational levels based on the theoretical state of the art in this area; then we proceed to present an exploratory but model-inspired case study in a software development center: a Brazilian subsidiary of a U.S. corporation. Based on the results of our case study and how it fits with the model developed, we develop and present strong practical implications for how organizations may better integrate their risk management approaches across different levels.

Next, we present the theoretical review and then the process integration proposal. In the subsequent section, the research method and case-study description, with practical implications of the integration proposed, are given. Finally, further considerations, suggestions for future studies, and research limitations are presented.

THEORETICAL REVIEW

In this section, we first discuss risk management, followed by a brief description of the most important characteristics in distributed software development that are likely to affect risk management approaches. The analysis of the literature shows that there is a strong need for an integrated approach to manage risk in distributed projects across different organizational levels; after this, we develop a proposal to such integration.

Risk Management

The word *risk* comes from the old Italian word *risicare*, which means "to dare" (Bernstein, 1997). Boehm (1991) examined risk through the spiral model: risk analysis in each iteration of software development. DeMarco and Lister (2003), on the other hand, define risk as a problem that has yet to occur, while a problem is a risk that has already materialized. Before it happens, a risk is just an abstraction. It is something that may affect a project, but it also may not.

Successfully managing risk requires more than good processes and the ability to think intuitively; it also requires discipline and method. DeMarco and Lister (2003) define risk management as the process of thinking out corrective actions before a problem occurs, while it is still an abstraction. The opposite of risk management is crisis management: trying to address a problem after it happens.

Risk management in software engineering should pervade the whole software life cycle. The risks cannot be thought of as add-ons to the project, but as the core of the business (Kerzner, 2000). In conclusion, risk management is proactive in preventing problems, and it is a continuous and concurrent process (many risks are managed at the same time).

Distributed Software Development

The software development process is a set of activities, methods, practices, and technologies used to develop and keep related software and products (Pressman, 2004). Commonly accepted premises are that the software quality is strongly dependent on the process quality used in its preparation and that the software process can be defined, managed, measured, and improved.

However, even using a well-defined development process, software development is not a simple task. To add to that, more than a decade ago, many organizations began to experiment with remotely located software development facilities (DSD), seeking lower costs and access to skilled resources. As part of the globalization efforts currently pervading society, software project teams have become geographically distributed (Damian & Moitra, 2006). The reasons for that are the possibility of exploring the market proximity advantages, including knowledge of customers and local conditions; the possibility to improve time to market using time-zone differences in round-the-clock development; and satisfaction of the need to have a global resource pool to successfully and cost-competitively have resources, wherever located.

Organizations search for competitive advantages in terms of cost, quality, and flexibility in the area of software development (Carmel & Tija, 2005; Karolak, 1998), looking for productivity increases as well as risk dilution (Prikladnicki, Yamaguti, & Antunes, 2004). Many times the search for these competitive advantages forces organizations to search for external solutions in other countries (offshore sourcing).

DSD profoundly impacts the way products are conceived, designed, constructed, tested, and delivered to customers (Damian & Moitra, 2006; Herbsleb & Moitra, 2001). Thus, the structure needed to support this kind of development is different from the one used in colocated environments. DSD has diverse effects on many levels, including on strategic, cultural, knowledge management, and technical issues.

Tools and technological environments have been developed over the last few years to help in the control and coordination of the development teams working in distributed environments. Many of these tools are focused on supporting procedures of formal communication such as automated document elaboration, processes, and other noninteractive communication channels.

Risk Management in Distributed IT Projects

Risk management in DSD is an important and necessary activity. Prikladnicki, Audy, & Evaristo (2003) suggested that effective risk management was an alternative to solving existent problems in distributed projects because it is hard to deploy, execute, and control projects in DSD environments, due not only to technical factors like specification, code, peer review, analysis, and so forth, but also to nontechnical factors such as social, cultural, behavioral, and political ones (Kiel, 2003). Other studies (e.g., Herbsleb & Grinter, 1999) present the same difficulties but also suggest additional technical factors such as the software development process, project management, project size, and complexity.

Therefore, risk management becomes important in projects developed by distributed teams (from the same or different organizations). Whether a project is developed globally or in the same city, having geographically dispersed teams using collaboration technologies and developing specific solutions to distributed projects adds risk factors to the projects.

The three categories of risks in DSD projects are organizational, technical, and communication risks (Karolak, 1998). Risks can belong to more than one category, and such shared risks are the most critical. Karolak also found that the risk management in DSD projects should occur not only at the project level, but also at the organizational level. First, the decision of whether globally dispersed teams can develop a particular project is difficult (strategic level). Moreover, the decision of where the project will be better developed can also be a problem (tactical level). A number of models are possible and appropriate under different circumstances to examine the cost-benefit ratio of such a decision. Morstead and Blount (2003) developed a road map and a risk management framework for reinventing the IT function into a low-cost, high-service organization leveraging remote resources.

Additionally, it is suggested that all the risks identified in the strategic and tactical levels should be reflected at the project development or operational level. The identified risks must then be passed to the project manager. Therefore, the project manager can plan response actions to these risks and add them to those of the whole project. In short, there are a set of inherent problems and challenges to software development. DSD, by adding factors like geographic and temporal dispersion as well as cultural differences, has accentuated such challenges and added new ones to the development process. As a result, the work in DSD environments is more problematic than in centralized ones, and the importance of effective risk management becomes even more critical.

PROCESS INTEGRATION PROPOSAL

Based on the literature review, we have developed a process integration proposal for risk management at the strategic, tactical, and operational levels. The need for this integration has been discussed in the literature (Karolak, 1998; Prikladnicki, Yamaguti, & Antunes, 2004; Sahay, Nicholson, & Krishna, 2003), but there has been no actual proposal on how to implement it. In our proposal, we develop activities with the purpose of enabling better and more effective risk management in distributed projects. The guiding principle is to document all decisions made at the higher levels, and to provide means to reflect such information at the operational level where the project will be executed and the risk management process based on the software development process will begin. As a result, a project manager at the operational level will also become aware of identified risks at the higher levels and he or she can plan response actions for these risks. The integration proposed is based on the reference model for distributed software development proposed by Prikladnicki and Audy (2006), and presented in Figure 1.

Two planning cycles can be identified for the management of DSD projects. The first is strategic planning, conducted by the organization headquarters to identify and prioritize new projects to be developed. The decision involves both projects from the organization as well as those requested by external clients. Moreover, the participants in this planning level are responsible for the strategic alignment between the perspectives and goals of each subsidiary and the headquarters.

The second cycle involves the tactical-operational planning level in each subsidiary. The overlapping of the two planning cycles occurs at the critical juncture of project allocation because those activities involve the planning and selection of projects that will be developed in each center and appropriate resource allocation. Project allocation involves the selection of projects that will be better developed in each subsidiary according to an allocation policy defined by the organization. The allocation must have as entry criteria the list of projects to be developed and the output of this step is the center or centers selected for each project. Global directors, vice presidents, or CIOs (chief information officers) are responsible for the strategic planning stage. The coordinators (directors and/or senior managers) in each subsidiary approve the tactical planning stage. The project managers are directly involved in the operational planning, which involves the software development project administration

Figure 1. The reference model for DSD (Prikladnicki & Audy, 2006)

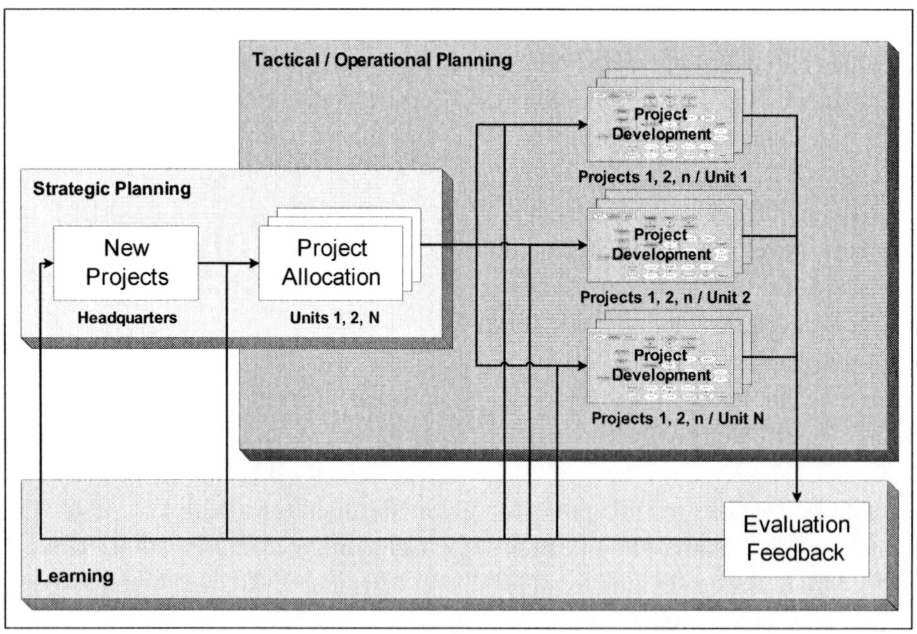

centered on the general coordination of the work between the collaborators, interfaces among teams, communication, contacts with clients, and conflict solving. At this time, the project risk management process is put in place. Finally, the last cycle proposed is the learning cycle, related to the evaluation and strategies adopted. This learning activity is reflected in future projects and allocation decisions.

The Integration Proposal

In order to integrate the risk management in the strategic, tactical, and operational levels, our integration proposal follows the same sequence as those in the reference model presented in Figure 1. First, the identification and prioritization of new projects to be developed is focused on the organizational road map for distributed development. Second, the project allocation involves the planning and selection of projects to be developed in each center and the resources allocated for each project. Finally, project development (project dimension and the risk management process itself) is the project execution activity, where the risk management process will run from the operational level point of view. The evaluation and feedback process (learning cycle) is the continuous learning process that needs to be included in all activities to reflect the decisions and experiences in the future selection of projects, project investigation, project and resource allocation, and project execution activities.

The last integration effort proposed is the creation of an artifact to document all risk analysis and risk assessment conducted at the strategic and tactical level. The purpose of this artifact is to give the project manager all information necessary from discussions with senior managers. The artifact is composed of two main sections: The first section describes details of the project investigation and the decision of which development center to use; the second session is related to the risk description, including strategies to reduce the risk, the exposure from the organizational point of view, and general comments. This document should be completed after the project selection decisions, and is sent to the project manager to be used as another input for the risk management process. Finally, when the project is finished, all lessons learned should be stored in the risk repository and used for new distributed IT projects to be developed.

CASE STUDY: EVALUATING THE INTEGRATION PROPOSED

This exploratory study was developed to initiate the evaluation of the risk management integration proposed. This study was performed in an offshore software development organization located in Brazil, a subsidiary of a U.S. corporation. Our main goal was to understand how the decision to develop a project using globally dispersed teams was made and how it was reflected in the project development itself. The risk management process was analyzed, as well as decisions at the strategic and tactical levels that impacted project development at the operational level.

Research Method

Qualitative methods are appropriate when studying state-of-the-art situations where practice precedes theory (Yin, 2001), which is the case here. The data collected were constituted of primary (individual interviews) and secondary sources (document reviews, business processes, meeting minutes, and software development process description). We interviewed two people from the U.S. headquarters (the director for global IT development and the global project management office [PMO] manager) and eight people from the Brazilian subsidiary (the director of the Brazilian software development center, three senior software development managers, three project managers, and the PMO manager). In the interviews, we asked about the risk management process in the context of a software development project, and the process to allocate a project to be developed in a subsidiary. Content analysis was developed for data analysis (Krippendorff, 2004; Oates, 2006).

We developed two questionnaires (Appendix 1), each considering a specific dimension to be explored. The first questionnaire explored the strategic and tactical decision levels, where the projects are analyzed and allocated (or not) to be developed in a subsidiary. The main respondents were the director for global IT development (Global Development Software Operations), the global PMO manager, one of the account managers responsible for some competence centers, the director of the Brazilian software development center, and three senior software development managers. The second questionnaire explored the operational decision level, were the project is executed (project development). The main respondents were the PMO manager for the Brazilian center, three senior software development managers, and three project managers. It is important to notice that all senior software development managers were interviewed in both dimensions because their role in participating in some decisions in the tactical level and as a supervisor in the operational level.

The Organization

The organization studied is an offshore software development center located in Brazil and owned by a multinational organization. In January of 2001, the company decided to transfer to Brazil part of its software development and created its first offshore software development center worldwide, with the goal of developing and improving the software applications that support the trade areas of the company around the world. When the research was conducted, the subsidiary had over 180 professionals working on many projects worldwide. Almost all projects were geographically distributed since

Figure 2. Global organizational structure

Figure 3. Brazilian subsidiary organizational structure

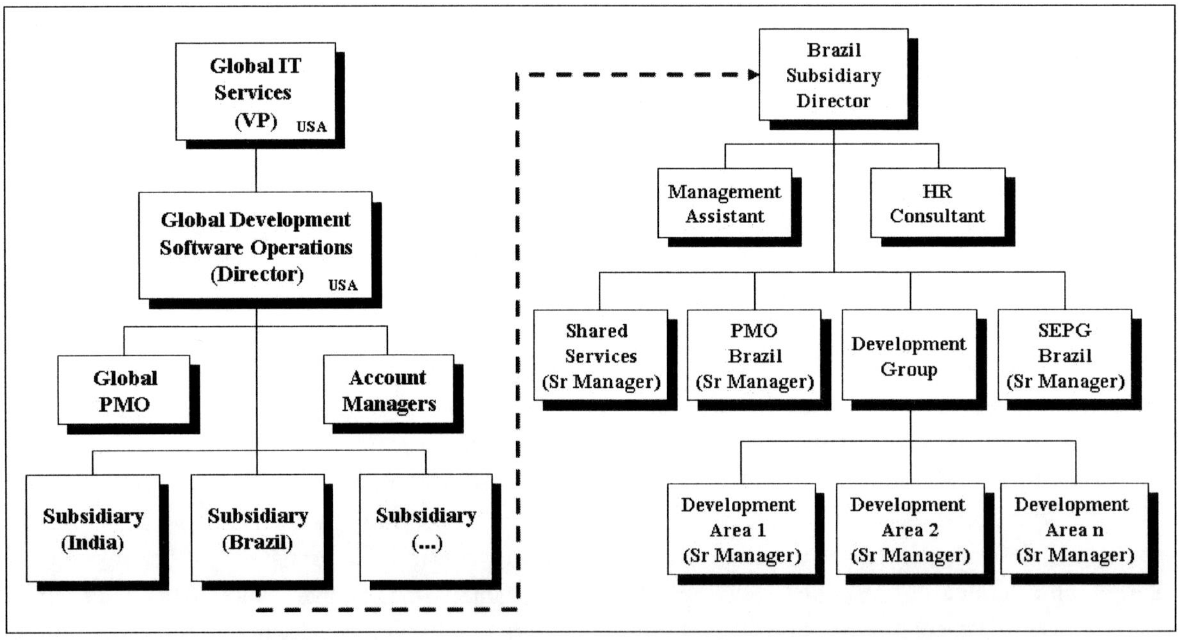

customers and users are located in the company's offices around the world. Figure 2 shows the global organizational structure, whereas Figure 3 shows the Brazilian organizational structure, highlighting its development areas (each managed by one senior manager, organized similarly to the organization departments in the United States), the shared services group, and the SEPG (software engineering process group), each managed by a senior manager.

Important strategic changes took place since the Brazilian center was created. In January 2002, the role of director for global IT development (Global Development Software Operations) was created in the company's head office in the United States to manage all orders of software projects of all trade areas in the company around the world. From then on, both the center in Brazil and the newly inaugurated center in India became directly subordinated to that director. Also, a global PMO team was created, with a PMO manager in each center to better control the projects during development. The software development process is based on the Microsoft Solutions Framework (MSF),[1] and on known methodologies such as Rational Unified Process (RUP),[2] and Project Management Institute (PMI).[3] Finally, the center studied has been an Capability Maturity Model for Software (SW-CMM)[4] Level 2 organization since 2003. As a result of the interviews conducted, we identified the main processes employed by the organization in order to manage risk in a distributed environment at all decision levels.

Project Allocation

In order to have better control of the distributed project allocation and planning, the organization created a set of activities to be implemented in all projects being developed in the several development centers of the organization (Figure 4). These activities include the selection of projects for offshore development through resource allocation. Interestingly, the director for global IT develop-

Figure 4. Project allocation to software development centers

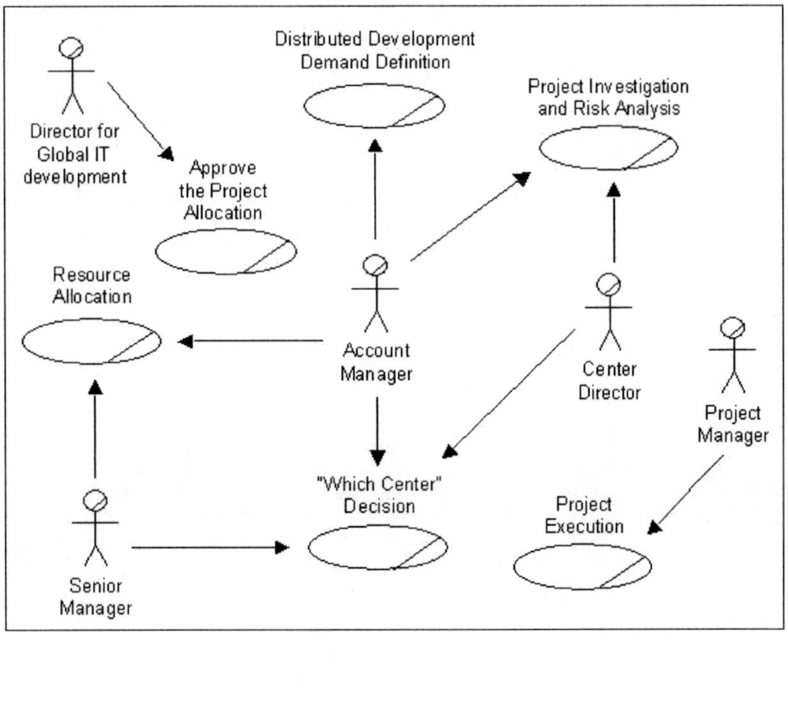

ment mentioned that many times the decision to develop a project in an offshore subsidiary was made based on his own intuition combined with the intuition of other directors and managers.

Once the project is planned and sent to the development centers, execution starts, following the defined software development process. These activities are part of a process called the software development distribution model (Figure 5), where all strategic and tactical decisions are made following a systematic process with well-defined steps (generally performed by senior managers, account managers, and development center directors, with the approval of the director for global IT development).

The risk analysis is represented in Step 2 where, after the distributed development road-map definition (Step 1), a risk and benefit analysis helps to decide whether the project can be allocated to a remote development center. The criteria guiding this process were based on a survey developed by the account manager interviewed. Account managers, senior managers, directors, and other people involved in this process identified what should be considered when deciding on an offshore development activity and location. They found that decisions are made considering the following criteria.

- Export compliance issues (if the project falls under export compliance restriction).
- Data privacy (if data need to move with the application, e.g., support, but on-site government regulation prohibits it).
- Intellectual property (if the application contains high value and the offshore centers are not stringent enough to protect its intellectual property).
- Physical environment availability (if the project requires an environment that is not the direction given by the organization, e.g., replicating a database outside the headquarters).
- Security constraints (if there is a concern that the security to the application will be compromised by moving offshore).
- The project size and cost of moving offshore.
- The level of documentation available for the offshore project team.
- Requirements clarity (if it is clearly defined and understood by the on-site team) and integration requirements (how many integration points the application would have).
- Technology risk (if the project is working with new or obsolete legacy technology).

Figure 5. Software development distribution model

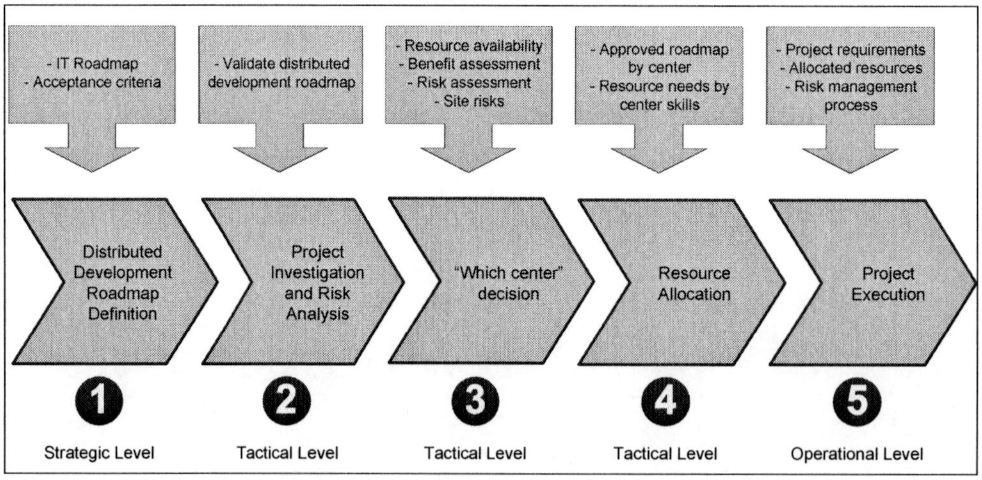

- On-site experience in offshore projects (how many projects the key members have participated in with offshore teams).
- Span of control (how many concurrent projects are managed by the on-site manager).
- Complexity of engagement (how many non-IT partners and sponsors will be engaged).
- Time frame of engagement (the lead time to ramp up the project).

Once the decision is made, a risk assessment is conducted in Step 3 to verify which center among all the organization's development centers can better develop each project. At this time, decisions are made considering the following criteria.

- Center capability (experience in each project activity in the context of current project requirements (coding, testing, modeling, etc.).
- Risks regarding knowledge (center's knowledge in application and technology domain required by the project).
- Risk concentration (based on the total combined risks of all projects an the country where the center is located).
- Resource availability (current level of bench strength of the center).
- Boarding time for new resources (current lead time to hire and train both employees and contractors).
- Physical space (availability of office space, desktops, phones, etc. to make a new resource productive).
- Turnover factor (current attrition rate).
- Language barrier (level of difficulty communicating with other teams).
- Time-zone implications (any hurdles the team might face to effectively collaborate due to difference in time zones).
- Transaction cost (any additional cost the center may have to consider due to moving

Figure 6. Risk management process based on the CMMI model

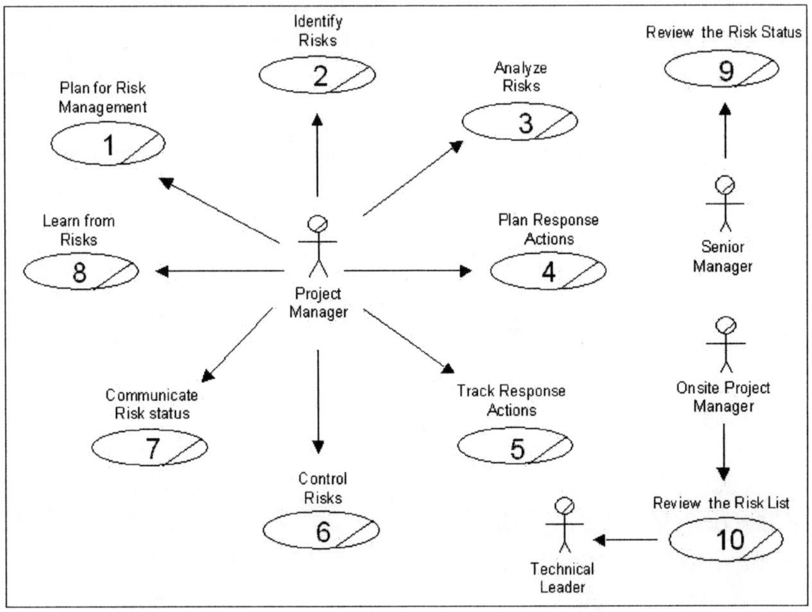

people across portfolios or projects, recruiting new people, and management overhead over usual and customer overhead).

Once the center is selected, the resource-allocation activity occurs in Step 4. Finally, in Step 5, project execution occurs following the organization's defined software development process.

Risk Management in the Software Development Process

Project execution involves all work concerning project development by the project team. This process includes risk management activities (performed by project managers). The risk management process in the operational level (project execution) was defined based on the CMMI model and has 10 activities (Figure 6). The organization uses *A Guide to the Project Management Body of Knowledge* (Project Management Institute, 2004) with input from the risk management process defined in the MSF. The risk management process was developed in order to be simple to implement and to improve the communication with all stakeholders involved.

The first activity involves planning and defining steps that will be used in the project's risk management, and identifying stakeholders. The project manager must define strategies to be used based on the strategies available in the organization software development process, which risks will be managed, when the risk status will be communicated, and so forth.

Afterward, all risks are identified. Risk identification includes searching for common risks and past risks in the risk repository and using the techniques planned in the first activity, and may involve all the project teams, including globally dispersed customers and/or users. Risk identification must consider as input project requirements, assumptions, and constraints. As an example, we can consider risks related to resource skills, late requirements, trust, cultural differences, the development environment, and so on. The technical leader and the on-site project manager review the risks identified in order to agree on the risk list (10th activity). Additionally, the risk identification can be based on a standard risk list that was created based on the most common risks from old projects. This list is maintained and updated periodically. Once the risks are identified, they have to be analyzed (third activity). Risk analysis involves determining the probability and impact of each risk and its exposure, which can be determined using a standard scale from 1 to 5 where 1 is the least probable to occur and the lowest impact, and 5 is the most probable and the highest impact in the project.

The fourth activity is one of the most important in the risk management process since it involves the planning of response actions for each risk. All actions must define the type of response (mitigation, avoidance, transfer, etc.), who is responsible for the action, and the timeline. All planned response actions must be tracked, as defined in the fifth activity. To track all response actions means to monitor if each response action is being implemented on time, and to track periodically the probability and impact of each risk. If a risk occurs, a contingency action (planned in the fourth activity) must take place, and the risk must be controlled. The risk control is defined in the sixth activity. All risks being managed in a project will have a status that must be communicated (seventh activity) to specific stakeholders as defined in the first activity. The senior manager usually reviews this status (ninth activity). Finally, when a project is finished, lessons learned should be registered in the risk repository for use in future projects (eighth activity).

Critical Analysis

Both the software development distribution model and risk management in the software development process have activities related to risk management. The software development distribution model implements a risk analysis and a risk assessment at the strategic and tactical levels in order to help in the decision to distribute development and the selection of the best center for a specific project.

On the other hand, risk management in the software development process involves risk management concerning the project itself at the operational level. However, a key point in the whole process is the integration of the risk analysis and assessment in the strategic and tactical levels (generally performed by senior managers, account managers, and development center directors with the approval of the director for global IT development) with the risk management process in the operational level (performed by project managers). The global PMO manager said, "After we established the process to make allocation decisions in offshore development, we realized that we also needed to integrate such decision with execution" because "integration would improve risk management, giving the whole picture of the project concerning all risks involved, the selection of projects to be developed in an offshore center, center and resource allocation and the project execution."

For this reason, despite the process to select appropriate centers to develop each project, and a process to manage risks when a project is being executed, the processes are not typically integrated, according to the interviews conducted in the organization studied. All interviews showed the need for a more integrated and improved risk management process.

The Brazilian software development center director stated the following:

Based on my experience with the Brazilian team, I start to see some projects with many difficulties, and some of them with the same or similar difficulty. From this point, I decided to discover where the main problem was. I was part of the project allocation decision process and at the same time I was managing the Brazilian team. Then when I start observing the project execution and talking to the project teams, I found that the risk management integration could be a good approach to minimize project issues, managed by the project manager, in the future.

One of the senior software development managers stated,

Most of the projects being developed in the offshore subsidiary were having problems considering risks that were identified in the project investigation and allocation. Since each one of us coordinates one or more project managers and their project teams, we lived some experiences where some risks were not considered in the project execution. The reason is that those risks were identified by the Brazilian center director, the global PMO manager or other people in a higher decision level, without a process or documented procedure to reflect in the project development. Risks such as requirements clarity, level of documentation, data privacy, and security constraints were the most common risks, in our opinion.
Finally, according to all project managers interviewed, they had many difficulties managing risks in a distributed environment without having a better knowledge of what had been done during the decision of allocating the project to their center, under their responsibility.

Ultimately, the case study supports a need for a shared understanding of all risks in the strategic and tactical levels to be reflected in the operational level through a systematic and documented procedure.

Analyzing the case-study results, the software development distribution model created by the organization and the risk management process defined in the operational level can be easily integrated, following the model proposed. Figure 7 identifies the processes.

Figure 7. The integration proposed analyzed with the case-study results

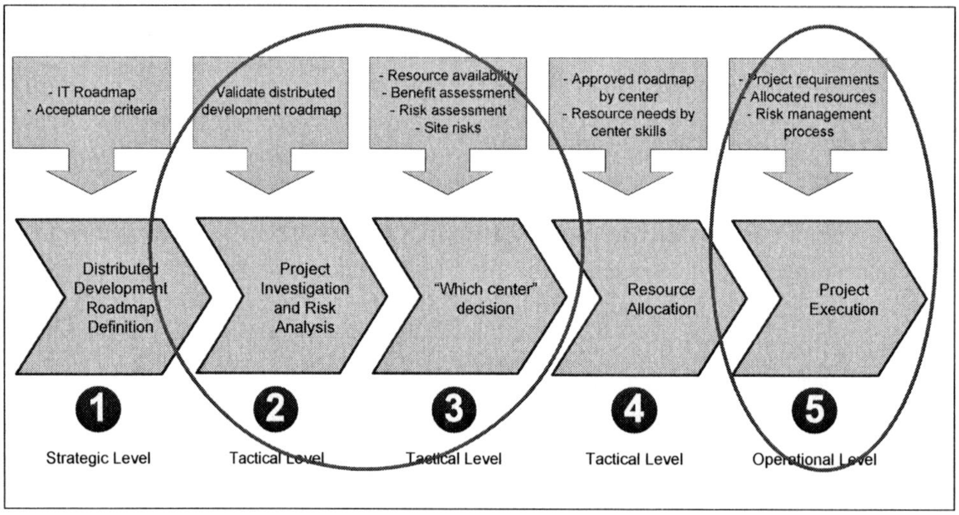

In some of the projects analyzed there was an integration effort, but without a formally documented procedure. The integration was dependent on who was working on a specific project and not on a rule or standard procedure to be followed in each decision level.

For this reason, all suggestions collected in the interviews can be implemented using the model proposed for risk management integration across decision levels. The last thing we think the organization can implement aiming at an improvement in the whole process is the learning cycle, that is, an evaluation of all decisions made to be reflected in future projects and decisions.

CONCLUSION

Software projects are dynamic and unique, leading to the existence of many risks that must be managed. However, one of the main reasons that risk management is inefficient or is not implemented in many organizations is the lack of documentation of both success and failures in projects. Knowledge about risk management alone is not sufficient.

The practice of learning from past experiences is mandatory to help senior managers and project managers to plan for and control risks (Kwak & Stoddard, 2004). DSD provides a good opportunity to exploit the benefits of knowledge management since the DSD model involves additional steps not found in traditional risk management models. Sometimes, risk management in this kind of project can take longer than in traditional projects because of the geographic dispersion and time-zone differences. Additionally, the DSD environment itself can be a risk for the projects and needs to be managed to avoid the common problems and challenges that Herbsleb and Moitra (2001) identified in their studies.

This chapter discussed the role of risk management in distributed IT projects, considering the strategic, tactical, and operational levels of an organization that has implemented DSD. From the point of view of the strategic and tactical levels, organizations can create a software development distribution model, where a risk analysis and a risk assessment can be performed in order to help in the offshore development decision. This can lead to the selection of the best center for a specific project. Additionally, from the operational-level point of view, the key point in the whole process is the integration of the risk analysis and assessment performed in the strategic and tactical levels with the risk management process executed in the operational level. Despite the process to select appropriate centers to develop each project, and a process to manage risks when a project is running, the processes must be integrated to achieve an efficient risk management process for DSD projects. One of the limitations of our study is that the case study was conducted on a particular type of distributed software development environment that admittedly is not the more common: The remote software development center was a wholly owned subsidiary of a U.S.-headquartered organization (Carmel & Agarwal, 2002; Subramanyam, 2004). However, we strongly believe that some of the issues surfaced would be there regardless of the type of ownership structure between the parties.

The integration followed activities proposed in a reference model for DSD (Figure 1), which helps in the identification of the main integration points. The integration itself is composed by simple activities in order to make clear the necessity of having procedures and artifacts to document the decisions taken in all levels and also the risks identified. The main contribution of this integration proposal is to help organizations improve their risk management in all DSD projects since it includes standard procedures to document all decisions made in the software development distribution model and is directly reflected in project execution in all levels. The risk management importance must be emphasized and its participation must be more decisive in this kind of project. Planned follow-up studies on this topic will analyze other organizations in the same situation (wholly owned subsidiaries) in order to improve this process integration.

REFERENCES

Bernstein, P. (1997). *Challenge to God: The risk history.* Rio de Janeiro, Brazil: Campus.

Boehm, B. (1991). Software risk management: Principles and practices. *IEEE Software, 8*(1), 32-41.

Boehm, B. (2006). A view of 20th and 21st century software engineering. In *Proceedings of the 28th International Conference on Software Engineering*, 12-29.

Carmel, E. (1999). *Global software teams: Collaborating across borders and time-zones.* Prentice Hall.

Carmel, E., & Agarwal, R. (2002). The maturation of offshore outsourcing of information technology work. *MIS Quarterly Executive, 1*(2), 65-77.

Carmel, E., & Tija, P. (2005). *Offshoring information technology: Sourcing and outsourcing to a global workforce.* UK: Cambridge.

Damian, D., & Moitra, D. (2006). Guest editors' introduction: Global software development: How far have we come? *IEEE Software, 23*(5), 17-19.

DeMarco, T., & Lister, T. (2003). *Waltzing with bears: Managing risks on software projects.* Dorset House Publishing.

Erickson, J., & Evaristo, J. R. (2005). *Risk factors in distributed projects.* Chicago: University of Illinois, College of Business Administration.

Evaristo, R., Scudder, R., & Desouza, K. (2004). A dimensional analysis of geographically distributed project teams: A case study. *Journal of Engineering and Technology Management, 21*(3), 175-189.

Herbsleb, J. D., & Grinter, R. (1999). Splitting the organization and integrating the code: Conway's law revisited. In *Proceedings of the International Conference on Software Engineering*, 85-96.

Herbsleb, J. D., & Moitra, D. (2001). Guest editors' introduction: Global software development. *IEEE Software, 18*(2), 16-20.

Karolak, D. W. (1998). *Global software development: Managing virtual teams and environments.* Los Alamitos: IEEE Computer Society Press.

Kerzner, H. (2000). *Project management: A systems approach to planning, scheduling, and controlling.* John Wiley & Sons.

Kiel, L. (2003). Experiences in distributed development: A case study. In *Proceedings of the International Workshop on Global Software Development at ICSE*, 44-47.

Krippendorff, K. (2004). *Content analysis: An introduction to its methodology.* CA: Sage Publications.

Kumar, R. L. (2002). Managing risks in IT projects: An option perspective. *Information & Management, 40*, 63-74.

Kwak, Y. H., & Stoddard, J. (2004). Project risk management: Lessons learned from software development. *Technovation, 24*(2004), 915-920.

Lanubile, F., Damian, D., & Oppenheimer, H. L. (2003). Global software development: Technical, organizational, and social challenges. *ACM SIGSOFT Software Engineering Notes, 28*(6).

Morstead, S., & Blount, G. (2003). *Offshore ready: Strategies to plan & profit from offshore IT enabled services.* ISANI Press.

Oates, B. J. (2006). *Researching information systems and computing.* Sage Publications.

Pressman, R. S. (2004). *Software engineering: A practitioner's approach* (6th ed.). McGraw Hill.

Prikladnicki, R., & Audy, J. L. N. (2006). A reference model for global software development: Findings from a case study. In *Proceedings of the First IEEE Conference on Global Software Engineering.*

Prikladnicki, R., Audy, J. L. N., & Evaristo, J. R. (2003). Global software development in practice: Lessons learned. *Journal of Software Process: Improvement and Practice, 8*(4), 267-281.

Prikladnicki, R., Yamaguti, M. H., & Antunes, D. (2004). Risk management in distributed software development: A process integration proposal. *In Proceedings of the 5th IFIP Working Conference on Virtual Enterprises*, 517-526.

Project Management Institute. (2004). *A guide to the project management body of knowledge* (3rd ed.). PA.

Robinson, M., & Kalakota, R. (2004). *Offshore outsourcing: Business models, ROI and best practices.* Mivar Press.

Sahay, S., Nicholson, B., & Krishna, S. (2003). *Global IT outsourcing: Software development across borders.* UK: Cambridge University Press.

Schmidt, R., Lyytinen, K., Keil, M., & Cule, P. (2001). Identifying software project risks: An international Delphi study. *Journal of Management Information Systems, 17*(4), 5-36.

Sengupta, B., Chandra, S., & Sinha, V. (2006). A research agenda for distributed software development. In *Proceedings of the 28th International Conference on Software Engineering*, 731-740.

Subramanyam, M. (2004). The impact of global IT outsourcing on IT providers. *Communications of the Association for Information Systems, 14*(25), 543-557.

Yin, R. K. (2001). *Case study research: Design and methods* (3rd ed.). CA: Sage Publications.

APPENDIX 1
QUESTIONNAIRES

Questionnaire 1: Strategic and Tactical Decision Levels

1. How is the organization organized in terms of roles and responsibilities for offshore development?
2. What is the process to allocate a project to an offshore subsidiary?
3. Who is involved in the project allocation decision?
4. What are the main risks involved in the project allocation decision?
5. How are the risks identified in the project allocation phase reflected to the project manager in the subsidiary that was selected to develop the project?
6. What are the main difficulties around the decision of developing a project in an offshore subsidiary?
7. What are the solutions you have adopted to solve the difficulties?

Questionnaire 2: Operational Decision Level

1. How do you receive the notification that a project will be developed by your team?
2. What are the main risks involved in the development of a project in a subsidiary?
3. What is your role in the project allocation decision?
4. When the project ends, how are the lessons learned reflected in other projects, specifically those lessons about the offshore development environment?
5. Where do you document the risks of your project?
6. What is the organization risk management process for the projects you manage?
7. What are the main difficulties around the development of a project in an offshore subsidiary?
8. What are the solutions you have adopted to solve the difficulties?

Chapter X
Project Management Issues in IT Offshore Outsourcing

Kathy S. Schwaig
Kennesaw State University, USA

Steve Gillam
Accenture, USA

Elke Leeds
State University, USA

ABSTRACT

Global partnerships are forming to take advantage of the cost savings associated with offshoring as well as other strategic benefits. Not all information technology offshoring projects, however, are successful. Cost overruns, increased complexity and defective code cause organizations to rethink their offshoring strategy and their methods for managing these projects. In this paper, project management issues associated with offshore information technology outsourcing projects are identified and specific recommendations for addressing these issues are presented.

INTRODUCTION

Globalization, increased competition, shorter cycle times and an uncertain economic climate are the primary motivators for firms to reexamine traditional methods of delivering information technology (IT) based services and systems. Offshore IT outsourcing is an increasingly popular option for many firms as they transfer a wide array of software development and customer service applications to countries overseas (Misra, 2004). The rate of growth in the global IT outsourcing market is projected to grow at a compound annual rate of 26%, increasing from \$10 billion in 2003 to an estimated \$31 billion in 2008. During the same period, savings in offshore projects are expected to increase from $6.7 billion to $20.9 billion (Behravesh & Klein, 2004). McKinsey & Company, for example, purports that every dollar offshored from the U.S. or the UK creates $1.45 to $1.47 of value, with an estimated $1.12 reinvested back into the UK or U.S. and the rest going to the vendor country (Financial Times Limited, April 2005).

The dominant firms sending software development offshore are U.S. and European, and the typical offshore destinations are India, China, and the Philippines. Both sides of the list are expected to grow, however, as firms from more nations become involved (Kedia, Lahiri, & Lovvorn, 2005). Currently, India is projected to receive about 70% of the world's software and business process offshoring business while China, Ireland, Singapore, Mexico, and the Philippines comprise the rest (Bharadwaj, 2004).

Firms that offshore their IT processes expect to benefit from cost savings, access to highly skilled labor, increased quality and reduced development times (Bharadway, 2004). Based on the touted benefits and the expected growth in the offshoring market, it is clear that offshore IT outsourcing will increasingly become a part of an effective strategy for gaining and sustaining competitive advantage (Nair & Prasad, 2004; Kedia, Lahiri, & Lovvorn, 2005). Unfortunately, many offshore IT outsourcing projects to date have not delivered the projected costs savings and the expected strategic benefit (Saran, 2004). Traditional, "onshore" IT projects are replete with both management and technical complexities that challenge even the best project managers (McDonald, 2001). Geographic distance and cultural differences, among other factors, make managing offshore IT projects even more complex (Kliem, 2004; Laplante, Costello, Pawan, Sudi, & Landon, 2004).

Sound project management practices and techniques are crucial to delivering a successful offshore project. In fact, poor project management is one of the major causes of problems in IT offshore outsourcing efforts (Ante, 2004). A recent study of 101 CIOs involved in offshore IT projects found that their greatest challenges dealt with project management issues such as managing team communication, addressing cultural differences, specifying internal work processes, and developing internal management skills (Ware, 2003). Each challenge can impact the functionality and quality of offshore projects and ultimately eliminate any expected financial or competitive benefits.

In this paper, project management issues encountered in an IT offshoring context are discussed and suggestions for addressing these issues are presented. The paper is organized as follows: the next section provides an overview of outsourcing, IT outsourcing, and offshore IT outsourcing. In the following section, a discussion of IT project management is presented along with the framework used to evaluate project management issues in an offshore context. The section afterwards, discusses offshore IT outsourcing challenges related to each project management area and provides insights for addressing each issue. The last section concludes the paper with a discussion of the paper's contributions as well as recommendations for future research.

OVERVIEW OF OUTSOURCING, IT OUTSOURCING AND IT OFFSHORE OUTSOURCING

Outsourcing

The term *outsourcing* refers to the relationship between a vendor and a client in which the vendor is responsible for one or more of the client's processes or functions (Rajkumar & Mani, 2001). Theoretically, the outsourcing firm achieves economies of scale by spreading the fixed costs of performing the service or producing the product over their entire customer base (Ang & Straub, 1998). The client firm benefits from the vendor's expertise and saves the production costs associated with developing the product or service. Added costs emerge, however, in that the client has to manage and coordinate the negotiation process as well as the relationship with the vendor during the life of the project. A tradeoff typically emerges in terms of production and coordination costs (Lacity & Hirschheim, 1993). As long as the coordination costs do not exceed the savings in production costs, the firm's cost saving goals are achieved. If the coordination costs exceed the production costs savings however, the client's overall costs of producing the product or providing the service are actually increased.

Additional forces beyond cost savings motivate firms to outsource certain parts of their business (Lacity & Hirschheim, 1993; Williams, 1997; Insinga & Werle, 2000). One major force has been the need for firms to focus on their core competencies; those key activities that the firm must do better than their competitors in order to effectively compete in the marketplace (Prahalad & Hamel, 1990; Lacity & Hirschheim, 1995). Firms using a core competency mindset assess whether or not a function or process is essential to their business. If the answer is "no", the activity is a candidate for outsourcing (Lokachari & Mohanarangan, 2001). Other forces driving firms toward outsourcing include access to scarce expertise or new technology and the need to have a scalable workforce (Lacity, Willcocks, & Feeny, 1996).

IT Outsourcing

Firms have outsourced select information technology functions since the early 1960's when clients allowed vendors to process accounting applications such as payroll and accounts payable. Wide scale IT outsourcing, however, had its beginning in late 1988 when Kodak engaged in a major outsourcing project (Lacity & Hirschheim, 1995). Using a core competency model, Kodak outsourced three components of its information technology division to three competing vendors, IBM, Businessland, and DEC. This event led to a *bandwagon effect* where hundreds of CEOs, dissatisfied with their IT department's failure to deliver quality systems on time and within budget, made similar decisions to outsource their IT functions (Fitzgerald & Willcocks, 1993). Since that time, much research has been conducted on the factors associated with successful IT outsourcing (Schary & Coakley, 1991; Fitzgerald & Willcocks, 1993; Hurley & Rundell, 1997; Lacity, Willcocks, & Feeny, 1996; Lacity & Willcocks, 1998; DiRomualdo & Gurbaxani, 1998; Grover, Teng, & Cheon, 1998; Willcocks & Lacity, 1998; Kern & Willcocks, 2002).

Even though the current trend in IT outsourcing began with Kodak in the late 1980's, two key transformations have occurred since that time (Qu & Brocklehurst, 2003). The first was a movement beyond outsourcing IT towards business process outsourcing such as customer service centers. The second transformation was the expansion of IT outsourcing from onshore to offshore (Qu & Brocklehurst, 2003).

Offshore IT Outsourcing

Khan and Fitzgerald (2004) define offshore outsourcing as "an activity where client firms outsource IT activities to external service providers in other countries...and excludes the global internal subsidiaries of client firms" (p. 37). IT offshoring began in the early '90s when ERP implementations created a need for companies in developed countries to access talent in developing countries in order to shorten development time (Bharadwaj, 2004). In the late '90s, Y2K and the booming e-commerce market greatly increased the need for IT professionals worldwide. The trend continued following the technology downturn in 2001. With the sluggish economy, firms continued to look for ways to cut cost while still delivering quality and timely IT projects (Bharadwaj, 2004).

More recently, interest in offshore IT outsourcing has increased dramatically (Khan & Fitzgerald, 2004). Vendors such as IBM, EDS, Computer Sciences Corporation (CSC), Accenture, and

Affiliated Computer Services, Inc. offer offshore outsourcing options to their clients. As mentioned earlier, expected benefits include lower cost for IT application maintenance and software development, faster time to market, access to global expertise, better allocation of resources, increased throughput and fewer labor restrictions (Apte et al., 1997). Concerns regarding the risks of IT offshore outsourcing exist as well. A few of the challenges include effectively managing cultural differences across geographically dispersed teams, maintaining security of data and systems, compliance to regulatory mandates, saving morale of onshore staff, establishing and maintaining communications, selecting an offshore an vendor, and negotiating and enforcing contracts across multiple legal and ethical systems (Kliem, 2004).

With the explosion in the IT outsourcing market in the past few years has come increased attention by the academic community. Some research focuses on the economic, social, and political impact of offshore outsourcing on specific countries such as the United States (Matloff, 2004), Indonesia (Setiadi & Kom, 2005) and India (Khan, Currie, Weerakkody, & Desai, 2002; Nair & Prasda, 2004; The Economist, 2005). One study analyzed the software development strategies for implementing short-cycle-time software development within Russia (Pries-Heje, Baskerville, & Hansen, 2005). Other work identifies the benefits, risks and challenges of offshoring (Apte et al., 1997; Rajkumar & Dawley, 1997; Khan et al., 2002; Kliem, 2004). Case-based research investigates the business models pursued by different firms and provides comparative results across multiple projects (Khan & Fitzgerald, 2004; Beulen, van Fenema, & Currie, 2005). Finally, Misra (2004) advocates the use of metrics to assure that the goals of all parties involved in an IT offshore outsourcing project are met.

The following section discusses the discipline of project management and its importance in successful offshore IT outsourcing projects. In addition, the framework used in this paper for assessing current project management practices and issues in the offshore IT outsourcing context is presented.

PROJECT MANAGEMENT

The history of project management crosses many disciplines and decades (Cleland, 2004). Project management is the application of expertise, skills, tools, and techniques to a set of interconnected activities in order to meet the requirements of the project (Grupe, Urwiler, Ramarapu, & Owrang, 1998). The goal of any project management effort is to manage the resources assimilated in such a way that the project is completed on time, within budget, and according to the functionality and quality expectations of the sponsor (McManus, 2004). Project management factors that lead to successful development and implementation of software have been identified (McConnell, 2001; Nidumolu & Subramani, 2003) as well as those factors that pose risk (Keil, 1995; Wiegers, 1998; Chang & Christensen, 1999; Abdel-Hamid, 1999; Conradi, Nguyen, Wang, & Liu, 2000). Numerous formal project management methodologies exist and many IT organizations formulate their own project management approaches for delivering successful systems (Cantor, 1998; Meredith & Mantel, 2000).

Unfortunately, the failure rate of IT projects is somewhat legendary (Lyytinen & Hirschheim, 1987; Keil & Robey, 2001). In 2002 alone, over $140 billion was lost on failed information technology projects (Larkowski, 2003). Over the past decade, the Standish Group published studies indicating that anywhere from 75% to 90% of IT projects are unsuccessful (Larkowski, 2003). Interestingly, research also indicates that the cause of failure is typically not attributed to the technology but rather to poor management (Keil & Montealegre, 2000). As the statistics indicate, managing traditional onshore IT projects presents challenges ranging from eliciting end user involvement to selecting and motivating a project team. The complexity of an offshore IT outsourcing project delivers many of the same challenges as well as new ones (Misra, 2004).

Some prior research exists examining project management issues in an IT offshore outsourcing context. Studies, for example, develop frameworks to help firms decide whether or not to outsource IT projects offshore and, if so, which locations are best (Smith, Mitra, & Narasimhan, 1996; Zatolyuk & Allgood, 2004; Nair & Prasad, 2004; Khan & Fitzgerald, 2004). Others examine the political reasons that offshore projects fail (Rost, 2004).

A few studies have looked specifically at managing the risks of offshore IT development projects. Kliem (2004), for example, noted that the benefits of IT offshore outsourcing will not be realized unless the associated risks are identified and managed. He developed a framework of risks associated with offshored projects and a process for developing appropriate controls to address the risks. Similarly, Ramarapu and Parzinger (1997) discussed the complexity and risks associated with offshoring beyond traditional IT outsourcing and warned that benefits of reduce costs, improved quality, and economies of scale will not be realized without careful assessment of the issues associated with offshoring.

The Project Management Institute (PMI) is a professional organization focused on the needs of project management professionals worldwide and across multiple disciplines and industries including information technology. As an international professional organization, PMI is recognized for its Project Management Body of Knowledge (PMBOK), which identifies and synthesizes important literature and research on project management into a coherent set of principles and standards for professional certification in project management (Zanoni & Audy, 2004). The PMBOK focuses on best practices in nine key areas of a project: scope, procurement, communication, time, human resource, quality, cost, risk, and integration (Zanoni & Audy, 2004; Schwalbe, 2004). Project managers deploy tools, techniques, and management processes specific to each of the nine areas with the intent of directing their projects to successful completion.

No prior research, however, has used the PMBOK model as a tool for identifying areas of project exposure on IT offshore software development projects. This paper uses the nine areas of the PMBOK as a framework for assessing project management issues in an offshore IT outsourcing context. While the PMBOK is one of many project management frameworks, it is used in the current study due to its simplicity and applicability to practice.

Table 1 presents and defines each PMBOK component. The following section identifies the offshore IT issues related to each PMBOK area and provides suggestions for addressing each issue.

PROJECT MANAGEMENT ISSUES IN AN OFFSHORE IT OUTSOURCING CONTEXT

Scope Management

Undertaking an offshore project involves considerable amounts of planning, coordination, and alignment with business strategy. Traditionally, a Work Breakdown Structure (WBS), identifying each unit of work required to complete a project, is constructed as one of the first steps in the project management process (Baar & Jacobson, 2004). In offshore outsourcing, with its added geographic and cultural complexity, a complete understanding of scope and impact is required prior to WBS creation.

Simply locating some IT functions overseas does not ensure improved company performance and competitiveness in the long run. Managers must have a clear understanding of the objectives and goals of the offshore project and how they relate to the overall business strategy of the firm (Kedia, Lahiri, & Lovvorn, 2005). For the most part, firms engage in IT offshore outsourcing in order to meet strategic business objectives such as cost savings and faster time to market. In order for real value to emerge from the project, the IT planning process must be closely aligned with the strategic planning process of the firm. Offshore projects that are selected for implementation because they are closely aligned and linked to the objectives of the business are much more likely to succeed (Straub, Weil, & Schwaig, 2004).

Critical in the scope planning and definition stage is the level of executive management support. Executive support is especially crucial in an offshore IT outsourcing project that has been identified as strategic for a firm and an enabler of competitive advantage. Most executives, unfortunately, have little or no exposure or experience with offshore IT outsourcing. Each offshore project needs an executive champion who will communicate the strategic and critical nature of the project throughout the organization and an executive sponsor who will assure adequate financial and human resources are available. These roles should be incorporated into the WBS to demonstrate firm commitment and executive management involvement. Strong executive support is essential to a successful, long-term offshore strategic initiative (Krell, 2004).

Executives and other supporters, however, should be cautious not to overstate the benefits of offshoring. Users and executive management may be too ambitious or optimistic in their appraisal of what an outsourcing arrangement can accomplish. In fact, inaccurate cost savings estimates are the primary reason why outsourcing arrangements are perceived as failures (Krell, 2004).

Identifying the scope of an offshore outsourcing project is key to answering the most important question regarding the decision to offshore or not: why do we want to offshore? Some firms choose to offshore processes that they see as non-core. Each business must evaluate its business model and judge whether or not offshoring all or part of their IT fits with their long-term strategy (Kedia, Lahiri, & Lovvorn, 2005).

Table 1. The nine knowledge areas of the project management body of knowledge (Schwalbe, 2004)

Knowledge Area	Definition
Scope Management	All processes involved in defining and controlling what is or is not included in a project.
Procurement Management	All processes required to acquire goods and services from an outside organization.
Communications Management	All processes required to ensure timely and appropriate generation, collection, dissemination, storage, and disposition of project information
Time Management	All processes required to ensure timely completion of a project.
Human Resource Management	All processes required to make the most effective use of all the people involved in a project.
Quality Management	All processes required to ensure that the project will satisfy the needs for which was undertaken.
Cost Management	All processes required to ensure that the project is completed within the approved budget.
Risk Management	All processes required to identify, analyze and respond to risk throughout the life of a project and in the best interests of achieving project objectives.
Integration Management	Includes the processes involved in coordinating all of the other project management knowledge areas.

Procurement Management

The success of offshore IT outsourcing projects is largely dependent upon the relationship established with the vendor firm. Vendor selection, therefore, is extremely important (Ramarapu & Parzinger, 1997). Traditional IT vendor selection processes include the following steps: identify potential vendors, request solution and cost proposals from vendors, check client references, analyze vendor proposals including financials, and select a vendor. All of these steps are essential in selecting an offshore vendor as well, but a few additional issues are raised.

Before selecting a vendor, it is extremely important to clearly articulate *what* business processes will be outsourced offshore and *why* an offshore solution is appropriate. Answering these two questions guides downstream decision-making and helps structure the offshoring project. Business and IT managers must be able to link the offshore initiative to their overall business goals before they can fully address *how* to best offshore a function.

Several other issues must be addressed in the vendor selection process. First, the political and economic conditions of the offshore country must be assessed (Zatolyuk & Allgood, 2004; Nair & Prasad, 2004). Factors that should be understood include the judicial system, international trade laws, financial stability, labor laws, information privacy standards, software piracy laws, as well as the ability of law enforcement agencies to enforce regulations (Kliem, 2004).

Second, the foreign destination must have a stable and appropriate national infrastructure. Issues associated with infrastructure include bandwidth scalability, real estate cost, power availability and reliability, cost of living, transportation, and degree of educated labor pool (Kliem, 2004).

Third, the client should assure that the vendor's goals, objectives, mission, and vision are consistent with its own. This is especially important if the client wants to sustain a long-term offshore relationship with the vendor. Switching vendors can be extremely costly especially in an offshore context. Assurance of the leadership and willingness of the vendor to adapt to the client's standards and processes, therefore, is extremely important.

Fourth, a detailed investigation of specific vendor processes and capabilities should be conducted. Issues such as their software development capacity (employee retention, turnover, skills) and maturity levels, technology infrastructure, conflict management process, project management capabilities, and redundancy plans should be assessed. The goal here is to assure that the vendor has the technology and management processes in place to assure business continuity.

Finally, special attention and detail should be given to the negotiation and contracting process. Contract terms should explicitly define financial commitments, payment methods, roles and responsibilities, schedules, quality, and functional expectations. Dispute management and early termination clauses should be clearly established. A firm must understand how different countries handle breaches of contract and criminal behavior. In the United States, for example, it is relatively easy to get an injunction. Canada and Ireland are similar to the U.S., while it is much harder in India and virtually impossible in China (Briggs, 2005). Ownership of information and software must be clearly stated in light of international intellectual property laws (Laplante et al., 2004). Increasingly important is the security and privacy of information exchanged. Security practices related to data exchange, access, use, and storage must be established.

Communications Management

Communications management in a project includes two related tasks: instituting and managing the reporting cycle, and managing and coordinating meetings with team members as well as upper management. As team members work on project deliverables, their progress must be monitored and controlled. A sound reporting cycle assures that team members and managers are aware of quality expectations and deliverable deadlines and also assures that accountability is assigned and problems are systematically addressed. Formal and informal meetings are used to report on deliverable progress, discuss project issues, resolve problems, and communicate any major or minor changes in the project. A major component of communications management is handling project documentation for all stages of the project.

Communication is made more complex in an offshore context due to dispersed team members, language, time zone, and cultural differences (Ramarapu & Parzinger, 1997; Kliem, 2004). Team members and management are typically located at both the client and vendor sites. Due to their remote geographic location, it is easy to exclude offshore team members from group meetings and communication. Special consideration must be made in order to prevent the organization from alienating the offshore IT members, which may result in poor decision-making and reduced productivity. It is equally important to prevent the offshore team from making decisions without the involvement of the onshore group. Management needs to be accessible to both the onshore and the offshore associates, ensuring that everyone is receiving the same message, communication, and direction (Gilliard & Joharisen, 2004).

Cultural differences can also be seen in project communication. Generally, a U.S. worker is more likely to voice an opinion, ask questions, or offer suggestions. Some cultures, however, view a subordinate expressing their opinion as a sign of disrespect (Nicholls & Ellement, 1997). A project manager must understand this dynamic and manage it in a culturally appropriate manner (Kliem, 2004). While cultural differences and their impact on communication can be seen in an "onshore" project equally, the impact of cultural differences and lack of awareness may impact offshore projects more significantly.

Client and vendor management both must be kept informed about project progress. Advances in communication technologies such as e-mail, internet/intranet access, instant messenger, discussions/chat rooms, online/offline work, video-conferencing, and so forth, help to facilitate communication to team members as well as management (Chidambaram & Jones, 1993; Xue, Sankar, &Mbarika, 2005). While the technology is used to organize and send information, the message is actually communicated when it is interpreted by another person (Gillard & Joharisen, 2004). The effect of cultural interpretation and norms, therefore, plays an important part in technologically based communication (Xue, Sankar, & Mbarika, 2005). Finally, during the contracting process with the vendor, it is imperative that the client and the vendor agree on the quantity and frequency of communication as well as on a standardized reporting procedure.

Time Management

The use of offshore IT resources for project development activities can impact the timing of deliverables and the ultimate schedule of the project. Defining activities, estimating durations, and identifying sequencing relationships among activities, as well as monitoring and controlling the schedule, are more complicated when managing geographically dispersed project activities. Finely developed communication, coordination, and integration skills are essential in avoiding project cost and schedule overruns.

A major benefit of offshoring is the ability to perform project work on a 24×7 schedule. Some firms incorporate the "follow-the-sun" model in order to achieve round-the-clock development, management, and support (Edwards & Sridhar, 2005). The time difference can be a challenge in some cases and a real benefit in others. In a typical offshore situation, the IT resources may be located thousands of miles away and generally 9-12 hours apart. In this situation, a client development team can work all day and, while they are off work, another offshore team can continue the work from the point where the client team left off. This scenario allows for continuous development, reduces deployment time, increases throughput, and lowers cost in terms of faster time to market. Conversely, 24x7 development can add costs in terms of the increased need for coordination, communication, and control of the development environment (Dube & Pare, 2001).

Projects involving web hosting, design, and other internet-related work, are candidates for shorter cycle times. When a shorter cycle time is required, project managers must use more adaptive project management methodologies and techniques (Pries-Heje, Baskerville, & Hansen, 2005). In other instances, such as the rollout of an enterprise resource planning system or a software migration, projects may take months or even years. Whether the cycle time is long or short, project managers on offshore projects must have a project management skill set flexible enough to adapt to the type of project they are leading.

Human Resource Management

The offshore vendor team is the key to the overall success of the project and represents an extension of the client firm. As discussed earlier, communication and cultural challenges can make it difficult to achieve synergy between the domestic and remote teams. As a general guideline, one liaison from the onshore team should be present for every 20 members of the offshore team (Laplante et al., 2004). Other challenges include managing employee turnover, acquiring appropriate skills, sharing knowledge, and addressing cultural differences. Each is discussed below.

Employee turnover is one of the primary considerations in managing offshoring projects. Due to the aggressive growth of the offshoring industry overseas, the job market is extremely competitive with job turnover rates climbing as high as 35% in some countries (Overby, 2003). To guard against potential schedule delays and cost overruns, the client firm should closely assess the hiring, retention, and training strategies of the vendor firm prior to signing the contract. In addition, the project manager should ensure that the vendor has incorporated appropriate reward and incentives systems to reduce the risk of turnover. The initiating firm should communicate this directly to their internal human resources (HR) group and solicit involvement throughout the project.

Assessing employees' skills is challenging in the onshore realm. Transitioning the assessment to offshore vendor managed employees exponentially increases the difficulty. Because the outsourcing firm's business practices and methods may differ from the foreign country's practices, poorly developed and tested systems may result. The project manager must make sure, before the contract is signed, that the vendor can provide the technical and business talent necessary for a successful project.

Depending on the function being outsourced, the firm may require that the vendor establish a "captive data center" or "pavilion" staffed with fulltime employees who are dedicated 100% to the firm's outsourced function. This builds a knowledge base within the vendor's staff regarding the client's business and enables staff loyalty to the client.

Sometimes it is necessary to transfer employees between two teams either to facilitate training, convey team values, or to encourage knowledge sharing. Face-to-face meetings among members can benefit the project by increasing team cohesiveness and providing a shared understanding of project goals and deliverables (Hollingshead & McGrath, 1993). The trips, however, typically involve great distances, are expensive, and take a great deal of project time. Depending upon the length of stay, obtaining travel and work visas in the client country may also be a challenge. The U.S. government, for example, capped the number of H1B visas from 195,000 per year in 2002 to 65,000 per year in 2003. Traveling between locations can be complicated to manage and should be planned and organized well in advance.

Cultural differences can significantly influence the success or failure of an offshore IT project (Ramarapu & Parzinger, 1997). Project managers must be sensitive to differences in terms of vacation expectations, work ethic, holidays, religious observances, communication styles, and business methods. For example, during 2003, the Manilan president declared three additional holidays not originally on the calendar. Adjustments had to be made to work schedules and project timelines. In other cultures, it is typical for workers to take extended vacations at specific times of the year. The project manager who is unaware of these customs and who fails to incorporate them into the initial schedule will have to deal with project delays and cost overruns (Dube & Pare, 2001).

Quality Management

Project quality management ensures that the project will meet functionality requirements and achieve the goals for which the project was originally undertaken. Quality management includes both quality assurance as well as quality control. Firms should have sound knowledge of their internal and external quality specifications prior to coordinating a quality guarantee with an offshore vendor.

Firms that outsource software development or programming to cut expenses often find that those savings never materialize due to poorly written and poorly tested code (Matloff, 2004). In some cases, firms had their own engineers rewrite the code from scratch. A recent study found that the median Indian project had 10% more code errors than did comparable U.S. projects (Ante, 2004).

To address this problem, it is crucial that clients select reliable and reputable vendors as well as implement a robust quality management plan on the client side. Details related to percentage error, applicability, and adherence to specifications will benefit firms in terms of project success, vendor assessment, and contract renegotiation. Large providers that have proven themselves with other clients are typically less risky and more likely to deliver quality products. Checking customer references and software-certification levels is essential. For example, many offshore sites have capability maturity model (CMM) certification, which recognizes process improvement and quality in software development environments. Firms choosing to outsource offshore should consider the compatibility of the vendor's CMM level to their own level of capability. A vendor operating at CMM level 5 may find it very difficult to service a client operating at level 1 or level 2. Conversely, client organizations should not rely solely on the vendor's CMM level when assessing quality. CMM assesses the outsourcer's management techniques, not the quality of their personnel or staff (Matloff, 2004). In India, for example, it is common to staff projects with young, inexperienced developers in order to control costs (Matloff, 2004). Clients must look at both the management processes in place to guide the project as well as the quality of their human resource processes. Stark differences may exist between the client and the vendor regarding their technical environments, their interpretation of quality, and development cultures causing additional project management challenges.

Firms must carefully choose which tasks and functions they move overseas. Tasks that involve repetition and predictability can be done effectively. Any function that requires strong specific language skills, in depth knowledge of a specific country's accounting rules or standards, or quick decision-making should be kept in the client's country (Ante, 2004).

Finally, an offshore vendor must have an extensive quality assurance process that effectively assesses the quality of all software products and applications. The vendor should demonstrate, through the software testing process, high standards of data integrity and security as well as functionality, performance, compatibility, and usability (Misra, 2004).

Cost Management

Budget, or cost management, includes all the processes required to ensure that the project is completed within budget. During the planning phase, the costs of activities are estimated and the contract is structured (von Branconi & Loch, 2003). During project execution, actual costs are compared to budget, estimates are revised, and budgets updated.

The total cost of an IT outsourcing project often exceeds expectations. Most companies find that their first-year savings from offshore outsourcing is 15%-20%, significantly below the 50%-60% often touted (Krell, 2004). While savings of 30% or more on labor costs do materialize, other hidden costs often emerge. When firms outsource to India, for example, they typically pay $8,000 to $12,000 per employee for facilities, telecommunications, and the technology infrastructure (Krell, 2004). In addition, the client firm may need to invest in upgrading their own technology infrastructure in order to interact with the offshore firm. Relationship management costs are typically not figured into expected costs savings projections on the front end of the project and can add an additional 8% to overall project costs. Finally, the costs of terminating domestic employees, leases, supporting expatriates, and legal and consulting expenses should also be included in the overall costs (Krell, 2004). It is imperative that these issues are addressed in the contract negotiation phase, and that top management of both the client and vendor firms are involved.

Risk Management

Project risk management involves identifying, analyzing and mitigating factors that could lead to project failure. Risks are inherent in any strategic business initiative but especially in those that involve two companies located in different geographical and political environments with diverse work forces (Beulen, van Fenema, & Currie, 2005; Briggs, 2005). Firms must have a risk management process that identifies those factors most likely to significantly and negatively impact the project, and have a plan to mitigate those risks (Kliem, 2004).

An overarching issue in managing risk is the level and degree of control that the client retains over its strategic business processes and the security of the data that supports these processes (Cocheo, 2004). Client organizations must assure that foreign operations are meeting the standards set by regulators in their own countries. For example, a Canadian bank that is offshoring development to India may need to embed an auditor in the foreign operation to assure that all Canadian banking regulations are met. In some cases, a firm may use *management interchanges* where a manager from the client facility spends time in the foreign facility to assure compliance with company standards and government regulations (Cocheo, 2004). In many regulated industries, compliance officers will want to make sure that adequate controls (technical, contractual, and procedural) are in place to assure the integrity of the data and accuracy of the development process. Assuring that the vendor has appropriate security procedures including disaster recovery and backup plans in place is essential to assuring the security of the firm's data. Failure to demonstrate data protection can result in excessive non-compliance fines (Cocheo, 2004).

One of the major mechanisms for addressing risk is the contract between the client and the offshore firm. Contract negotiations can often become contentious as both sides flesh out all the terms and conditions of the deal. According to Krell (2004), the success of an IT outsourcing relationship depends on each party's ability to address three areas: complexity, governance and flexibility. Each is discussed below.

- **Complexity:** Negotiating large offshore outsourcing deals is multi-faceted and often results in detailed contracts 500-1,500 pages long. Because so much is at stake, inexperienced clients often need to hire specialists in the following areas: technology transaction outsourcing, employment and labor law, tax law, and real estate law.
- **Governance:** Failure to establish an appropriate governance structure poses the largest risk to a successful project. Too often, clients focus too much on vendor selection and not enough on monitoring service levels once the deal is in place. Typically, governance of offshoring projects involves a steering committee tasked with designing and implementing a conflict resolution process. Also, firms need to retain 7-12% of their displaced IT staff to manage the relationship with the vendor and to assist the organization with assimilating business process changes that result from moving large portions of IT outside the organization (Krell, 2004).
- **Flexibility:** Organizations can undergo several rounds of strategic changes during the course of a multi-year offshoring agreement. Business and market forces cause firms to change their business strategies and objectives resulting in a need to reposition their existing offshoring arrangement. Clients need to build into their contract mechanisms for renegotiating or even terminating the relationship.

Managing these three primary risk factors is vital in assuring a successful IT offshore outsourcing project. Managers must continually balance the benefits of a project against its inherent risk (Kliem, 2004).

Integration

The final project management area is integration, which deals with the manager's ability to coordinate and synchronize the many phases and processes of an IT project. Integration ensures that all aspects of the project are mutually intertwined and working together to achieve the project goals and objectives. This is especially critical in an IT offshore outsourcing context, where physical and cultural separation occurs between project components and isolation becomes an increased risk.

While integration is a separate area in the project management body of knowledge, properly managing integration on any IT project naturally encompasses the other eight areas. A change in scope, for example, should be accounted for and integrated into the project schedule, project budget, quality management plan, and risk management plan. Assuring that communication processes and control mechanisms are in place to fully synchronize all the activities and issues in a project is the domain of project integration. Successful project integration is the domain of an experienced project manager that understands the complexities and connectedness of the environment in which they manage. Specific challenges associated with each of the other eight domains relate to the notion of integration, and since they have already been discussed, they will not be reiterated here.

Summary

Project management issues associated with offshore information technology outsourcing projects have been presented. Table 2 summarizes the previous

Table 2. Project management implications of offshore IT outsourcing

Knowledge Area	Challenge	PM Implications
Scope Management	• What are the strategic and organizational benefits of engaging in a specific IT outsourcing project?	• Link IT offshore outsourcing to strategic objectives of the firm • Articulate why an offshore solution is appropriate • Assure strong and effective executive management involvement • Adopt realistic expectations for offshore IT outsourcing success
Procurement Management	• Given the political, social, and cultural complications associated with IT offshore outsourcing, how does the firm identify, select and contract with the best vendor for a specific project?	• Understand the vendor's socio-political climate, legal and criminal system, national infrastructure, and privacy and data protection standards • Articulate a consistent mission, vision, and objective with vendor management • Negotiate a flexible contract
Communications Management	• How are the cultural, language, and technology barriers addressed in order to insure effective and productive communications in a largely virtual environment?	• Establish specific communication processes that address cultural and language differences • Clearly state communication procedures and vehicles • Include both onshore and offshore team members in communications processes
Time Management	• How is the time difference managed in a way that positively impacts the project schedule as well minimizes coordination costs?	• Recognize the benefits and risks associated with time zone differences • Prepare for the increased need and costs for coordination, communication, and control
Human Resource Management	• How can a client assure that the best employees are assigned to their project and that vendor employee turnover is managed?	• Employ appropriate client to vendor ratios • Assess the quality of the work environment, pay scale, etc to assure that vendor retention rates are being managed. • Monitor the skill sets of offshore employees assigned to your projects
Quality Management	• How is quality management and control addressed in an offshore project given the geographic distance?	• Assess the quality of the offshore vendor's development processes and its fit with your environment. • Establish realistic mechanisms for assessing quality • Offshore only those tasks that are appropriate for a given vendor
Cost Management	• What mechanisms are used to assure that realistic costs savings are achieved and that coordination costs are minimized?	• Conduct financial feasibility studies that realistically estimate cost savings • Anticipate additional costs associated with the project (i.e. upgrading communications infrastructure)

Table 2. continued

Knowledge Area	Challenge	PM Implications
Risk Management	• How are risks addressed in an IT offshoring project?	• Implement a risk management process that identifies, analyzes and mitigates risks associated with offshoring • Establish governance mechanisms that control and monitor the offshoring relationship • Hire specialist to address risks specific to a certain technology, vendor, or country • Contract for renegotiation and termination clauses
Integration Management	• How do changes in one aspect of a project impact other aspects?	• Coordinate and synchronize the various phases and components of the project • Employ experienced project managers on more complex offshore projects • Respond to changes in the project environment by addressing multiple issues in a synergistic and consistent manner

discussion by highlighting the challenges and the project management implications associated with each PMBOK area. The following section discusses the limitations and contributions of the current paper as well as recommendations for future research.

LIMITATION, CONTRIBUTIONS, AND IMPLICATIONS FOR FUTURE RESEARCH

Limitations

While this paper makes some initial observations regarding project management issues in an IT offshore outsourcing context, some limitations of the work are noted. First, the PMBOK framework, while grounded in the practice of project management, is not strongly used in academic project management research. Second, the PMBOK framework is general and not exhaustive. All possible project management issues associated with IT offshore outsourcing are not identified with a cursory application of the framework. Finally, while the current study integrates both practioner and academic literature, the results are neither empirically derived nor examined. The paper, however, provides a great deal of foundational information and insights that can be used to formulate important research question that can guide future empirical investigations. These research opportunities and the contributions of the paper are presented next.

Contributions and Future Research

The research is useful to both practioners and academicians alike. First, the PMBOK model has been shown to be an appropriate lens for examining IT offshore projects. A myriad of issues were identified and appropriate solutions were given. In the future, project managers can use the PMBOK framework to identify important issues in the context of their specific IT offshore project. In addition, the PMBOK framework is not country-specific. It can be applied across borders and across cultures. For academicians, the paper provides an impetus for future research into issues impacting the success of IT offshore projects. First, an empirical validation of the framework and its applicability in a variety of IT offshore outsourcing contexts is warranted. Second, research is needed to identify the "best practices" used by successful firms in terms of addressing issues in each of the nine PMBOK areas. Finally, research is needed to determine what organizational characteristics best influence successful project management of IT offshore projects.

CONCLUSION

This paper used an existing project management framework in an effort to gain a better understanding of the project management issues associated with an IT offshore outsourcing project. Most firms want the results of their offshore outsourcing effort to appear seamless and transparent to those ultimately impacted by the project, while simultaneously reaping potential cost savings and strategic benefits. Yet despite the potential benefits, IT offshoring projects face all of the same challenges and risks of onshore projects as well as some very unique challenges due to geographical, cultural and social differences. Project managers must recognize that offshore IT projects need to be managed with these special challenges in mind. Otherwise, the potential savings and strategic benefit of offshoring will not be realized and the chance of failure will increase.

REFERENCES

Abdel-Hamid, T. K. (1999). The impact of goals on software project management: An experimental investigation. *MIS Quarterly, 23*(4), 531.

Ang, S., & Straub D. W. (1998). Production and transaction economies and IS outsourcing: A study of the U.S. Banking Industry. *MIS Quarterly, 22*(4), 535.

Ante, S. (2004). Shifting work offshore? Outsourcer beware; quality and security woes can eat expected savings. *Business Week*, No. 3865, p. 36.

Apte, U. M., Sobal, M. G., Hanaoka, S., Shimada, T., Saarinen, T., & Versalainen, A. P. J. (1997). IS outsourcing practices in the USA, Japan, & Finland: A comparative study. *Journal of Information Technology, 12*, 289.

Baar, J. E., & Jacobson, S. M. (2004). The keys to forecasting—*Work breakdown structure. Cost Engineering, 46*(3), 12.

Behravesh, N., & Klein, L. (2004). *The comprehensive impact of offshore IT software and services outsourcing on the U.S. economy and the IT industry.* Report prepared by Global Insight of the Information Technology Association of America. Arlington, VA.

Beulen, E., van Fenema, P., & Currie, W. (2005). From application outsourcing to infrastructure management: Extending the offshore outsourcing service provider. *European Management Journal, 23*(2), 133.

Bharadwaj, S. S. (2004). Offshore outsourcing: Path to new efficiencies in IT and business processes. *Journal of Information Technology Cases and Applications, 6*(3), 55.

Briggs, B. (2005). Offshore outsourcing poses risks. *Health Data Management, 13*(2), 68.

Cantor, M. R. (1998). *Object-oriented project management with UML.* Wiley Computer Publishing.

Chang, C. K., & Christensen, M. (1999). A net practice for software project management. *IEEE Software, 16*(6), 80.

Chidambaram, L., & Jones, B. (1993).Impact of communication medium and computer support on group perceptions and performance: A comparison of face-to-face and dispersed meetings. *MIS Quarterly Executive, 17*(4), 465.

Cleland, D. I. (2004). The evolution of project management. *IEEE Transactions on Engineering Management, 51*(4), 396.

Cocheo, S. (2004, May). Regulators keep a risk-based eye on offshore outsourcing. *ABA Banking Journal*, p. 64.

Conradi, R., Nguyen, M. N., Wang, A. I., Liu, C. (2000). Planning support to software process evolution. *International Journal of SEKE, 10*(1), 31.

DiRomualdo, A., & Gurbaxani, V. (1998). Strategic intent for IT outsourcing. *Sloan Management Review, 39*(4), 67.

Dube, L., & Pare, G. (2001). Global virtual teams. *Communications of the ACM, 44*(12), 71.

The Economist. (2005). Special report: The Bangalore paradox—Outsourcing and IT in India: India's IT and outsourcing industries. *The Economist, 375*(8423), 82.

Edwards, H. K., & Sridhar, V. (2005). Analysis of software requirements engineering exercises in a global virtual team setup. *Journal of Global Information Management, 13*(2), 21.

Financial Times Limited. (2005). Special reports: Outsourcing—Making offshoring happen. *Accountancy, 135*(1340), 55.

Fitzgerald, G., & Willcocks, L. (1993). *Information technology outsourcing practice: A U.K. survey.* London: Business Intelligence.

Gillard, S., & Joharisen, J. (2004). Project management communication: A systems approach. *Journal of Information Science, 30*(1), 23.

Grover, V., Teng, J., & Cheon, M. J. (1998). Towards a theoretically-based model of information systems outsourcing. In L. P. Willcocks, & M. Lacity (Eds.), *Strategic source of information systems: Perspective and practices.* New York: Wiley.

Grupe, F. H., Urwiler, R., Ramarapu, N. K., & Owrang, M. (1998). The application of case-based reasoning to the software development process. *Information and Software Technology, 40*(9), 493.

Hollingshead, A. B. & McGrath, J. E. (1993). Group task performance and communication technology. *Small Group Research, 24*(3), 307.

Hurley, M., & Rundell, J. (1997). *The IT outsourcing imperative: Intelligently managing the inevitable decision* (KPMG Report). Australia.

Insinga, R. C., &. Werle, M. J. (2000). Linking outsourcing to business strategy. *Academy of Management Executive, 14*(4), 59.

Kedia, B., Lahiri, S., & Lovvorn, A. (2005). Seeking competitive advantage on distant shores. *European Business Forum, 21*, 37.

Keil, M. (1995). Pulling the plug: Software project management and the problem of project escalation. *MIS Quarterly, 19*(4), 421.

Keil, M., & Robey D (2001). Blowing the whistle on troubled software projects. *Communications of the ACM, 44*(4), 87.

Keil, M., & Montealegre, R. (2000, Spring).Cutting your losses: Extricating your organization when big project goes awry. *Sloan Management Review, 396-311*, 55.

Kern, T., & Willcocks, L. P. (2002). Service provision and the net: Risky application sourcing? In R. Hirschheim, A. Heinzl, & J. Dibbern (Eds.), *Information systems outsourcing: Enduring themes, emergent patterns, and future directions.* Heidelberg: Springer

Khan, N., Currie, W. L., Weerakkody, V., & Desai, B., (2002). Evaluating offshore IT outsourcing in India: Supplier and customer scenarios. In *IEEE Proceedings of the 36th Hawaii International Conference on System Sciences.*

Khan N., & Fitzgerald (2004). Dimensions of offshore outsourcing business models. *Journal of Information Technology Cases and Applications, 6*(3), 35.

Kliem, R. (2004). Managing the risks of offshore IT development projects. *Information Systems Management, 21*(3), 6.

Krell, E. (2004). The total cost of IT outsourcing. *Business Finance Magazine, 10*(3), 48.

Lacity, M., & Hirschheim R. (1993). *Information systems outsourcing myths, metaphors, and realities.* New York: John Wiley.

Lacity, M., & Hirschheim, R. (1995). *Beyond the information systems outsourcing bandwagon: The insourcing response.* Chichester: John Wiley.

Lacity, M. C., & Willcocks, L.P. (1998). An empirical investigation of information technology sourcing practices: Lessons from experience. *MIS Quarterly, 22*, 363.

Lacity, M. C., Willcocks, L. P., & Feeny, D. F. (1996). The value of selective IT sourcing. *Sloan Management Review, 37*(3), 13.

Laplante, P. A., Costello, T., Pawan, S., Sudi, B., & Landon, M. (2004). The who, what, why, where, and when of IT outsourcing. *IT Professional, 6*(1), 19.

Larkowski, K. (2003). *Latest Standish Group CHAOS report shows that projects success rates have improved by 50%.* West Yarmouth, MA: Standish Group International.

Lokachari, P. S., & Mohanarangan, M. (2001). Outsourcing of information technology services: A decision-making framework. In *Proceedings of the Portland International Conference on Management of Engineering and Technology (PICMET)* (Vol. 1, p. 411).

Lyytinen, K., & Hirschheim, R. (1987). Information systems failures—A survey and classification of the empirical literature. In *Oxford Surveys in Information Technology.* Oxford University Press.

Matloff, N. (2004). Globalization and the American IT worker. *Communications of the ACM, 47*(11), 27.

McConnell, S. (2001). Common sense. *IEEE Software, 18*(4), 5.

McDonald, J. (2001). Why is software project management difficult? And what that implies for teaching software project management.*Computer Science Education, 11*(1), 55.

McManus, J. (2004). A stakeholder perspective in software project management. *Management Services, 48*(5), 8.

Meredith, J. R., & Mantel, S. J. (2000). *Project management: A managerial approach* (4th ed.). New York: John Wiley & Sons.

META Group, Inc. (2003). Retrieved from http://www.metagroup.com

Misra, R. B. (2004). Global IT outsourcing: Metrics for success of all parties. *Journal of Information Technology Cases and Applications, 6*(3), 21.

Nair, K. G. K., & Prasad, P. N. (2004, Summer). Offshore outsourcing: A SWOT analysis of a state in India. *Information Systems Journal*, p. 34.

Nicholls, C., & Ellement, G. (1997). *Ellen Moore (A): Living and working in Korea* (Richard Ivey School of Business Case #9A97G029) (p. 1-20). Richard Ivey School of Business.

Nidumolu, S. R., & Subramani, M. R. (2003). The matrix of control: Combining process and structure approaches to managing software development. *Journal of Management Information Systems, 20*, 159.

Overby, S. (2003). *Bringing I.T. back home. CIO.*

Prahalad, C. K., & Hamel, G. (1990). The core competence of the corporation. *Harvard Business Review, 68*(3), 79.

Pries-Heje, J., Baskerville, R., & Hansen, G. I., (2005). Strategy models for enabling offshore outsourcing: Russian short-cycle-time software development. *Information Technology for Development, 11*(1), 5.

Qu, Z., & Brocklehurst, M. (2003). What will it take for China to become a competitive force in offshore outsourcing? An analysis of the role of transaction costs in supplier selection. *Journal of Information Technology, 18*, 53.

Rajkumar, T. M., & Dawley, D. L. (1997). Problems and issues in offshore development of software. In *Strategic sourcing of information systems: Perspectives and practices* (pp. 369-386). John Wiley & Sons.

Rajkumar, T. M., & Mani, R. V. S. (2001). Offshore software development: The view from Indian suppliers. *Information Systems Management*, 63-73.

Ramarapu, N., & Parzinger, M. J. (1997). Issues in foreign outsourcing. *Information Systems Management, 14*(2), 27.

Rost, J. (2004). Political reasons for failed software projects. *IEEE Software, 21*(6), 104.

Saran, C. (2004). Badly-managed offshore software development costs firms millions. *Computer Weekly*, p. 5.

Schary, P. B., & Coakley, J. (1991). Logistics organization and the information system. *International Journal of Logistics Management, 2*(2), 22.

Schwalbe, K. (2004). *Information technology project management*. Boston: Thomson Learning.

Setiadi, R., & Kom, S. (2005, February 16). Implementing IT outsourcing in Indonesia: A smart decision or a big mistake. In *Proceedings of Issues and Challenges in Global Development Conference*, The University of Melbourne, Victoria, Australia.

Smith, M. A., Mitra, S., & Narasimhan, S. (1996). Offshore outsourcing of software development and maintenance. *Information and Management, 31*(3), 165.

Straub, D., Weil, P., & Schwaig, K. S. (2004). *Strategy and the control of the IT resource: A test of the strategic control model*. Working Paper Series, Georgia State University.

von Branconi, C., & Loch, C. H. (2004). Contracting for major projects: Eight business levers for top management. *International Journal of Project Management, 22,* 119.

Ware, L. (2003). Weighing the benefits of offshore outsourcing. *CIO.com.* Retrieved from http://www.cio.com

Wiegers, K. (1998). Know your enemy: Software risk management. *Software Development, 6*(10), 38.

Williams, K. (1997). Has your company outsourced yet? *Management Accounting, 78*(9), 14.

Willcocks, L., & Lacity, M. (1998). *Strategic sourcing of information systems: Perspectives and practices*. New York: John Wiley & Sons.

Xue, Y., Sankar, C. S., & Mbarika, V. W. A. (2005). Information technology outsourcing and virtual team. *The Journal of Computer Information Systems, 45*(2), 9.

Zanoni, R., & Audy, J. (2004). Project management model: Proposal for performance in a physically distributed software development environment. *Engineering Management Journal, 16*(2), 28.

Zatolyuk, S., & Allgood, B. (2004). Evaluating a country for offshore outsourcing: Software development providers in the Ukraine. *Information System Journal, 28.*

This work was previously published in International Journal of e-Collaboration, Vol. 2, Issue 4, edited by N. Kock, pp. 53.73, copyright 2004 by IGI Publishing, formerly known as Idea Group Publishing (an imprint of IGI Global).

Section III
Emerging Issues and Debate

Chapter XI
Hacker Wars:
Cyber Warfare Previews

Richard Baskerville
Georgia State University, USA

ABSTRACT

This chapter develops an analytical framework for new forms of information warfare that may threaten commercial and government computing systems by using e-collaboration in new ways. The framework covers (a) a strategic model, (b) a strategic arena, (c) e-collaboration, and (d) ethics and law. The framework is then used to compare two recorded instances of major hacker wars that erupted in the shadow of kinetic conflicts. In both cases, the hacker war appears to have been a grassroots collaborative enterprise led by loosely organized civilians, with neither government control nor permission. Collaborating across networks to coordinate their attacks, such hacker wars can attack both government and commercial computer networks without warning. The analysis shows how hacker wars demonstrate characteristics found in the frameworks, and that there are forms of e-collaboration that represent a potentially difficult new source of threat for globalized information systems.

INTRODUCTION

The collaborative use of computing, or e-collaboration, uses computers to support the coordination and cooperation of groups of people in order to perform a task or solve a problem (Bafoutsou & Mentzas, 2002). Building on work in virtual teams, the development of e-collaboration represents advances in virtual reality in the sense that virtual workplaces for work groups are often involved (Rutkowski, Vogel, Genuchten, Bemelmans, & Favier, 2002). The application of e-collaboration in most circumstances is a constructive activity: Teams of people using technology to develop work products, coordinate their activities, and communicate their knowledge. The use of information and communications technologies for e-collaboration extends beyond the workplace and into the public arena.

The widespread public availability of ICT makes it possible for grassroots and voluntary e-collaboration to make myriad positive contributions to the welfare of people anywhere in the world. Some computer conferencing tools are widely and near-freely available, such as NetMeeting and BSCW. Trends to make this technology available for public service are in sight. For example, organizations and the general public used ICT in many formal and informal ways to coordinate the relief efforts for the tsunami disaster of 2004 (Hempel, 2005).

We should not overlook, however, the dark-side potential of voluntary and public e-collaboration when used, however well intentioned, for coordinating and collaborating in attacks upon computing resources belonging to others. There are myriad sources of threats for commercial information systems today. These wellsprings of hazards include natural disasters, criminals, vandals, and that most human of all threats, human error (Baskerville, 1996; Im & Baskerville, 2005). With the advent of widespread public networking (the Internet), all of these threat sources have become real-time threats. Many, if not most, information systems are vulnerable through their network connections to all of these threat sources. Information security risk managers must appraise the risks to their systems from each of these sources. The task is growing more complex and extensive as our networks and computer systems grow more complex and extensive (Cronin & Crawford, 1999).

The value of information within the context of warfare is certainly nothing new, and has been recognized for millennia:

Thus, what enables the wise sovereign and the good general to strike and conquer, and achieve things beyond the reach of ordinary men, is foreknowledge. (Sun Tzu, 1995, p. 90)

However, electronic improvements to the use of information within warfare have been evolving over the past century. The original version was known as electronic warfare and regarded the control of the electromagnetic spectrum in military action (Armistead, 2004). To a degree, information warfare arises from electronic warfare and entails actions to protect, exploit, corrupt, deny, or destroy information or information resources in order to achieve a significant advantage over an adversary (Alger, 1996). Information warfare may be critical in the opening phases of any military offensive (Bishop, 2006). Because information warfare encompasses perception management, it is closely related to the struggle to win the hearts and minds of the population. Brutal experience has shown that the information war can be more significant for victory than the armed confrontation that accompanies it (Sturges, Katjihingua, & Mchombu, 2005).

Warfare and terrorism currently lie on the distant horizon as a source of threat to commercial information systems. The central focus of concern for warfare as a source of risks for commercial systems have been directed mostly at those commercial systems concerned with national critical infrastructures. Risk planners are assuming that warfare or terrorist strategies would attack commercial computer systems only as a means to disrupt essential services such as energy, transportation, communications, and so forth (The President's Commission on Critical Infrastructure Protection, 1997). Little concern has been expressed for warfare or terrorism strategies directed at the destruction or disruption of commercial computing per se (Furnell & Warren, 1999).

In this chapter, we explore risks that arise from the use of e-collaborative technologies for the purposes of warfare and terrorist strategies aimed at disrupting or destroying commercial computing capacity as an end rather than as a means. We will explore cases that involved both random and strategically formulated attacks on widespread commercial and government computing facilities. We select perhaps the most interesting and well-known cases of such destruction: the eruption of hacking warfare between nations in the shadow of military confrontations, shooting wars, and other material conflicts.

The common usage of the term *hacker* refers to people who break into computer systems, usually via network access, obtaining privileges on these systems that have not been assigned by the system owners. This common usage might be more properly qualified as "black hat hacker" or just "black hat." Black hat hackers do not have permission from system owners, and may vandalize or abuse systems once access has been acquired illegitimately. "White hat" denotes someone ethically opposed to computer abuse who aims to help secure a system by hacking it to discover its security flaws. Teams of professional white hats employed by system owners for the purposes of identifying flaws for correction are called tiger teams (Wayne, 2006; "White Hat," 2007).

Hacking is also categorized into types by goal. Criminal hacking is illegal hacking for the purpose of fraud, identity theft, vandalism, and so forth. Cyber terrorists invoke significant damage from hacking to coerce or intimidate a country or region. Cyber espionage involves hacking into systems for purposes of intelligence gathering. Cyber warfare may involve hacking for the purposes of information warfare (Vatis, 2006). The alignment of white hats and black hats with morally good and bad hacking becomes obscure and less useful in warfare settings because black-hat hacking can be a component of offensive military operations.

For our purposes, we shall also distinguish between cyber wars and kinetic wars. We shall define cyber wars as computer- and computer-network-based conflicts. In opposition, we define kinetic wars as material wars that lead to the direct physical injury of people (Hall, 2003).

This distinction is not entirely discriminating since destruction of commercial computing capacity can sometimes physically injure people. For example, breakdowns in control systems could lead to physical injuries, such as clinical medication systems in hospitals or air traffic control computers. However, even when such physical injuries are the intentional effect of a cyber war, the result is indirect. In a kinetic war, such injuries are a direct effect.

Hacker wars are a form of cyber war, and these have erupted between nations in parallel with kinetic conflicts. While the ultimate effects of such hacker wars have been minimal so far, and are overshadowed by the drama of the kinetic wars, these suggest the possibility that future cyber wars might be seen as acceptable political alternatives to kinetic wars. Because of ICT and many collaborative computing technologies, the world is increasingly becoming a globalized village. It can be said that, largely, this world universally abhors resort to kinetic warfare for dispute settlement. When kinetic warfare becomes the last resort and political negotiation the first resort, where do nations turn when such negotiations break down? A middle ground of resort may include the international legal justice system and nonkinetic conflicts such as trade sanctions and cyber warfare. Grassroots cyber warfare, waged by the public and seemingly beyond the control of the government, is a particularly problematic prospect (Baskervillel 2005).

Given this prospect, security managers and risk planners in today's commercial environments may need to consider the long-term risks that are evolving from cyber warfare. Thus far in the brief history of direct cyber conflict, government and commercial computers have been targeted either randomly or strategically for political and commercial purposes, as well as for critical infrastructure disruptions. This suggests that all Internet-connected information systems, including retail, electronic mail, marketing, and other systems, are potential targets in national cyber war conflicts.

What makes such conflicts different from existing threats is the resource base of cyber warfare attackers. With increasingly widespread availability of collaborative technologies, cyber attacks are likely to be better organized, more persistent, and technologically advanced. This is especially the case where such national backing as military or political organizations is made available through back channels. Security penetrations are likely to be faster, better coordinated, and more destructive than those of casual vandals or solitary criminals (Hinde, 2001).

To illustrate the threat of public e-collaboration in the waging of cyber warfare on commercial systems, we compare two cyber conflicts that have erupted as hacker wars between nations. We shall briefly describe each of these conflicts before analyzing them in terms of a cyber warfare framework. This framework, to be described below, includes various strategies and concepts in information warfare, such as perception management and offensive and defensive cyber warfare. We will also apply various military strategies that have been developed in kinetic warfare and applied in information warfare. These concepts and strategies arise in a theoretical military arena sometimes called the revolution in military affairs (RMA). These ideas include asymmetrical warfare, network warfare, and various information-driven tactical maneuvers such as "zap" and "swarm." These concepts and strategies are particularly dependent on tight communication and collaboration, and are powerfully enhanced by computer collaboration tools where public is drawn into the activities. There are parallel strategies to these activities in the realm of cyber warfare and each of the incidents of hacker wars can be analyzed from this part of the framework (Jones, Kovacich, & Luzwick, 2002). Finally, we will consider each of these cyber warfare incidents from the perspective of ethics and international law.

ANALYTICAL FRAMEWORK

Figure 1 illustrates the analytical framework used in our analysis. The framework consists of a number of nested concepts that interact in various ways. For example, all of the concepts in the figure are encapsulated by ethics and international or domestic law. This encapsulation is represented by the outermost area labeled ethics and law.

Figure 1. Cyber war analytical framework

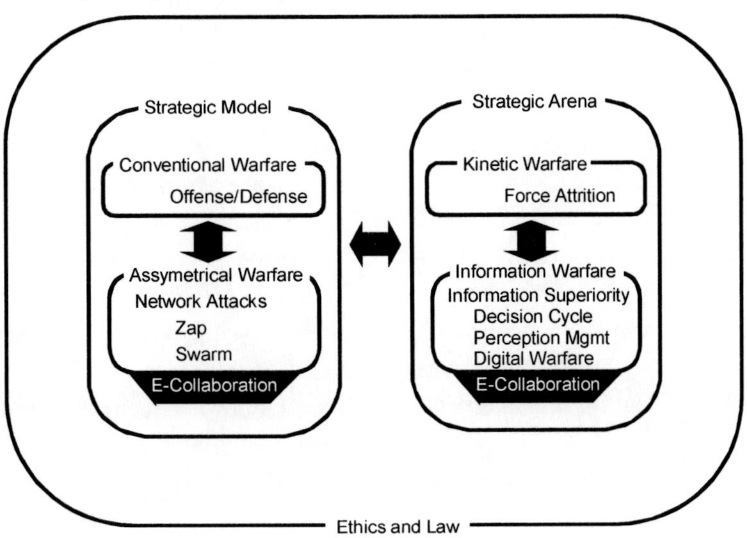

Within the environment defined by ethics and law are two interacting concepts. On the left are the various kinds of strategic model and on the right are the various kinds of strategic arena. A double-sided black arrow designates the interactive nature of these two ideas.

Strategic models regard the theoretical framework used to conceptualize warfare actions and operations. Within the strategic-model area are examples of two different kinds of strategic models. At the top is the conventional warfare model, and at the bottom is the asymmetric warfare model. The two-sided arrow between the two models indicates their interactive nature. The conventional model is older and regards traditional theories of military strategy. There are various elements in the conventional warfare model; however, the most important of these is the concept of offensive and defensive operations. The asymmetrical warfare model is newer and also consists of a number of elements. For our purposes, the most important of these is the idea of network attacks. Two models of network attack that are highly relevant for cyber warfare are zap and the swarm.

To of the right of the strategic model is the second major conceptual area of our framework: the strategic arena. Whereas the strategic model defines the kind of warfare activities, the strategic arena defines the conceptual location of the activities. Two of these locations include kinetic warfare and information warfare. These two strategic arenas interact and the two-headed arrow between them indicates this interaction. Kinetic warfare (shooting wars) is beyond our scope, but one primary goal of kinetic warfare is shown for illustrative purposes; this is force attrition. The most common goal in the area of information warfare is the attainment of information superiority. This superiority can involve better information capabilities in three different areas. These areas are the decision cycle, perception management, and cyber warfare.

For the purposes of our analysis, the two most important conceptual areas are asymmetric warfare as a strategic model and information warfare as a strategic arena. Both of these areas depend upon collaboration using ICT. While conventional warfare and kinetic warfare are related ideas, for the purposes of this chapter, we will exclude them from our boundaries. Asymmetric warfare and information warfare will be discussed in more detail below.

Strategic Model: Asymmetrical Warfare

Asymmetrical warfare leverages particular friendly strengths against particular opposition weaknesses. It involves orchestrating confrontations such that these occur in the exact arenas where the opponent has specific disadvantages in force, politics, or society. Such asymmetry is complete when these weaknesses in the opposition situation are arranged against highly advantageous strengths among friendly elements. Asymmetrical warfare is usually irregular, unmatched to existing defenses, unusual, and capitalized on surprise. Asymmetrical warfare does not usually focus on attrition (the destruction of opposition resources) as the means to compel an opponent; it more often is used in the information warfare arena (Hall, 2003).

Network Warfare

Network attacks as a strategic model in asymmetric warfare do not necessarily refer to computer or data networks. Network attacks regard the synchronization and coordination of forces that are dispersed either over close distances or far distances. A network attack is somewhat similar in nature to a virtual organization. Just as it is the case with virtual teams, network warfare elements can be switched on and off using ICT and e-collaboration technology in an instantaneous and momentary fashion to take advantage of opportunities (whether accidental or intentional). Consequently, network warfare often does depend on excellent data and computer networks, but the exact term *network* in this instance refers to a network of force elements rather than an information network (Kovacich, Jones, & Luzwick, 2002).

Zap

A zap is a particular kind of network attack that relies closely on information being acquired and delivered over a long distance. This information enables the exercise of highly focused and precise force across this very long range. Such zaps use information to enable a minimum amount of force to have a maximum impact on the opposition. For example, a zap attack could involve a large number of hackers collaborating at a specific point in time to attack and disrupt a very specific, key network node that in turn would disable the widest possible network service area. A zap is a longer range attack (Berkowitz, 2003) for which coordination is essential, and these attacks are increasingly dependent on some form of e-collaboration.

Swarm

A swarm is a particular kind of network attack that relies on widely dispersed small elements that use information to coordinate and assemble at a precise time and place in such a way as to overwhelm an opponent. It works well when the opponent has a similarly wide dispersal of elements. For example, a distributed denial-of-service attack (or DOS attack, one in which a large number of cloned "zombie" computers are illicitly used to overwhelm a target computer with senseless traffic) executes with sudden and widespread force, and is most likely to crash a target. The swarm is a close-range attack (Berkowitz, 2003). Again, coordination and collaboration are essential elements of this kind of network attack. To be effective in a purposeful way, e-collaboration is necessary for installing and triggering the clone computers.

E-Collaboration in Asymmetrical Warfare

Because of its dependence on communication and coordination, modern asymmetrical warfare is necessarily a creature that, like effective management, is dependent on e-collaboration. This dependence holds whether the strategic arena is kinetic warfare or information warfare.

Strategic Arena: Information Warfare

There are many definitions of information warfare, and these overlap and conflict to some extent. It refers to actions or operations that degrade opposition information (and information systems) or that defend friendly information (and systems) to achieve a significant advantage, usually in times of crisis or conflict (Alger, 1996; Denning, 1999).

Information Superiority

Information superiority is the central objective of information warfare. It regards an overwhelming command of both friendly and opposing decisions, perceptions, and information communication and processing. In a world that is increasingly dependent on information and technology, superior information in warfare is essential. Real information superiority is not usually demonstrated by a crashing and momentous event of victory. Information superiority is very often nuanced: a matter of small but important advantages (Berkowitz, 2003; Hall, 2003).

Decision Cycle

The decision cycle regards the quality and the length of time required to collect the necessary information and to make well-formed decisions. In military strategy, the decision cycle consists of multiple stages, such as to observe, orient, decide, and act. Control over this cycle can be used to improve the quality and speed of friendly decision cycles. Control over an adversary's cycle is also used to degrade opposition quality and speed of decisions. Information is obviously necessary for observation and orientation, but the ability to decide and act is also dependent on the ability to communicate information and coordinate activities. Hence, e-collaboration technologies are essential for causing ineffective decision cycles. Information superiority over decision cycles means that control over both friendly and opposition decision cycles is held (Hall, 2003).

Perception Management

Perception management refers to actions or operations that affect the perception of both friendly and adversary groups or individuals in order to influence their emotions, reasoning, decisions, and actions (Denning, 1999). Perception management is often related to psychological operations and propaganda. Data networks and e-collaboration technologies provide vehicles for manipulating and communicating propaganda and psychologically "spun" information (Crilley, 2001).

Cyber Warfare

Cyber warfare is the digital side of information warfare and includes the protection (defensive operations) and compromise (offensive operations) of computer-based technology that bears information. Whereas perception management regards the manipulation of the information content of systems or other communications media, digital information warfare regards the manipulation of the computers and data networks that process and carry such information.

Ethics and Law

Is it legal or ethical to pervert the use of e-collaboration technologies for warfare or terrorism? Ethics and the subsequent international and domestic legal frameworks provide an overarching environment for the status of cyber warfare against commercial systems. When computer systems are attacked across national borders, a number of complex legal issues develop. Depending on the precise situation of the attack, it might be considered to be criminal, terrorism, or cyber warfare.

The status of cyber warfare is not yet tested by courts nor protected by specific treaties. Military theorists, however, are generally treating information and cyber warfare as a field in which the capabilities of information warfare systems are to be regarded as a weapon system available for use during armed conflict. Military organizations are asserting, in a case-by-case fashion, the right to use force in cyberspace. This interpretation is probably consistent with Article 51 of the United Nations (UN) charter, which protects inherent rights of states to self-defense and also to use force against potential aggression. The broad language of this article reaches out to encompass future advances in technology and new forms of weaponry such as those being employed in information warfare. Accordingly, customary principles of warfare must guide the use of cyber warfare weapons between nations. These principles are summarized in Table 1, which details six common principles in warfare and illustrates each principle with a brief description of lawful conflict along with ways in which cyber warfare might be illegal as a result of the violation of these principles. Of particular interest when considering commercial targets in cyber warfare are principles of discrimination and humanity. Principles of discrimination require warfare to target military elements and not civilian elements. Attacks on national critical information infrastructures probably violate this principle. Principles of humanity suggest that cyber attacks should cause no unnecessary suffering, and attacks on critical information infrastructures probably also violate this principle. Other principals such as proportionality and necessity require cyber warfare acts to have clear military objectives. Indiscriminate attacks on commercial computing facilities probably violate these principles (Darnton, 2003).

Cyber warfare weaponry has a surprisingly long history of terrorist incidents. Terrorism usually involves a conflict between a state and a nonstate adversary. Such adversaries may be internal groups that seek to overthrow a government (revolutions). Terrorism usually depends upon a terror cycle that aims to divide the government and the populace. The cycle iterates violent acts on the populace with repressive security responses by the government. With repeated violence, the government increases repression. The duel effects of seeming government helplessness and increasing government repressiveness steadily undermine the faith and support of the populace until the civil government finally collapses (Baker, Friedman, & Miller, 1983).

Where terrorists are not supported or condoned by any state, the legal status of these acts is defined by criminal law, both international and domestic. Importantly, the UN General Assembly Resolution 2625 prohibits a state from condoning or operating with "cloaked" agents. The resolution requires states to refrain from organizing or assisting terrorism or civil strife in other states. States must not even acquiesce in permitting these activities within their territories. Where agents involved in cyber warfare can be shown to operate with the acquiescence of a state, principles of international warfare apply. Where agents and states are definitely separated, criminal law applies (Delibasis, 2002). In our discussion of hacker wars below, the issue of separation between agent and state becomes extremely significant for existing cyber warfare attacks on commercial information systems.

CASES OF HACKER WARS

There are at least four recorded instances of major hacker wars erupting in the shadow of kinetic conflicts. These hacker wars involve geographically dispersed individuals who collaborate to conduct coordinated attacks on computer systems in highly strategic fashions. These individuals are members of the public acting with a sense of civic and political determination. It is likely that most of these individuals have never met face to face and do not belong to any particular organization. In each of the four cases, the hacker war appears to have been a grassroots enterprise by civilians, without government control or condoning. Vatis (2001) provides an excellent summary of all four of these incidents. These incidents include the 1999 Yugoslavia-U.S. hacker war in which Yugoslav hackers successfully flooded multiple NATO (North Atlantic Treaty Organization) sites with viruses and e-mail. In concert, Serbian, Russian, and Chinese hackers launched Web site

Table 1. Generally accepted principles of legal warfare (adapted from Darnton, 2003)

Principle	Lawful Conflict	IW Violations
Discrimination	Target military, not civilians	Scenarios targeting critical infrastructures
Proportionality	Not excessive for objective	Scenarios lacking clear military objective
Lawfulness	Respects treaties and international law	Disruption and spoofing of international communications
Necessity	Needed for objective	Scenarios lacking clear military objectives
Humanity	No unnecessary suffering	Scenarios targeting critical infrastructures
Neutrality	Harmless to neutral countries	Difficult to confine IO from neutral facilities

defacement attacks on numerous U.S. commercial sites. The Indo-Pakistani hacker war began in 2001, with myriad groups engaging in Web site defacement. Hundreds of sites were defaced on both sides, with more than 500 attacks in 2003. The 2000 Israeli-Palestinian hacker war and the 2001 China-U.S. hacker war will be discussed below. Of these attacks, the latter two have been selected as comparative examples to illustrate the analytical framework.

2000 Israeli-Palestinian Hacker War

According to Vatis (2001), such hacker wars emerge in the shadows of kinetic conflicts. In this case, shortly after several Israeli soldiers were kidnapped, pro-Israeli hackers created a Web site called Wizel.com. This site was used as a host for "FloodNet" attacks, DOS attacks that effectively disable Web sites with a flood of reloads. The attack disabled six Hezbollah sites, Hamas.org, and other Palestinian informational sites (Gentile, 2000). Pro-Palestinian hackers retaliated with an attack on Wizel.com, the Bank of Israel, Tel Aviv, and a wide variety of government sites. Disabled government sites included the Israeli parliament, the defense forces, and the foreign ministry. According to a pro-Palestinian group called Unity, the Palestinian cyber attack involved a four-phase strategy, disabling (a) Israeli government sites, (b) the Bank of Israel and the Tel Aviv Stock Exchange, (c) Lucent, Golden Lines, and NetVision, and (d) Israeli e-commerce Web sites.

- **Strategic model:** Both sides operated with asymmetric strategic models using network attacks. Pro-Israel hackers used a zap maneuver precisely directed at several high-visibility Web sites. The pro-Palestinian strategy was rather more sophisticated—more of a swarm attack with an escalating approach. The attack on high-visibility government computers could undermine confidence in the government to protect its infrastructures. This was followed by a financial attack that was inherently directed at a critical information infrastructure. Similarly, the planned attack on the communications network providers is another critical information infrastructure attack, appropriately staged after the financial damage was to be complete. Finally, the commercial attacks were random vandalism that reinforced the helplessness of the government to protect its population.
- **Strategic arena:** Both sides were operating with cyber warfare techniques. The technology used during the war included DOS, distributed denial of service (DDOS), worms, Trojan horses, and break-ins. Information superiority was sought with sheer offense as the hackers tried to dominate by denying information assets to the adversaries. The effects were not that successful. Most of the damage was embarrassment and public Web sites that were temporarily rendered unavailable.
- **E-collaboration:** There is no evidence of the use of specific, advanced e-collaboration technologies. It is likely that e-mail, chat rooms, and private Web sites supported the e-collaboration. Essentially, the attackers operated with a form of ad hoc intranet cobbled together from preexisting communication settings that provided a basic social and technical infrastructure.
- **Ethics and law:** There are no firm indications in the Israeli-Palestinian hacker war (or any of the hacker wars thus far) that the attacks on either side have been state sponsored. No assumptions that these hackers were cloaked agents can be justified. It appears that the hacking is criminal rather than a conflict between states. This hacker war does seem to qualify under some definitions of terrorism since it appears to be a conflict between a state and a nonstate adversary. In particular, in the Israeli-Palestinian hacker

> war, the strategies clearly aimed at attacking government Web sites and national infrastructures. In this case, however, it is not clear that the nonstate adversary is an organized body of any permanence. It would appear that the hacking agents were irregular, informal, and assembled ad hoc without clear centralized direction or leadership. The activity seems to have been a form of swarm that occurred as a collective decision of dispersed individuals: a temporary self-organization rather than a carefully organized, planned, and closely controlled group.

There is little record of concentrated law enforcement activity in pursuit of the hacking perpetrators. However, this might not be seen as that unusual since the damage resulting from the hacking was not significant on either side.

If state sponsorship could be established, it is unlikely the hacker wars would emerge as legal under common principles of warfare. Referring to Table 1, the attacks were mostly on political organizations, critical infrastructures, and commercial Web sites. These are clearly civilian targets with no clear military significance. It is unclear whether any military objective would have been achieved with the strategies exercised by either side in this hacker war. Thus, if the hackers were state sponsored, the states violated the principles of discrimination, proportionality, necessity, and humanity.

2001 China-U.S. Hacker War

Like the Israeli-Palestinian hacker war, the China-U.S. war emerged in the shadow of a rather small but nevertheless kinetic conflict in which a midair collision between Chinese and U.S. warplanes led to the Chinese seizure of a U.S. spy plane. At the height of the tensions during the following month, Chinese hackers seemed to have assembled as a self-organized group using Internet relay chat and bulletin boards. The campaign was styled as "Hack USA Week" and was publicized in the Chinese press. The resulting campaign is estimated to have damaged nearly 1,200 U.S. Web sites including those of the White House, the Air Force, the Navy, the National Institutes of Health, the California Department of Energy, the U.S. Department of Labor, the Department of Energy, and many commercial companies (Lemos, 2001).

U.S. hackers retaliated by breaking into hundreds of Chinese Web sites, leaving hate messages, pornography, and pro-drug-abuse messages. U.S. hackers attacked Chinese government sites, university sites, and nonprofit organizations randomly. In general, it appeared American hackers chose targets based on their vulnerability rather than content or ownership (Wallace, 2001). More than 700 Chinese Web sites were damaged (Radcliff, 2001).

- **Strategic model:** Both sides appeared to operate with simple symmetric models using swarm network attacks. The attacks did not appear to follow any specific strategies other than to compromise as many systems that were highly visible and highly vulnerable as possible. Both sides were operating with cyber warfare techniques. The Chinese hackers used DDOS attacks or defaced Web sites through common intrusions. Attackers from both countries covered their tracks by bouncing attacks through servers in Korea, Russia, and elsewhere (Radcliff, 2001).
- **Strategic arena:** As in the Israeli-Palestinian hacker war, information superiority was sought through sheer offensive strategies. The objective seemed to be the defacement or service denial of the most sites. The vandalism was very nearly random on government, university, and commercial Web sites. If there was any goal beyond the targeted vandalism, it would seem to be to reinforce the helplessness of the respective governments to protect its population and

establish the technical supremacy of one state over the other.

- **E-collaboration:** Similar to the previous example, there is no evidence of the use of specific, advanced e-collaboration technologies. In addition to e-mail, chat rooms, and private Web sites that supported the e-collaboration, there was some usage of the press. As previously, it appears that the attackers operated with a form of ad hoc intranet.
- **Ethics and law:** Similar to the Israeli-Palestinian hacker war, there is no clear evidence of state sponsorship from either China or the United States. However, some authorities did conclude at the time that the Chinese government had made no attempt to investigate the Chinese criminal activity and therefore had tolerated the attacks (Cohen as quoted in Lemos, 2001; Vatis, 2001). These are significant accusations because such cloaked agents would escalate the incident into a state-sponsored cyber war. Press reports indicate that civilian watchdog agencies reacted promptly in both countries (Anonymous, 2001; Krebs, 2001). In the United States, law enforcement agencies were prominently involved in investigating the incidents, including the FBI (Federal Bureau of Investigation; Radcliff, 2001). In China, however, the criminal penalties are dramatically higher; criminal hacking in China is a capital offense.

Because the attacks were directed against states by ostensibly unsponsored criminal groups, the China-U.S. hacker war should probably be qualified as a form of cyber terrorism (claims to the contrary notwithstanding). If reviewed under the principles of warfare, this hacker war would violate the principles of proportionality and necessity because of the random nature of the vandalism. The attacks were also indiscriminate and failed the principle that seeks to avoid civilian targets when targeting military targets in lawful conflict.

DISCUSSION

In comparison, these two examples demonstrate that the impact of cyber conflict has proven thus far to be entirely mild. While there were defacements by the thousands, the activity was not particularly extraordinary. While CERT, the FBI, and commercial security analysts were concerned, the activity levels were rather normal (Krebs, 2001). According to Berkowitz (2003, p. 159) about the China-U.S. hacker war, "...none of this amounted to a hill of beans. It was a food fight, not a geopolitical death match. The sites that 'got owned' during Hack the USA Week were just web sites that various organizations use to provide information to the public."

In other words, the hacker wars thus far have not had serious consequences. The results might be best summed by chauvinistic claims that "our hackers beat the daylights out of their hackers" from one or both combatants (McConnell, 2004, p. 13). However, based on analysis with the framework above, these wars do raise some interesting issues. The Israeli-Palestinian hacker war followed a comparatively sophisticated strategy with evidence of both the zap and swarm network attacks—forms of asymmetrical warfare that have a more clear relevance to strategic modeling than the more recent China-U.S. hacker war. While both hacker wars operated in a strategic arena seeking information superiority through cyber warfare, the China-U.S. war simply sought offensive superiority by maximum vandalism of random sites. The Israeli-Palestinian hacker war sought information superiority by threatening national critical infrastructures.

Is it likely that this food fight will become a geopolitical battleground? There are indications that it will. At least 25 countries are believed to have the ability to carry out significant cyber warfare attacks (Vatis, 2006). Among these, China's army may have developed a cadre of professional hackers skilled at delivering a disinformation program in the opening phases of a military offensive. In the presence of growing tensions between Taiwan and China, in June 2006, hackers electronically sent a series of press releases that seemed to appear as if they had come from Taiwan's Ministry of Defense (Bishop, 2006).

From the viewpoint of e-collaboration, it is important to recognize that both hacker wars arise as examples of self-organizing terrorist activities. Thus, criminal law rather than laws surrounding the international use of force would appear to be most relevant. The contemporary accusations that the Chinese government condoned (if not indeed sponsored) the Chinese side of the hacker war are very interesting. Although the charges appear to be groundless posturing, they represent an attempt to escalate a relatively small-scale, grassroots, and largely uncontrollable terrorist action into a class of state-sponsored warfare. Serious law enforcement agencies worked to deflate perception of the incidents (Radcliff, 2001).

The analysis shows that strategic models, strategic arenas, and principles of warfare apply and suggest that e-collaboration technologies have a clear future role in this arena. While the hacker wars seem to have been somewhat benign in impact, the concern about escalation drawn from the framework suggests that e-collaboration technologies could have a role as an escalation tool in cyber warfare. Current conflict escalations involve such mild dispute resolution avenues as diplomacy, legal resort, or trade sanctions. The gulf between these and (potentially lethal) kinetic warfare is large. Cyber warfare and e-collaboration technology might fulfill a role as a median dispute resolution tool (a next-to-the-last resort) that is more serious than trade sanctions, but short of kinetic warfare. The mild impact thus far simply enhances the value and attractiveness of such a conflict vehicle. There is also potential for operating a cyber war in concert with kinetic conflict. Cyber warfare is a new and operational conflict alternative, and its dependence on e-collaboration is notable.

The framework also draws out the importance of civilian targets in cyber warfare and the dark side of e-collaboration. Like other technologies, e-collaboration can be applied for better or worse purposes. It can enhance commerce, spread knowledge, and advance civilization. It is also an equally useful tool for terror and war. Indeed, the concern of basic defensive vehicles such as law enforcement authorities and commercial security analysts is justified. Any future experiments with cyber warfare will likely involve e-collaboration technologies and ought to include agencies and organizations from both the private and the public sector (Adams, 2002).

CONCLUSION

The analytical framework illustrates the role of e-collaboration technologies in supporting electronic warfare and cyber terrorism. While it highlights how ineffective hacker wars have proven thus far, it suggests the greater potential of state-sponsored cyber war where e-collaboration technologies might be better developed and applied. It is clear that there is potential for the use of cyber warfare in national conflicts, but it is not exactly clear how such a war can be waged within the boundaries of current principles that govern lawful warfare and common criminal law. Hacker wars thus far illustrate how theories and principles of warfare apply in cyberspace, and illustrate the role of e-collaboration. Today, they are, at the most, crude and inconsequential previews of future cyber warfare.

REFERENCES

Adams, J. (2002). Virtual defense. *Foreign Affairs, 80*(3), 98-112.

Alger, J. (1996). Introduction. In W. Schwartau (Ed.), *Information warfare. Cyberterrorism: Protecting your personal security in the electronic age* (2nd ed., pp. 8-14). New York: Thunder's Mouth Press.

Anonymous. (2001). *Authority warns of hacker war.* Retrieved February 17, 2004, from http://fpeng.peopledaily.com.cn/200105/04/eng20010504_69242.html

Armistead, L. (Ed.). (2004). *Information operations: Warfare and the hard reality of soft power.* Dulles, VA: Brassey's.

Bafoutsou, G., & Mentzas, G. (2002). Review and functional classification of collaborative systems. *International Journal of Information Management, 22*(4), 281-305.

Baker, D., Friedman, R., & Miller, D. (1983). *Intelligence warfare.* New York: Crescent Books.

Baskerville, R. (1996). A taxonomy for analyzing hazards to information systems. In S. Katsikas & D. Gritzalis (Eds.), *Information systems security: Facing the information society of the 21st century* (pp. 167-176.). London: Chapman-Hall.

Baskerville, R. (2005). Hacker wars: E-collaboration by vandals and warriors. *International Journal of e-Collaboration, 2*(1), 1-16.

Berkowitz, B. (2003). *The new face of war: How war will be fought in the 21st century.* New York: The Free Press.

Bishop, M. W. (2006). China's cyberwarriors. *Foreign Policy, 156*, 93.

Crilley, K. (2001). Information warfare: New battle fields, terrorists, propaganda and the Internet. *Aslib Proceedings, 53*(7), 250-264.

Cronin, B., & Crawford, H. (1999). Information warfare: Its application in military and civilian contexts. *The Information Society, 15*, 257-263.

Darnton, G. (2003). Information warfare and the laws of war. *2nd European Conference on Information Warfare and Security*, 107-116.

Delibasis, D. (2002). The right of states to use force in cyberspace: Defining the rules of engagement. *Information & Communication Technology Law, 11*(3), 255-268.

Denning, D. E. (1999). *Information warfare and security.* Reading, MA: Addison-Wesley.

Furnell, S. M., & Warren, M. J. (1999). Computer hacking and cyber terrorism: The real threats in the new millennium? *Computers & Security, 18*(1), 28-34.

Gentile, C. J. (2000). Hacker war rages in Holy Land. *Wired News.* Retrieved February 7, 2004, from http://www.wired.com/news/print/0,1294,40030,00.html

Hall, W. M. (2003). *Stray voltage: War in the information age.* Annapolis, MD: Naval Institute Press.

Hempel, J. (2005, January 31). A corporate peace corps catches on. *Business Week*, p. 14.

Hinde, S. (2001). Incalculable potential for damage by cyber-terrorism. *Computers & Security, 20*, 568-572.

Im, G., & Baskerville, R. (2005). A longitudinal study of information system threat categories: The enduring problem of human error. *The Database for Advances in Information Systems, 36*(4), 68-79.

Jones, A., Kovacich, G. L., & Luzwick, P. G. (2002). Everything you wanted to know about information warfare but were afraid to ask: Part 1. *Information Systems Security, 11*(4), 9-20.

Kovacich, G. L., Jones, A., & Luzwick, P. G. (2002). Global information warfare: How businesses, governments, and others achieve objectives and attain competitive advantages. *Security Management Practices, 11*(5), 15-23.

Krebs, B. (2001). *Experts yawn over US-China hacker war.* Retrieved February 17, 2004, from http://www.findarticles.com/cf_dls/m0NEW/2001_May_1/74015540/p1/article.jhtml

Lemos, R. (2001). *Defacements rise in China hacker war.* Retrieved January 17, 2004, from http://news.com.com/2100-1001-256732.html?legacy=cnet

McConnell, M. (2004). Cyber terrorism and espionage is real and growing. *Bulletin of Northwest Public Power Association, 58*(4), 13-14.

The President's Commission on Critical Infrastructure Protection. (1997). *Critical foundations protecting America's infrastructures.* Washington, DC.

Radcliff, D. (2001). We have met the enemy and he is us. *Computerworld.* Retrieved February 17, 2004, from http://www.computerworld.com/industrytopics/retail/story/0,10801,60466,00.html

Rutkowski, A. F., Vogel, D. R., Genuchten, M. v., Bemelmans, T. M. A., & Favier, M. (2002). E-collaboration: The reality of virtuality. *IEEE Transactions on Professional Communication, 45*(4), 219-230.

Sturges, P., Katjihingua, M., & Mchombu, K. (2005). Information in the national liberation struggle: Modelling the case of Namibia (1966-1990). *Journal of Documentation, 61*(6), 735-750.

Sun Tzu. (1995). *The art of war.* London: Hodder & Stoughton.

Vatis, M. A. (2001). *Cyber attacks during the war on terrorism: A predictive analysis.* Retrieved December 6, 2004, from http://www.ists.dartmouth.edu/library/analysis/cyber_a1.pdf

Vatis, M. A. (2006). The next battlefield. *Harvard International Review, 28*(3), 56-61.

Wallace, R. (2001). *It's an all-out cyber war as U.S. hackers fight back at China.* Retrieved February 17, 2004, from http://www.foxnews.com/story/0,2933,19337,00.html

Wayne, H. (2006). Computer security firm uses hacker's instincts, skills. *Knight Ridder/Tribune Business News*, p. 1.

White hat. (2007). *Wikipedia.* Retrieved February 23, 2007, from http://en.wikipedia.org/wiki/White_hat

Chapter XII
Effects of Leadership Style and Anonymity on Arguments and Intentions Related to Acting Unethically

Surinder S. Kahai
State University of New York at Binghamton, USA

Bruce J. Avolio
University of Nebraska, USA

ABSTRACT

A laboratory experiment was conducted with 42 student groups to evaluate the effects of transactional vs. transformational leadership styles and anonymity when groups supported by an electronic meeting system (EMS) discussed the ethical issue of copying copyrighted software. A confederate leader displayed either transformational or transactional behaviors. Transformational leaders motivate effort by raising the awareness of followers to make them aspire to higher order needs and values, and developing them to fulfill their aspirations. Transactional leaders motivate effort by highlighting the contractual exchange involved in a relationship. Participants working with a transformational confederate were more likely to make arguments that challenged the copying of copyrighted software than those working with one who was more transactional. These arguments in turn caused groups exposed to such arguments to have greater deviation among its members in intentions to copy the software. Participants working with a transactional confederate were more likely to make arguments in favor of copying copyrighted software. These arguments in turn caused groups exposed to such arguments to have a greater mean of intentions to copy the software. Implications for practice and future research on ethics and leadership are discussed.

EFFECTS OF LEADERSHIP STYLE AND ANONYMITY ON ARGUMENTS AND INTENTIONS RELATED TO ACTING UNETHICALLY

Despite increasing governmental regulation, the problem of unethical behavior persists and imposes a significant cost on organizations (Verschoor, 2006). Additionally, the increasing digitization of intellectual property combined with increasing deployment of electronic networks is making it more likely for organizational workers to engage in unethical acts that compromise the rights of owners of digital content (Stead & Gilbert, 2001). Consequently, it is not surprising that public- and private-sector establishments, universities, and professional societies are interested in understanding how workers and students can be developed to reason ethically (Giacalone & Thompson, 2006). Among the issues that these entities are focusing on is the linkage between a leader's behavior and the ethical intentions and behavior of the leader's followers (Brown & Trevino, 2006; Cuilla, 2004). Underlying this focus is a widespread belief that leaders set the tone for ethical behavior in organizations.

Due to an increasing reliance on computer-mediated communication and the geographic spread of an organization's workforce, organizations are also focusing today on the use of online communication media for the development of ethical reasoning (French, 2006; Painter-Morland, Fontrodona, Hoffman, & Rowe, 2003). The relevance of online communication media for the development of ethical reasoning was apparent in the emergence of Web-based discussions of hundreds of hypothetical ethical scenarios not too long ago. These discussions broke out after it was revealed by a prestigious university in March 2005 that it had turned down over 100 applicants who snooped into an online database to find out the status of their applications (Zeller, 2005). Visitors to one such discussion at Collegeconfidential.com, for instance, included students, school teachers, and parents who engaged each other about how they would act and presented reasons for why one action would be more appropriate or inappropriate than another.

Due to the potential of online discussions to shape the ethical attitudes and behaviors of today's and future personnel, organizations would benefit from an understanding of the effects of interventions introduced in such discussions. Interventions could take the form of the introduction of a leader who facilitates the discussion and tries to influence the tone of the discussion, and the manipulation of features of online communication media, such as anonymity in providing input. Motivated by these considerations, we decided to examine the connection between leadership style and ethical reasoning in an electronic context.

Specifically, we compared the effects of transactional and transformational styles displayed by confederate leaders introduced in groups to facilitate discussion on the issue of copying copyrighted software in groups. Members of these groups communicated anonymously or nonanonymously via an electronic meeting system (EMS). We also studied the effects of different types of arguments raised on the intentions within the groups regarding action on the issue. Though prior literature has examined the relationship of leadership to ethics (e.g., Brown & Trevino, 2006; Grojean, Resick, Dickson, & Smith, 2004; Turner, Barling, Epitropaki, Butcher, & Milner, 2002), the authors are not aware of any empirical research that has examined the relationship between leadership styles and ethical reasoning in electronic contexts.

Our focus on transformational and transactional styles was motivated, in part, by the impact that these leadership styles have on group processes and outcomes across a broad range of organizations (see Judge & Piccolo, 2004) and settings, including electronic settings (Kahai, Sosik, & Avolio, 2003; Sosik, Avolio, & Kahai, 1997). Both transactional and transformational leadership styles are likely to influence ethical decision making, though they may do so by evoking different values. Transformational leaders rely upon end values to raise the standards that their followers use to make their decisions (Burns, 1978). Transactional leaders, on the other hand, rely more on what Burns called modal values. Modal values include such things as fairness in an exchange, while end values represent such things as liberty or justice. Transformational leaders may also influence ethical behavior by role modeling ethical behavior and by articulating and stressing clear ethical guidelines for their followers to live up to (Trevino, Brown, & Hartman, 2003). As noted by Cuilla (2004, p. 316), "transformational leadership has become almost synonymous with ethical leadership."

Our focus on anonymity was motivated by the common use of electronic media, especially EMS, to enable anonymous communication and the ability of leadership to interact with anonymity to influence group processes and outcomes (Kahai et al., 2003). We focused on the copying of copyrighted software partly because illegal duplication of software and of digital intellectual property, in general, has received considerable attention in ethics-related research and education and also as part of a global debate regarding intellectual property rights (e.g., Cronan, Foltz, & Jones, 2006; Moores & Chang, 2006).

General Theoretical Model

Our model of the effects of leadership styles and anonymity within an electronic context is based on a key premise of Street, Douglas, Geiger, and Martinko's (2001) cognitive elaboration model of ethical decision making. According to this model, which synthesizes Jones' (1991) model of ethical decision making and the elaboration likelihood model (Petty & Cacioppo, 1981), an individual's intentions about an ethical issue are determined by the level of cognitive elaboration by that individual on the issue. When an individual performs a high level of cognitive elaboration, that individual is using central information processing to more deeply analyze the ethical dilemma. With a deeper level of processing, we expect the individual to recognize the moral aspects of an ethical issue before establishing an intention how to act. Conversely, when an individual performs a low level of cognitive elaboration, that individual is using peripheral information processing. Here the individual is less likely to recognize the moral aspects of the issue and may rely on unethical considerations, such as self-interest, to establish an intention on how to act.

We suggest that in a group discussion supported by an EMS, the style of a leader facilitating the discussion and anonymity in providing input are two determinants of cognitive elaboration by individuals within the group. Leadership style influences the motivation to think about an issue and present one's ideas to the group. For instance, a leader displaying transformational behaviors may relate action on an issue to one's self-concept of doing the right thing, thereby making it personally relevant and motivating deeper thinking about the issue. By encouraging consideration for each individual's unique perspective, a leader displaying transformational behavior is also likely to help create an environment in which group members feel comfortable presenting their ideas to others.

Anonymity influences the motivation and ability of group members to elaborate cognitively. For instance, anonymity within a group may increase the motivation to think about an issue by creating a safe environment in which one cannot be associated with her or his ideas and thus avoid sanctions (Nunamaker, Dennis, Valacich, Vogel, & George, 1991). It may also influence the ability to cognitively elaborate by reducing distraction due to attention to others and what the others may be saying (Kahai, Avolio, & Sosik, 1998). As discussed later in the development of hypotheses, anonymity may also influence the effect of a leader's behavior.

In Figure 1 we present a model of the effects of leadership style and anonymity on group discussion of an ethical issue in an EMS context. Leadership style and the level of anonymity influence the ideas that group members think about and present to others via the EMS. The ideas presented, in turn, influence the intentions of members about how to act in the ethical situation at hand as well as further elaboration on the issue. Further elaboration may occur due to support and stimulation provided by others.

In the next section, we develop hypotheses about the effects of transformational vs. transactional leadership styles and anonymity on arguments expressed within a group and on group members' intentions about how to act in an ethical situation. When considering arguments presented within a group, we focused on (a) arguments in favor of copying software, (b) arguments against copying software based on ethical considerations, and (c) arguments against copying software based on consequences of copying.

We also focused on the change in the state of intentions within a group to copy the copyrighted software as a result of group discussion. We considered two indicators of the state of intentions before and after the discussion: the group mean of intentions to copy and the deviation in intentions within a group to copy. The group mean of intentions to copy represents the overall inclination of the group to take the action of copying the copyrighted software. It does not presuppose homogeneity in intentions to copy but reflects a summarization of the intentions, such as that which may occur as a result of a vote. The presence of deviation in intentions (to copy the software) within a group of students is significant because it would indicate lack of conformity to a norm among the student population that illegal copying is acceptable (Cohen & Cornwell, 1989) and stimulate debate. In a recent survey of students, only 29.8% of the respondents regarded software piracy as wrong (Hinduja, 2003).

Figure 1. General model of the effects of leadership style and anonymity on group process and outcomes for discussion of ethical issues

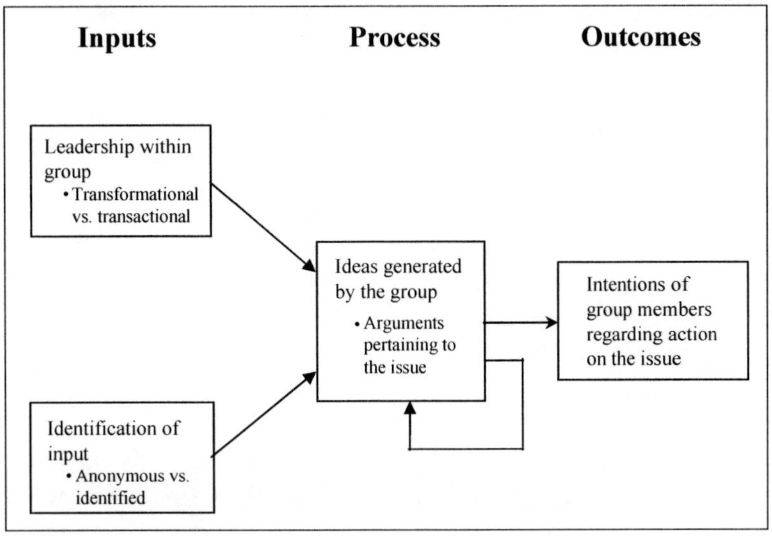

Hypotheses

Effect of Leadership Style on Arguments

Both transformational and transactional leadership behaviors are likely to motivate thinking about the ethical issue at hand and presenting one's ideas to the group (hereafter, thinking about the issue and presenting one's ideas are jointly referred to as elaboration).

Transformational leaders display inspirational motivation (IM) and promote intellectual stimulation (IS) and individualized consideration (IC). Transformational leaders raise the awareness of followers to make them aspire to higher order needs and values and develop them to fulfill their aspirations (Bass, 1985; Bass & Steidlmeier, 1999; Burns, 1978). The transformational confederate leader in the current study indicated to the participants that they would learn from their discussion (IM). The linking of the discussion to participants' learning was expected to engage their self-concepts and intrinsically motivate them to elaborate on the issue at hand (Shamir, House, & Arthur, 1993). The leader highlighted the collective efficacy of the group by indicating that by working together, they could arrive at a better set of ideas. This IM behavior, in turn, was expected to increase participants' self-efficacy and motivate them to elaborate more on the issue at hand (Shamir et al.).

A transformational leader encourages group members to challenge and reframe each other's ideas (IS) and also show understanding and consideration for each other's opinions (IC). By clarifying the leader's expectations, these behaviors motivate participants to stimulate others and express understanding and support for their ideas. Group members elaborate more when others force them out of their molds by challenging them and reframing their ideas. They also elaborate more when others display understanding and support for their ideas.

Transactional leaders motivate effort by highlighting the contractual exchange involved in a relationship (Bass, 1985). In this study, the transactional confederate leader was expected to motivate elaboration by highlighting the extrinsic outcomes that would result in exchange for participants' effort (i.e., a long list of ideas, a feeling of satisfaction for performing that task as agreed to, and course credit given by the course instructor) and building effort-accomplishment expectancies. The transactional leader strengthened these expectancies and the motivation to elaborate by clarifying the task (i.e., generate and exchange as many ideas as possible about the ethical issue) and rewarding participants by expressing satisfaction with their progress.

As per the above discussion, both leaders were expected to motivate elaboration. However, the transformational and transactional leaders were also expected to influence perceptions of the ethical issue's personal relevance and moral intensity differently and cause the nature of elaboration to differ (Street et al., 2001). In the transformational condition, the leader highlighted that not copying copyrighted software is simply the right and ethical thing to do and linked it to the important value of creating a greater good for the society. The leader thus made the behavior of not copying the software meaningful and personally relevant by linking it to important personal values of (a) being ethical by doing the right thing and (b) helping society. It is expected that by emphasizing end values, the leader motivated participants to rise above their individual interests and argue against copying based on ethical considerations.

In the transactional condition, the leader was expected to influence the perceptions of the issue's moral intensity by highlighting the magnitude of consequences of one's action (Jones, 1991). Here the leader indicated that if consumers pay for the software, the producer gets a fair rate of return and continues to provide innovative products, which is in consumers' own best interest and simply the more effective transaction for consumers.

By taking such an exchange-oriented perspective in which there is a payback for what you do, a leader is likely to motivate participants to argue against copying based on the harms or negative consequences of copying. Based on our discussion, we hypothesize the following:

Hypothesis 1. In groups discussing the issue of copying copyrighted software, the nature of arguments raised will differ depending on the style of the confederate leader. Specifically,

a. There will be a greater incidence of arguments against copying based on ethical considerations in groups led by a transformational vs. a transactional leader.
b. There will be a greater incidence of arguments against copying based on negative consequences in groups led by a transactional vs. a transformational leader.

By taking an exchange-oriented perspective in which there is payback for what you do, a transactional confederate leader is also likely to motivate economic- and individual-oriented thinking in which the potential for self-interested behavior rises (Bass, 1985). In the case of the issue of copying copyrighted software, self-interest favors copying the software (Hinduja, 2003). Accordingly, we hypothesize the following:

Hypothesis 1c. There will be a greater incidence of arguments in favor of copying in groups led by a transactional vs. transformational confederate leader.

Effect of Leadership Style and Anonymity on Arguments

We expected the anonymity of participants to influence the effect of leadership style on arguments raised during group discussion. We argued above that both transformational and transactional styles displayed by confederate leaders will motivate participants to elaborate on the ethical issue. Anonymity is expected to influence the effect of these leadership styles on the motivation to elaborate.

As argued in the development of Hypothesis 1, transformational behaviors increase the motivation to elaborate by (a) highlighting the collective efficacy of the group, (b) encouraging group members to challenge and reframe each others' ideas, and (c) encouraging understanding and consideration for each others' opinions. Anonymity is expected to enhance all these effects. By disconnecting individual input from individual identities, anonymity takes attention away from individual identities and interpersonal differences (Kahai et al., 1998). This has the effect of enhancing the salience of a group and increasing focus on the content of ideas rather than on who is presenting them (Jessup, Connolly, & Galegher, 1990). When the salience of a group is enhanced, efforts by a transformational leader to highlight a group's efficacy are likely to have a more powerful effect on motivating elaboration by group members (Sosik, et. al., 1997). The increase in focus on the content being presented rather than on who is presenting it is also likely to enhance participants' understanding and consideration of ideas encouraged by a transformational leader (Kahai et al., 2003). Additionally, anonymity creates an environment in which group members feel safe to question, challenge, and reframe each others ideas by reducing the possibility of identification and sanctions. In such an environment, the encouragement by a transformational leader to challenge and reframe others' ideas is likely to have a stronger effect.

Transactional behaviors motivate participants to elaborate by clarifying what the participants were expected to do and rewarding them by expressing satisfaction on their progress. Anonymity is expected to interfere with the rewarding behavior of a transactional confederate leader. In the absence of association of individual identities with input, participants may perceive the leader's rewarding behavior as less directed toward their individual input (Kahai et al., 2003), which, in turn, may reduce the potency of the rewarding behavior and interfere with a leader's effect on motivating participants to elaborate (Sosik et al., 1997).

Based on the above discussion, we expect to see a greater incidence of the arguments encouraged by a transformational confederate leader and a lower incidence of the arguments encouraged by a transactional confederate leader in the anonymous condition relative to the identified condition, leading us to the following hypothesis,

Hypothesis 2. In groups discussing the issue of copying copyrighted software, anonymity will moderate the effect of leadership style on the incidence of (a) arguments against copying based on ethical considerations, (b) arguments against copying based on negative consequences, and (c) arguments in favor of copying.

Effect of Arguments on Intentions

We focused on the effects that the incidence of arguments of different types referred to in the above hypotheses had on the mean of and deviation in postdiscussion intentions to copy within groups. According to persuasive-arguments theory (Vinokur & Burnstein, 1974), a group member forms her or his intentions about the ethical issue based on the arguments that she or he is aware of. As others present their arguments, the group member becomes aware of new arguments, which in turn influence her or his intentions. In groups with a higher incidence of arguments for copying, members are more likely to be introduced to new arguments for copying the software, which would lead them and, consequently, the group mean of intentions to copy to shift more in favor of copying. Likewise, as the incidence of arguments against copying based on ethical considerations or negative consequences increases within groups, members are likely to be introduced to a greater number of new arguments against copying, which would lead them and, consequently, the group mean of intentions to copy to shift more against copying. Therefore, we hypothesize the following:

Hypothesis 3. In groups discussing the issue of copying copyrighted software:

a. A higher incidence of arguments against copying based on ethical considerations will cause the group mean of intentions to copy to shift more against copying.
b. A higher incidence of arguments based on negative consequences will cause the group mean of intentions to copy to shift more against copying.
c. A higher incidence of arguments in favor of copying will cause the group mean of intentions to copy to shift more in favor of copying.

Students consider the copying of copyrighted software as acceptable and normative behavior (Cohen & Cornwell, 1989; Glass & Wood, 1996; Hinduja, 2003). When the incidence of arguments in favor of copying within such a group increases, group members are likely to be exposed to a greater number of arguments that support their position and, consequently, are likely to converge to similar intentions to copy the software at the end of their discussion. However, a greater incidence of arguments against the copying based on ethical consideration or negative consequences is likely to introduce participants to perspectives that they had not considered earlier and, consequently, should create greater deviation in intentions to copy the software at the end of discussion. This leads to the following hypothesis.

Hypothesis 4. In groups discussing the issue of copying copyrighted software,

a. A higher incidence of arguments against copying based on ethical considerations will lead to a greater deviation in intentions to copy.

b. A higher incidence of arguments based on negative consequences will lead to a greater deviation in intentions to copy.

c. A higher incidence of arguments in favor of copying will lead to a lower deviation in intentions to copy.

Method

Participants

Two hundred undergraduate students enrolled in an introductory information systems course participated in a laboratory experiment for course credit. Participants were randomly assigned to 42 groups, the majority of which (34) consisted of five members each while the rest consisted of four members each. These groups had no history of prior interaction, nor any expectation of future interaction.

Overview of Experimental Session and Task

Participants used Ventana Corporation's GROUPSYSTEMS in a decision room setting (Dennis, George, Jessup, Nunamaker, & Vogel, 1988). A facilitator greeted the participants as they entered the decision room and then randomly assigned them to designated terminals. Before the start of the experimental session, the facilitator introduced participants to other members of their group and the confederate who led their group. Each experimental session consisted of three phases and lasted for approximately 90 minutes. Phase I was a 15-minute training session designed to acquaint participants with the electronic brainstorming system (EBS) tool in GROUPSYSTEMS V. Phase II involved using the idea organizer (IO) and the EBS tools to comment on and discuss the software-copying scenario presented below. In Phase III, participants answered a questionnaire pertaining to their IO and EBS sessions in Phase II.

In Phase II, each group was presented the following ethical scenario to comment on and discuss.

Your roommate just purchased a new software package at a price of $250. The software would enable you to complete your MIS assignments. Having the software on your computer will eliminate the need to go to public computer facilities on campus. The software is not copy protected (i.e., it can easily be copied), however it is copyrighted (i.e., the producer has the exclusive legal right to make copies).

If your roommate said it was okay to copy, would you copy the software?

Each group's task was made up of three parts. First, participants used the IO tool for 10 minutes to privately indicate whether or not they would copy the software, and to state supportive arguments. Second, participants used the EBS tool for 20 minutes to discuss the issue anonymously or nonanonymously with group members by presenting and arguing for their positions. Consistent with common usage of the EBS tool, only electronic discussion was permitted (Jessup & Tansik, 1991; Kahai et al., 1998; Kahai et al., 2003; Valacich, Dennis, & Connolly, 1994). Third, participants used the IO tool for 10 minutes to privately indicate whether or not they would copy the software and state supporting arguments. Anonymity effects examined in this study pertain to group discussion.

Research Design

We employed a 2x2 factorial design (identified vs. anonymous discussion and transactional vs. transformational leadership). The experimental groups were randomly assigned across anonymity and leadership conditions. Depending on the anonymity condition, participants were informed that their comments were anonymous or not anonymous from other group members and the leader. The EBS tool tagged comments with the author's name only in the identified condition.

Female confederates led the group discussions. The instructions provided to participants described the leaders as having undergone (a) an independent study on the topic that the participants will be discussing and (b) training in facilitating group work in face-to-face and EMS-supported situations. The confederates displayed transactional or transformational behaviors. Introductory messages and scripts consisting of eight behavioral comments representing transactional and transformational leadership styles were created based on a comprehensive training program (Bass & Avolio, 1994). Before participants discussed the software-copying scenario, the EBS displayed an introductory message from their confederate leader. During the discussion, the confederates typed their scripted comments into the EBS at times held constant across conditions.

The introductory messages in the transactional and transformational conditions had a common beginning. The common beginning highlighted that software can be recognized as property and that it is up to the consumer to take responsibility for upholding the contractual agreement with the producer regarding certain rights and privileges for use and distribution that come with purchasing the software. Beyond this common beginning, the transactional confederate's introductory message conveyed to participants there are clear economic benefits for consumers and producers when consumers comply with software copyright laws. The transformational confederate's introductory message conveyed to participants that honoring copyright protection of software is simply the right and ethical thing to do.

The transactional confederate's comments during the discussion emphasized what the group was expected to do and the rewards (e.g., feelings of satisfaction, course credit) it would receive upon achieving the expected outcomes, and expressed satisfaction with the group's progress. The transformational confederate's comments during the discussion emphasized understanding and appreciating different needs and viewpoints within the group, stimulating each other's efforts to be creative by questioning assumptions, reframing problems, and approaching traditional situations in new ways. The transformational confederate also provided meaning and challenge to participants' discussion while motivating them to work together on an important issue.

Measures

Manipulation Checks. Individual-level perceptions pertaining to anonymity and leadership style were employed for manipulation checks. To obtain these perceptions, questionnaire items based on five-point scales (1 for *strongly disagree*; 2, *disagree*; 3, *neither*; 4, *agree*; 5, *strongly agree*) were employed. Three items were employed to measure perceptions of anonymity ($\alpha = .85$). These reverse-coded items assessed whether participants believed that other members of their group, their leader, or the experimenters will be able to trace their discussion comments to them.

Leadership manipulation was checked using items adapted from the *Multi-Factor Leadership Questionnaire: Form 5X* (Bass & Avolio, 1996). Two scales were employed to assess perceptions of transactional behavior by the leader during discussion. One scale, based on four items (e.g., "My group leader clarified the goals of this exercise" and "My group leader indicated that we will receive course credit for completing this exercise"), assessed perceptions of the leader clarifying goals and reward contingencies ($\alpha = .70$), and the other scale, based on two items ("My group leader indicated that we were meeting her expectations" and "My group leader recognized us for meeting her expectations"), assessed perceptions of the leader providing contingent rewards ($\alpha = .78$).

Three scales were employed to assess perceptions of transformational behavior by the leader during discussion. One scale, based on four items (e.g., "My group leader encouraged us to rethink our ideas"), assessed the extent to which participants perceived their leader exhibiting intellectual stimulation ($\alpha = .81$). A second scale, based on six items (e.g., "My group leader helped us to develop an appreciation for each others' views"), assessed the extent to which participants perceived their leader exhibiting individualized consideration ($\alpha = .88$). A third scale, based on six items (e.g., "My group leader indicated the importance of the issue that we were discussing" and "My group leader emphasized that we could build synergy by working together"), assessed the extent to which participants perceived their leader exhibiting inspirational motivation ($\alpha = .77$).

In addition, three items were employed to check perceptions of what the leader conveyed in the introductory message to participants. The items focused on whether she conveyed that (a) honoring a software producer's copyright is simply a part of the contractual agreement that each consumer enters into, (b) honoring copyright protection of software is simply the right and ethical thing to do, and (c) there are clear economic benefits for consumers and producers when consumers comply with software copyright laws.

Dependent Variables. The frequencies of arguments of different types were obtained by parsing and coding arguments within EMS transcripts of group discussions. Arguments were coded as per the scheme in Appendix A, which is largely based on McDonald and Pak's (1996) synthesis of various cognitive frameworks that individuals use when they consider ethical scenarios. Four raters who were blind to the study's conditions and hypotheses examined each transcript after they were trained by the first author. Two raters examined 12 transcripts and the other two raters examined the rest. After examining the transcripts individually, the raters met to reconcile their differences and produce reconciled transcripts that were used for measurement. The first two raters' demonstrated 91.6% agreement and the other two demonstrated 97% agreement.

The incidence of arguments against copying based on negative consequences was measured by adding the number of times group members made arguments against copying based on utilitarianism and self-interest (see Appendix A for definitions of these and other categories referred to later in this paragraph). The incidence of arguments against copying based on ethical considerations was measured by adding the number of times group members made arguments against copying based on categorical imperative, duty, justice, legality of action, and religious, philosophical, or ethical convictions. The incidence of arguments in favor of copying was measured by adding the number of times group members made arguments in favor of copying based on self-interest, light of day, and neutralization.

The intentions of group members to copy the software, which were used to compute the mean of intentions and deviation in intentions within a group before and after the discussion, were obtained by the EMS immediately before and after the discussion as a response to the question "Will you copy the software?" (1 for *definitely not*; 2, *probably not*; 3, *don't know*; 4, *probably yes*; 5, *definitely yes*). The standard deviation of responses within each group was employed as an indicator of deviation.

Analysis

All hypotheses tests were performed with group-level data. MANCOVA was employed to test Hypotheses 1 and 2, which focused on how leadership and anonymity affect the frequencies of arguments. The independent variables for MANCOVA included leadership (L) style (transactional vs. transformational), anonymity (A; identified vs. anonymous discussion), and LxA. Anonymity was employed as an independent variable because it can affect the participation of group members and, hence, the incidence of arguments of various types by reducing evaluation apprehension (Jessup et al., 1990). Group size was used as a covariate to control for variation in group sizes. The mean of prediscussion intentions to copy within a group, indicative of how the group as a whole was inclined toward copying the software, was also employed as a covariate because it can influence the incidence of arguments for or against the copying of software.

Partial correlations were employed to test Hypotheses 3 and 4, which focused on the relationship between frequencies of arguments of various types and the changes in the mean of and deviation in the intentions to copy the software within any group. To determine the relationship between the incidence of arguments of any type (e.g., arguments against copying based on negative consequences) and change in the mean of and deviation in the intentions, the relationships of intentions with arguments of the remaining two types (e.g., arguments against copying based on ethical considerations and arguments in favor of copying) were partialled out. The correlation analysis also partialled out the relationships of leadership style, anonymity, and the interaction of leadership style and anonymity to postdiscussion intentions beyond the relationships via arguments because leadership style and anonymity can affect the normative influence within a group beyond the cognitive effect via arguments by making the group and its norms salient (Kahai et al., 1998; Sosik et al., 1997).

Results

Tables 1 and 2 present means, standard deviations, sample sizes, and Pearson product-moment correlations among the measures used for manipulation checks and hypotheses testing.

Manipulation Checks

Perceptions of anonymity were significantly stronger in the anonymous vs. the identified condition (M = 3.33 vs. 2.49; F(1,198) = 37.64, p

Table 1. Means, standard deviations, sample size, and correlations among measures for manipulation checks

Variable	M	SD	n	1	2	3	4	5	6	7	8
1. Perceived anonymity	2.89	1.05	200	**.85**							
2. Perceived transformational nature of introductory message	3.73	1.13	200	.03	--						
3. Perceived transactional nature of introductory message	3.62	1.15	200	-.05	.27	--					
4. Perceived intellectual stimulation by leader	3.30	0.83	200	.01	.14	.01	**.81**				
5. Perceived individualized consideration by leader	3.23	0.81	200	-.07	.10	-.02	.71	**.88**			
6. Perceived inspirational motivation by leader	3.39	0.69	200	-.06	.09	-.00	.72	.83	**.77**		
7. Perceived clarification of goals and contingencies by leader	3.11	0.84	200	.01	-.04	-.00	.08	.13	.23	**.70**	
8. Perceived contingent rewarding by leader	3.66	0.90	200	-.02	.02	.12	.30	.31	.43	.34	**.78**

Note. Boldfaced diagonal elements are Cronbach's alphas for multi-item scales. Underlined correlations are significant at p ≤ .05.

Table 2. Means, standard deviations, sample size, and correlations for measures of study's variables

Variable	M	SD	n	1	2	3	4	5	6	7
1. Leadership	1.50	0.51	42	1						
2. Anonymity	1.48	0.51	42	.00	1					
Incidence of arguments										
3. against copying based on ethical considerations	2.05	2.57	42	.38	.00	1				
4. against copying based on negative consequences	1.52	1.78	42	-.19	.15	.19	1			
5. in favor of copying	21.93	7.59	42	-.49	-.14	-.28	-.04	1		
Intentions within group										
6. Change in mean of intentions to copy	0.10	0.22	42	.18	-.07	-.15	-.16	.24	1	
7. Change in deviation in intention	0.03	0.24	42	-.10	.05	.26	.08	-.07	-.34	1

Note. Underlined correlations are significant at p ≤ .05

< .01). Because the introductory messages of both confederate leaders highlighted the consumer's responsibility for upholding the contractual agreement with the producer, participants in both leadership-style conditions registered that the leader conveyed this message (M = 3.87 in both conditions; these means were significantly greater than 3 at p = .00, which represents neither agreement nor disagreement with the statement that the leader conveyed the message in question). Perceptions that the introductory message conveyed that honoring copyright protection of software is simply the right and ethical thing to do were significantly higher in the transformational vs. the transactional condition (M = 3.89 vs. 3.73; $F(1,198) = 4.08$, $p = .04$). Perceptions that the leader's introductory message conveyed that there are clear economic benefits for consumers and producers when consumers comply with software copyright laws were significantly higher in the transactional vs. the transformational condition (M = 3.83 vs. 3.40; $F(1,198) = 7.26$, $p < .01$).

Perceptions of the leader (a) clarifying goals and reward contingencies and (b) providing contingent rewards during discussion were significantly greater in the transactional vs. the transformational condition (for the act of clarifying, M = 3.52 vs. 2.71; $F(1,198) = 59.30$, $p < .01$; for contingent rewarding, M = 3.89 vs. 3.43; $F(1,198) = 13.53$, $p < .01$). Perceptions of the leader displaying intellectual stimulation, individualized consideration, and inspirational motivation during the discussion were significantly greater in the transformational vs. the transactional condition (for intellectual stimulation, M = 3.61 vs. 2.98; $F(1,198) = 33.46$, $p < .01$; for individualized consideration, M = 3.59 vs. 2.87; $F(1,198) = 48.22$, $p < .01$; for inspirational motivation, M = 3.65 vs. 3.13; $F(1,198) = 31.91$, $p < .01$).

Process and Outcome Effects

Table 3 shows descriptive statistics for the study's process and outcome variables in each cell in the study's design. Tables 4 and 5 show summary results for the study's hypotheses.

Hypothesis 1 was supported. The MANCOVA for arguments of different types showed a significant effect of leadership (multivariate $F(3,34) = 6.56$, $p = .00$), thereby permitting univariate

Table 3. Cell means, standard deviations, and sizes for dependent variables

Variable	Identified			Anonymous		
	M	SD	n	M	SD	n
Incidence of arguments against copying based on ethical considerations						
Transactional leadership	1.09	1.81	11	1.10	1.37	10
Transformational leadership	3.00	3.55	11	3.00	2.49	10
Incidence of arguments against copying based on negative consequences						
Transactional leadership	1.55	1.44	11	2.20	2.97	10
Transformational leadership	1.00	1.00	11	1.40	1.17	10
Incidence of arguments in favor of copying						
Transactional leadership	27.27	9.85	11	23.7	4.76	10
Transformational leadership	18.55	6.74	11	18.00	3.20	10
Change in mean of intentions						
Transactional leadership	0.09	0.25	11	0.03	0.18	10
Transformational leadership	0.14	0.26	11	0.14	0.18	10
Change in deviation in intentions						
Transactional leadership	-0.01	0.17	11	0.13	0.16	10
Transformational leadership	0.04	0.16	11	-0.04	0.40	10

Table 4. Analysis of variance results

	Effects								
Variables	L			A			L x A		
	F	df	p	F	df	p	F	df	p
Hypotheses 1 - 2 (Effects of L and L x A on arguments)									
Multivariate Analysis	6.56	(3,34)	.00	0.53	(3,34)	.67	0.41	(3,34)	.74
Univariate Analysis									
Incidence of arguments against copying based on									
(a) ethical considerations	6.80	(1,36)	.01	.05	(1,36)	.83	0.11	(1,36)	.74
(b) negative consequences	3.97	(1,36)	.05	.99	(1,36)	.33	0.85	(1,36)	.36
Incidence of arguments in favor of copying	11.11	(1,36)	.00	.79	(1,36)	.38	0.32	(1,36)	.58

Table 5. Partial correlation results

Incidence of arguments	Intentions to copy					
	Change in group mean			Change in deviation within group		
	r	df	p	r	df	p
Hypotheses 3 – 4 (Relationship of arguments with post-discussion intentions to copy)						
Incidence of arguments against copying based on						
(a) ethical considerations	-.20	35	.24	.33	35	.05
(b) negative consequences	-.03	35	.90	-.06	35	.72
Incidence of arguments in favor of copying	.35	35	.04	-.07	35	.67

analyses. Consistent with Hypothesis 1a, the incidence of arguments against copying based on ethical considerations was higher in the transformational vs. the transactional condition (M = 3.00 vs. 1.10; F(1,36) = 6.80, p = .00). The incidence of arguments against copying based on negative consequences was higher in the transactional vs. the transformational condition (M = 1.86 vs. 1.19; F(1,36) = 3.97, p = .05), thereby supporting Hypothesis 1b. In accord with Hypothesis 1c, the incidence of arguments in favor of copying was higher in the transactional vs. the transformational condition (M = 25.57 vs. 18.29; F(1,36) = 11.11, p = .00).

Hypothesis 2 was not supported. The interaction of leadership style and anonymity did not influence the incidence of arguments against copying based on ethical considerations (F(1,36) = 0.11, p = .74), the incidence of arguments against copying based on negative consequences (F(1,36) = 0.85, p = .36), or the incidence of arguments in favor of copying (F(1,36) = 0.32, p = .58).

The incidence of arguments against copying based on ethical considerations or negative consequences was not related to the shift in the group mean of intentions to copy (r = -.20, p = .24 and r = -.03, p = .90), thereby indicating lack of support for Hypothesis 3a and Hypothesis 3b. Consistent with Hypothesis 3c, a higher incidence of arguments in favor of copying caused the group mean of intentions to copy to shift more in favor of copying (r = .35, p = .04).

As per Hypothesis 4a, a higher incidence of arguments against copying based on ethical considerations led to an increase in the deviation within groups in intentions to copy the software (r = .33, p = .05). However, the incidence of arguments against copying based on negative consequences did not influence the deviation in intentions with the group (r = -.06, p = .72) nor did the incidence of arguments in favor of copying (r = -.07, p = .67), thereby indicating lack of support for Hypothesis 4b and Hypothesis 4c.

DISCUSSION

The leadership style of the confederate leaders affected the incidence of arguments for and against copying as expected. Contrary to expectations, however, anonymity did not influence this effect of leadership style. This absence of interaction between leadership style and anonymity may be due to the high level of consensus among participants on the ethical issue presented to them; 82.5% of the students who participated in our study indicated prior to their discussion that they would copy copyrighted software. In the development of Hypothesis 2, we indicated that anonymity may enhance the transformational confederate's effect on (a) highlighting the collective efficacy of the group, (b) encouraging group members to challenge and reframe each others' ideas, and (c) encouraging understanding and consideration for each others' opinions. Anonymity was also expected to interfere with a transactional confederate's effect on motivating elaboration by making the leader's rewarding behavior appear less personal.

The high level of consensus among participants may have taken attention away from individual identities and enhanced (a) the salience of the group and (b) focus on the content of ideas and, therefore, their understanding and consideration as well. In the presence of consensus, participants may have been less apprehensive about challenging and reframing each others' ideas to begin with. Consequently, there may have been less opportunity for anonymity to contribute to (a) taking attention away from individual identities and enhancing the salience of the group and (b) reducing apprehension about challenging and reframing each others' ideas. Also, in the presence of consensus, participants may have found the support from their peers as sufficient and may not have cared enough about the positive feedback from the transactional confederate.

A higher incidence of arguments against copying based on ethical considerations caused a greater deviation in intentions to copy within groups but did not have a significant effect on the group mean of intentions to copy. Perhaps the incidence of arguments against copying based on ethical considerations was sufficient to induce a deeper level of processing in some of the participants and led them to deviate from others, but it needed to be higher to cause the group mean to move away from favoring copying. It is also possible that the discussion time was not sufficient to cause more members of the group to be influenced by arguments based on ethical considerations and enable the group mean to move away from favoring copying.

The incidence of arguments against copying based on negative consequences did not influence the group mean of or deviation in intentions to copy. The incidence of such arguments was low and may not have been sufficient to induce a deeper processing in any of the participants and lead them to deviate from others or cause the group mean to move away from favoring copying.

The incidence of arguments in favor of copying was sufficiently high to cause the group mean to change toward favoring copying. However, with the level of consensus high at the outset, there may have been little room for the deviation within groups to reduce. Accordingly, we may not have seen the incidence of arguments in favor of copying influencing deviation in intentions to copy within groups.

CONCLUSION

Recent ethics scandals have led to greater urgency in developing ethical reasoning among today's and future personnel. Organizations of various kinds, including universities, are interested in understanding how leadership influences ethical reasoning and behavior (Brown & Trevino, 2006; Ritter, 2006). Highlighting the relationship of leadership to ethics, Cuilla (2004, p. 302) stated, "In leadership we see morality magnified, and that is why the study of ethics is fundamental to our understanding leadership." Organizations are also interested in learning about the efficacy of using the online communication medium for developing ethical reasoning among personnel and students (French, 2006; Painter-Morland et al., 2003). These considerations motivated our examination of the effects of transactional vs. transformational leadership styles and anonymity in groups discussing an ethical issue electronically.

In a relatively short interaction with a confederate leader, we observed that the transformational leadership style led to a greater incidence of arguments against copying based on ethical considerations. These arguments led to greater deviation in intentions to copy within groups. Though the transactional leadership style led to a greater incidence of arguments against copying based on negative consequences, it also led to a greater incidence of arguments in favor of copying. A greater incidence of arguments in favor of copying led to stronger intentions to copy software within groups. These preliminary results support the idea that transformational leadership would promote ethical thinking in which one rises above one's self-interest, whereas transactional leadership would promote an exchange-oriented perspective that not only suggests that there is payback for what you do but may also encourage self-interested behavior. Our confidence in the transformational leader's ability to promote ethical thinking is bolstered by the difficult test that we created to assess the transformational leader's efficacy. University students have a high proclivity to illegally copy software (Hinduja, 2003). That a transformational confederate was able to motivate ethical thinking in spite of a high barrier is indeed noteworthy.

The possibility that transformational leadership motivated deeper processing and caused participants to focus on end values is supported by earlier work by Jung and Avolio (1999), in which they compared transformational and transactional leadership in a task in which business school students were asked to generate ideas to enhance their school's application for AACSB reaccreditation. Whereas students under a transactional leader generated a greater number of ideas, those under a transformational leader generated higher quality ideas. Jung and Avolio concluded that transformational leadership appeared to motivate a deeper level of processing that resulted in fewer but higher quality ideas.

Additional evidence that transformational leadership may have led to ethical thinking by influencing participants' values comes from a study by Brown and Trevino (2006), who found that socialized charismatic leadership style, which overlaps with transformational leadership, was negatively associated with deviant behavior in work groups. According to the authors, measures of deviant behavior overlap considerably with measures of unethical organizational behavior. Brown and Trevino also found that the effect of socialized charismatic leadership on interpersonal deviance was mediated by the congruence of values between a leader and her or his followers. These findings offer support to our idea that transformational leadership may have led to ethical arguments through their effect on followers' values.

Study's Limitations

Generalizability of the study's results is limited. It is limited to situations in which ad hoc groups of university students come together in the same place to role-play and discuss an ethical scenario over an EMS for a relatively short period of time. Such situations, while limiting the study's generalizability, are likely to be relevant in the context of ethical training at universities and in organizations. For instance, universities employ role-playing scenarios to develop ethical reasoning in their students. Students in a class may be put together into ad hoc groups and made to role-play an ethical scenario for a relatively short period of time (see Sanyal, 2000). Furthermore, the role-playing may be facilitated by assistants acting as group leaders and may be carried out over an EMS. EMSs similar to the one employed in this study have been implemented at various universities and are being used for classroom learning (e.g., Ready, Hostager, Lester, & Bergmann, 2004; Vogel, van Genuchten, Lou, Verveen, van Eekout, & Adams, 2001).

The generalizability of the study's results may also extend beyond university boundaries to other organizational settings. Paralleling the use of confederate leaders who facilitated the discussion of the ethical scenario in our study, ethical training in many organizations may be facilitated by leaders to whom participants have no reporting relationship (Tyler, 2005). Students in this study came together in groups consisting of other students whom they saw in their classes but did not necessarily work with in a group. This is similar to what may happen in organizational training sessions when organizational members from different parts of an organization sign up for a session. These members are likely to have seen each other within the organization (e.g., in the hallways or the cafeteria) but have not necessarily worked together.

IMPLICATIONS AND FUTURE RESEARCH

We are living in an era in which technology that enables free and illegal copying of intellectual property and cultural artifacts such as music and movies is always a step ahead of legal methods to curtail such copying. Commentators are suggesting that perhaps the only recourse we have is to appeal to our citizens that what they are doing is not right (Henninger, 2005). Like these commentators, our results suggest that in groups discussing an ethical issue and having a proclivity to act unethically, a leader should act transformationally. By operating at a higher level of moral reasoning and appealing to followers to do the right thing (Avolio, 1999; Turner et al., 2002), a transformational leader promotes ethical reasoning, which, in turn, promotes deviation in intent among group members. A transformational leader also discourages reasoning that promotes unethical intent.

From the viewpoint of leadership development and ethics management in organizations, the implications of our results reinforce those that follow from the work of Brown and Trevino (2006). Specifically, our results provide an additional motivation for organizations to encourage transformational leadership behaviors among managers. Our results also suggest that organizations should consider combining leadership training with ethics training. Currently, leadership training and ethics training are separate programs in most organizations. Given the close connection between transformational leadership and ethical reasoning and behavior, combining leadership training with ethical training may be a reasonable move for organizations (Brown & Trevino).

In organizational settings today, members often operate from geographically dispersed locations via technology. Such a setting creates a challenge for a leader to enhance the identification of their followers with the group and organization. We provide some evidence that indicates that a transformational leader motivated followers to think deeper about an ethical issue. By getting followers to think deeper about their own views and how their views might be related to the mission of the organization or group of which they are a part, a transformational leader may be able to enhance followers' identification with their organization or their group.

The geographical dispersion of workers is also creating a question for organizations about whether they can deliver ethics training online (French, 2006). Even universities are facing this issue as they attempt to reach out to more students via distance-learning technology. Results of our study offer evidence about the ability of a facilitator to influence the thinking of a group of individuals who are meeting online to role-play an ethical scenario. From a practical viewpoint, our results also suggest that a facilitator should emphasize transformational behaviors and de-emphasize transactional behaviors if she or he is seeking to (a) encourage ethics-based consideration of the issue (i.e., consideration of justice, one's duty, the legality of one's actions, and one's religious, philosophical, or ethical conviction), and (b) discourage self-interest-based consideration of the issue. If the facilitator is seeking deviation in opinions on an ethical issue within a group after a discussion, she or he should inject arguments based on ethical considerations.

Extending beyond the study's conditions offers opportunities for future research. Future research should examine the role-playing of ethical scenarios in intact groups where the leader is a member of the group and has credibility established from prior interactions. In such a situation, the effects of leadership on group process and outcomes in an EMS context may be stronger than what we have found. It is also likely that intact groups with significant history show different effects than what we observed here. For instance, Pinsonneault, Barki, Gallupe, and Hoppen (1999) reported that members of groups with history appeared to hold back ideas during electronic brainstorming based on the assumption that other members of the group already knew what they would say.

Longitudinal research designs can also be used to examine whether the present study's results are reproduced over longer periods of time because of the potential of group processes to change (Walther, 1995). Groups may employ technology over time in ways that cause leadership, and hence its effects, to change over time.

Future studies could also examine tasks where participants have to make a collective decision for an ethical issue. In the current study, participants discussed an ethical issue and indicated their individual intentions, but they were not required to make a collective decision. Additionally, leaders in the present study exhibited behaviors that were predominantly transactional or transformational. The combination of these styles is often optimal for performance (Bass & Avolio, 1994) and should be examined.

The groups in the current study interacted via a single computer-mediated communication system that allowed text messages only. Group members working at a distance from one another may use several computer-mediated communication systems, some rich and some lean, to support their work. Future studies should examine the effects when combinations of communication systems are used.

Future research should explore the effects of leadership style and anonymity when group members are discussing an ethical issue in which there is less consensus among participants. We believe that anonymity did not influence the effect of leadership style on argumentation because of the nature of the ethical issue discussed. The high level of consensus among students about the appropriateness of copying may have caused anonymity to not have any effect.

We believe this study and previous ones in electronic contexts (e.g., Kahai, Sosik, & Avolio, 1997; Kahai et al., 2003; Kahai, Sosik, & Avolio, 2004; Sosik et al., 1997; Sosik, Kahai, & Avolio, 1998) indicate that exploring the effects of leadership and its interaction with features of electronic communication technologies remains a promising area for future studies. Due to the increasing level of electronic communication in organizations around the world, continued research seems timely and warranted.

REFERENCES

Avolio, B. J. (1999). *Full leadership development: Building the vital forces in organizations.* Thousand Oaks, CA: Sage Publications.

Bass, B. M. (1985). *Leadership and performance beyond expectations.* New York: Free Press.

Bass, B. M., & Avolio, B. J. (1994). *Improving organizational effectiveness through transformational leadership.* Thousand Oaks, CA: Sage Publications.

Bass, B. M., & Avolio, B. J. (1996). *Multifactor leadership questionnaire manual.* Palo Alto, CA: Mindgarden.

Bass, B. M., & Steidlmeier, P. (1999). Ethics, character, and authentic transformational leadership behavior. *The Leadership Quarterly, 10*(2), 181-217.

Brown, M., & Trevino, L. (2006). Socialized charismatic leadership, values congruence, and deviance in work groups. *Journal of Applied Psychology, 91*(4), 954-962.

Burns, J. M. (1978). *Leadership.* New York: Harper & Row.

Cohen, E., & Cornwell, L. (1989). College students believe piracy is acceptable. *CIS Educator's Forum, 1*(3), 2-5.

Cronan, T., Foltz, C., & Jones, T. (2006). Piracy, computer crime, and IS misuse at the university. *Communications of the ACM, 49*(6), 85-90.

Cuilla, J. (2004). Ethics and leadership effectiveness. In J. Antonakis, A. T. Cianciolo, & R. J. Sternberg (Eds.), *The nature of leadership* (pp. 302-328). Thousand Oaks, CA: Sage.

Dennis, A., George, J., Jessup, L., Nunamaker, J., & Vogel, D. (1988). Information technology to support electronic meetings. *MIS Quarterly, 12*(4), 591-624.

French, W. (2006). Business ethics training: Face-to-face and at a distance. *Journal of Business Ethics, 66*(1), 117-126

Giacalone, R., & Thompson, K. (2006). From the guest co-editors: Special issue on ethics and social responsibility. *Academy of Management Learning & Education, 5*(3), 261-265.

Glass, R. S., & Wood, W. A. (1996). Situational determinants of software piracy: An equity theory perspective. *Journal of Business Ethics, 15*(11), 1189-1198.

Grojean, M. W., Resick, C. J., Dickson, M. W., & Smith, D. B. (2004). Leaders, values, and organizational climate: Examining leadership strategies for establishing an organizational climate regarding ethics. *Journal of Business Ethics, 55*(3), 223-241.

Henninger, D. (2005, April 1). Wonder land: Can Justice Scalia solve the riddles of the Internet? *Wall Street Journal*, p. A10.

Hinduja, S. (2003). Trends and patterns among online software pirates. *Ethics and Information Technology, 5*(1), 49-61.

Jessup, L., & Tansik, D. (1991). Decision making in an automated environment: The effects of anonymity and proximity on group process and outcome with a GDSS. *Decision Sciences, 22*(2), 266-279.

Jessup, L. M., Connolly, T., & Galegher, J. (1990). The effects of anonymity on GDSS group process with an idea-generating task. *MIS Quarterly, 14*(3), 313-321.

Jones, T. M. (1991). Ethical decision making by individuals in organizations: An issue-contingent model. *The Academy of Management Review, 16*(2), 366-395.

Judge, T. A., & Piccolo, R. F. (2004). Transformational and transactional leadership: A meta-analytic test of their relative validity. *Journal of Applied Psychology, 89*(5), 755-768.

Jung, D. I., & Avolio, B. J. (1999). Effects of leadership style and followers' cultural orientation on performance in group and individual task conditions. *Academy of Management Journal, 42*(2), 208-218.

Kahai, S., Avolio, B., & Sosik, J. (1998). Effects of source and participant anonymity and difference in initial opinions in an EMS context. *Decision Sciences, 29*(2), 427-460.

Kahai, S., Sosik, J., & Avolio, B. (1997). Effects of leadership style and problem structure on work group process and outcomes in an electronic meeting system environment. *Personnel Psychology, 50*(1), 121-146.

Kahai, S. S., Sosik, J. J., & Avolio, B. J. (2003). Effects of leadership style, anonymity, and rewards on creativity-relevant processes and outcomes in an electronic meeting system context. *The Leadership Quarterly, 14*(4-5), 499-524.

Kahai, S., Sosik, J., & Avolio, B. (2004). Effects of participative and directive leadership in electronic groups. *Groups & Organization Management, 29*(1), 67-105.

Luthans, F., & Avolio, B. J. (2003). Authentic leadership: A positive developmental approach. In K. S. Cameron, J. E. Dutton, & R. E. Quinn (Eds.), *Positive organizational scholarship* (pp. 241-258). San Francisco: Berrett-Koehler.

McDonald, G., & Pak, P. C. (1996). It's all fair in love, war, and business: Cognitive philosophies in ethical decision making. *Journal of Business Ethics, 15*(9), 973-996.

Moores, T., & Chang, J. (2006). Ethical decision making in software piracy: Initial development and test of a four-component model. *MIS Quarterly, 30*(1), 167-180.

Nunamaker, J. F., Dennis, A. R., Valacich, J. S., Vogel, D. R., & George, J. F. (1991). Electronic meeting systems to support group work: Theory and practice at Arizona. *Communications of the ACM, 34*(7), 40-61.

Painter-Morland, M., Fontrodona, J., Hoffman, W. M., & Rowe, M. (2003). Conversations across continents: Teaching business ethics online. *Journal of Business Ethics, 48*(1), 75-88.

Petty, R. E., & Cacioppo, J. T. (1981). *Attitudes and persuasion: Classic and contemporary approaches.* Dubuque, IA: Brown.

Pinsonneault, A., Barki, H., Gallupe, R. B., & Hoppen, N. (1999). The illusion of electronic brainstorming productivity: Theoretical and empirical issues. *Information Systems Research, 10*(4), 378-380.

Ready, K. J., Hostager, T. J., Lester, S., & Bergmann, M. (2004). Beyond the silo approach: Using group support systems in organizational behavior classes to facilitate student understanding of individual and group behavior in electronic meetings. *Journal of Management Education, 28*(6), 770.

Ritter, B. (2006). Can business ethics be trained? A study of the ethical decision-making process in business students. *Journal of Business Ethics, 68*(2), 153-164.

Sanyal, R. N. (2000). An experiential approach to teaching ethics in international business. *Teaching Business Ethics, 4*(2), 137-189.

Shamir, B., House, R., & Arthur, M. (1993). The motivational effects of charismatic leadership: A self-concept based theory. *Organization Science, 4*(4), 1-17.

Sosik, J., Avolio, B., & Kahai, S. (1997). Effects of leadership style and anonymity on group potency and effectiveness in a GDSS environment. *The Journal of Applied Psychology, 82*(1), 89-103.

Sosik, J. J., Kahai, S. S., & Avolio, B. J. (1998). Transformational leadership and dimensions of creativity: Motivating idea generation in computer-mediated groups. *Creativity Research Journal, 11*(2), 111-121.

Stead, B., & Gilbert, J. (2001). Ethical issues in electronic commerce. *Journal of Business Ethics, 34*(2), 75-85.

Street, M., Douglas, S., Geiger, S., & Martinko, M. (2001). The impact of cognitive expenditure on the ethical decision-making process: The cognitive elaboration model. *Organizational Behavior and Human Decision Processes, 86*(2), 256-277.

Trevino, L. K., Brown, M., & Hartman, L. P. (2003). A qualitative investigation of perceived executive ethical leadership: Perceptions from inside and outside the executive suite. *Human Relations, 56*(1), 5-37.

Turner, N., Barling, J., Epitropaki, O., Butcher, V., & Milner, C. (2002). Transformational leadership and moral reasoning. *Journal of Applied Psychology, 87*(2), 304-311.

Tyler, K. (2005). Do the right thing. *HR Magazine, 50*(2), 99.

Valacich, J. S., Dennis, A. R., & Connolly, T. (1994). Idea generation in computer-based groups: A new ending to an old story. *Organizational Behavior and Human Decision Processes, 57*(3), 448-467.

Verschoor, C. C. (2006). Surveys show ethics problems persist. *Strategic Finance, 88*(4), 19-21.

Vinokur, A., & Burnstein, E. (1974). Effects of partially shared persuasive arguments on group-induced shifts: A group-problem-solving approach. *Journal of Personality and Social Psychology, 29*(3), 305-315.

Vogel, D. R., van Genuchten, M., Lou, D., Verveen, S., van Eekout, M., & Adams, A. (2001). Exploratory research on the role of national and professional cultures in a distributed learning project. *IEEE Transactions on Professional Communication, 44*(2), 114-125.

Walther, J. B. (1995). Relational aspects of computer-mediated communication: Experimental observations over time. *Organization Science, 6*(2), 186-203.

Zeller, T., Jr. (2005, March). Not yet in business school, and already flunking ethics. *New York Times*, p. C3.

APPENDIX A

Coding-Scheme Categories for Arguments

Arguments Against Copying the Software

Utilitarianism
Arguments in which the principal concern is to act so as to produce the greatest ratio of good over bad for everyone (e.g., "I would not copy the software because we all would lose from it")

Self-Interest Based
Argument based on the concern that it is not in the best interest of the individual or group to copy the software (e.g., "I would not copy the software because I would not be able to get support from the software company if I had any problems")

Categorical Imperative
Arguments based on the principal that an action is either right or wrong regardless of consequence (e.g., "I would not copy the software because it is not the right thing to do")

Duty
Arguments that involve consideration of whether an action is consistent or inconsistent with any prescribed rules of duty, such as those of a member or student organization, or those of an American citizen (e.g., "I would not copy the software because it is against my duty as a member of the MIS association to promote software piracy")

Justice
Arguments based on whether the action leads to an unjust distribution of benefits and burdens among all those concerned due to their age, sex, religion, interests, income, personal characteristics, or social or occupational position (e.g., "Just because the producer of software is making a lot of money, it does not mean that we are entitled to copying the software")

Legal
Arguments based on whether an action is legal or not (e.g., "I would not copy the software because it is illegal to do so")

Religious, Philosophical, or Ethical Conviction
Arguments based on one's religious or philosophical conviction (e.g., "I would not copy because that is against my religion")

Arguments for Copying the Software

Self-Interest Based
Arguments in which the principal concern is to promote personal interests or satisfaction (e.g., "I would copy the software because it would make it easy for me to complete my assignments")

Light of Day

Arguments based on whether the action will see the light of day or, if it does, what the possible consequences are (e.g., "I would copy because there is a very little chance of getting caught")

Neutralization

Arguments whereby an individual feels that the norms they are violating do not apply to their particular instance (e.g., "I would copy because I do not intend to keep the software after the course")

Chapter XIII
The Role of Culture in Knowledge Management: A Case Study of Two Global Firms

Dorothy Leidner
Baylor University, USA

Maryam Alavi
Emory Univeristy, USA

Timothy Kayworth
Baylor University, USA

ABSTRACT

Knowledge management approaches have been broadly considered to entail either a focus on organizing communities or a focus on the process of knowledge creation, sharing, and distribution. While these two approaches are not mutually exclusive and organizations may adopt aspects of both, the two approaches entail different challenges. Some organizational cultures might be more receptive to the community approach whereas others are more receptive to the process approach. Although culture has been widely cited as a challenge in knowledge management initiatives and many studies have considered the implications of organizational culture on knowledge sharing, few empirical studies address the influence of culture on the approach taken to knowledge management. Using a case-study approach to compare and contrast the cultures and knowledge management approaches of two organizations, the study suggests the ways in which organizational culture influences knowledge management initiatives as well as the evolution of knowledge management in organizations. Whereas in one organization the KM effort became little more than an information repository, in the second organization, the KM effort evolved into a highly collaborative system fostering the formation of electronic communities.

INTRODUCTION

Knowledge management (KM) efforts are often seen to encounter difficulties from corporate culture and, as a result, have limited impact (DeLong & Fahey, 2000; O'Dell & Grayson, 1998). An Ernst and Young study identified culture as the biggest impediment to knowledge transfer, citing the inability to change people's behaviors as the biggest hindrance to managing knowledge (Watson, 1998). In another study of 453 firms, over half indicated that organizational culture was a major barrier to success in their knowledge management initiatives (Ruggles, 1998). The importance of culture is also evident from consulting firms such as KPMG, who report that a major aspect of knowledge management initiatives involves working to shape organizational cultures that hinder their knowledge management programs (KPMG Management Consulting, 1998). These findings and others (Hasan & Gould, 2001; Schultze & Boland, 2000) help demonstrate the profound impact that culture may have on knowledge management practice and the crucial role of senior management in fostering cultures conducive to these practices (Brown & Duguid, 2000; Davenport, DeLong, & Beers, 1998; DeLong & Fahey, 2000; KPMG Management Consulting; Gupta & Govindarajan, 2000; Hargadon, 1998; von Krogh, 1998).

While studies have shown that culture influences knowledge management and, in particular, knowledge sharing (DeLong &Fahey, 2000; Jarvenpaa & Staples, 2001), there is little research on the broader aspects of the nature and means through which organizational culture influences the overall approach taken to knowledge management in a firm. The purpose of this research is to examine how organizational culture influences knowledge management initiatives. We use a case-study methodology to help ascertain the relationship of the organizational culture to the knowledge management approaches within two companies. The following section discusses knowledge management approaches and organizational culture. The third presents the methodology. The fourth section presents the two cases, and the fifth discusses the case findings, the implications, and the conclusion.

LITERATURE REVIEW

Knowledge Management

Knowledge can be defined as a form of high-value information (either explicit or tacit) combined with experience, context, interpretation, and reflection that is ready to be applied to decisions and actions (Davenport et al., 1998). While all firms may have a given pool of knowledge resources distributed throughout their respective organization, they may be unaware of the existence of these resources as well as how to effectively leverage them for competitive advantage. Therefore, firms must engage in activities that seek to build, sustain, and leverage these intellectual resources. These types of activities, generally characterized as knowledge management, can be defined as the conscious practice or process of systematically identifying, capturing, and leveraging knowledge resources to help firms compete more effectively (Hansen, Nohria, & Tierney, 1999; O'Dell & Grayson, 1998).

Approaches and strategies for managing knowledge have been conceptualized in various ways. One early conceptualization of KM approaches distinguished between the process and practice approaches (Hansen et al., 1999). The process approach attempts to codify organizational knowledge through formalized controls, processes, and technologies (Hansen et al.). Organizations adopting the process approach may implement explicit policies governing how knowledge is to be collected, stored, and disseminated throughout the organization. The process approach frequently involves the use of information technologies, such as intranets, data warehousing, knowledge repositories, decision support tools, and groupware (Ruggles, 1998), to enhance the quality and speed of knowledge creation and distribution in the organizations. The main criticisms of this process approach are that it fails to capture much of the tacit knowledge embedded in firms and that it forces individuals into fixed patterns of thinking (Brown & Duguid, 2000; DeLong & Fahey, 2000; Hargadon, 1998; von Krogh, 2000).

In contrast, the practice approach to knowledge management assumes that a great deal of organizational knowledge is tacit in nature and that formal controls, processes, and technologies are not suitable for transmitting this type of understanding. Rather than building formal systems to manage knowledge, the focus of this approach is to build the social environments or communities of practice necessary to facilitate the sharing of tacit understanding (Brown & Duguid, 2000; DeLong & Fahey, 2000; Gupta & Govindarajan, 2000; Hansen et al, 1999; Wenger & Snyder, 2000). These communities are informal social groups that meet regularly to share ideas, insights, and best practices.

More recent work on approaches to KM has suggested three KM strategies: knowledge hierarchies, knowledge markets, and knowledge communities (Dennis & Vessey, 2004). These three strategies reflect differing approaches to handling the three primary knowledge processes: knowledge creation, knowledge development, and knowledge reuse (Dennis & Vessey). Knowledge hierarchies are a strategy in which knowledge is treated as a formal, organizational resource in which one person's knowledge is substituted for another. As such, this strategy suggests knowledge repositories (or expert systems or templates) that are directed toward a specific subset of users rather than to the organization as a whole (Dennis & Vessey). The structural requirements for a knowledge hierarchy are significant: A staff must be available to isolate appropriate knowledge, prepare it for consumption by others, and identify the others who might benefit from such knowledge. Knowledge markets represent another extreme in which organizational efforts are focused more on capturing and storing knowledge than in systematically creating and developing it for consumption by specific others (Dennis & Vessey). Use of a knowledge market is unlikely to be required, and depending upon the enthusiasm of potential contributors, may be overpopulated,

Table 1. The process vs. practice approaches to knowledge management

	Process Approach	Practice Approach
Type of Knowledge Supported	Explicit knowledge codified in rules, tools, and processes	Mostly tacit knowledge unarticulated and not easily captured or codified
Means of Transmission	Formal controls, procedures, and standard operating procedures with heavy emphasis on information technologies to support knowledge creation, codification, and transfer	Informal social groups that engage in storytelling and improvisation
Benefits	Provides structure to harness generated ideas and knowledge Achieves scale in knowledge reuse	Provides an environment to generate and transfer high-value tacit knowledge Provides spark for fresh ideas and responsiveness to changing environment
Disadvantages	Fails to tap into tacit knowledge; may limit innovation and forces participants into fixed patterns of thinking	Can result in inefficiency; abundance of ideas with no structure to implement them
Role of Information Technology	Heavy investment in IT to connect people with reusable codified knowledge	Moderate investment in IT to facilitate conversations and transfer of tacit knowledge

or underpopulated, with knowledge. The knowledge market is a laissez-faire approach in which the supply and demand of knowledge are left to reach some equilibrium with little pressure from managers. The knowledge-community strategy lies somewhere in between the focused, specific knowledge-hierarchy approach and the unstructured, unfocused knowledge-market approach. Knowledge communities organize knowledge around communities in which norms of participation, contribution, and reuse are established by each community. An organization might have several such communities. Less controls exist in the development of the knowledge, and hence the quality is lower than in a hierarchy, but more flexibility exists in reuse, with attention focused on augmenting users' knowledge rather than replacing it (Dennis & Vessey). This latter strategy recognizes the time dependence of knowledge and that complete substitution of knowledge across time and contexts might be unwise, even while learning from the previous knowledge and context might be quite valuable.

Organizational Culture

Schein (1985) defines organizational culture as a set of implicit assumptions held by members of a group that determines how the group behaves and responds to its environment. At its deepest level, culture consists of core values and beliefs that are embedded tacit preferences about what the organization should strive to attain and how it should do it (DeLong & Fahey, 2000). These tacit values and beliefs determine the more observable organizational norms and practices that consist of rules, expectations, rituals and routines, stories and myths, symbols, power structures, organizational structures, and control systems (Bloor & Dawson, 1994; Johnson, 1992). In turn, these norms and practices drive subsequent behaviors through providing the social context through which people communicate and act (DeLong & Fahey). Putting this into the context of knowledge management, organizational culture determines the social context (consisting of norms and practices) that determines "who is expected to control what knowledge, as well as who must share it, and who can hoard it" (Delong & Fahey, p. 118). Figure 1 illustrates this conceptual linkage between culture and knowledge management behavior.

As Figure 1 depicts, the social context (consisting of norms and practices) is the medium for transmission of underlying values and beliefs into specific knowledge management behaviors. While Figure 1 is useful to explain the conceptual link-

Figure 1. The impact of organizational culture on knowledge management behaviors

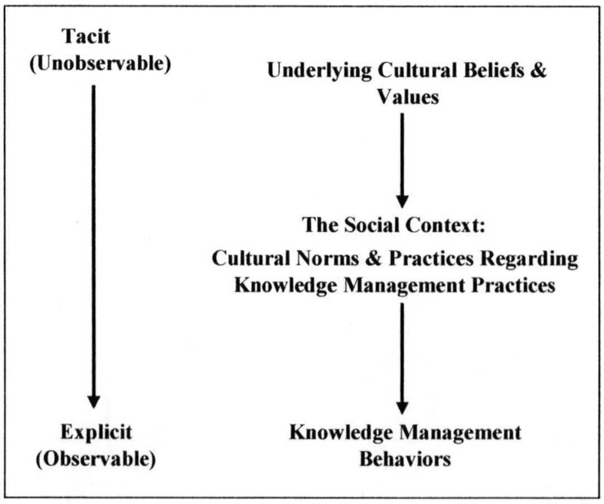

age between culture and knowledge management behavior, further explanation is needed to inform our understanding of the types of cultures that exist within organizations.

A number of theories have attempted to define culture at the organizational level. Wallach (1983) conceptualizes organizational culture as a composite of three distinctive cultural types: bureaucratic, innovative, and supportive. In bureaucratic cultures, there are clear lines of authority and work is highly regulated and systematized. Innovative cultures are characterized as being creative, risk-taking environments where burnout, stress, and pressure are commonplace. In contrast, supportive cultures are those that provide a friendly, warm environment and where workers tend to be fair, open, and honest. From Wallach's standpoint, any given firm will have all three types of culture, each to varying levels of degree. Wallach's cultural dimensions were developed based upon a synthesis of other major organizational culture indices. Wallach's cultural dimensions were applied by Kanungo, Sadavarti, and Srinivas (2001) to study the relationship between IT strategy and organizational culture. Part of the attractiveness of Wallach's dimensions, in comparison with other commonly used cultural indices, such as the organizational culture profile scale (O'Reilly, Chatman, & Caldwell, 1991), the competing values framework (Quinn & Rohrbaugh, 1983), or the organizational value congruence scale (Enz, 1986), is that it is highly intuitive. Managers can readily identify with the descriptions of the three general culture types. Consistent with Kanungo et al., we will employ Wallach's approach to describe organizational cultures.

Organizational Culture and KM

While not extensive, there is an early stream of work developing in the area of culture and KM. One focus has been on the effect of culture on knowledge-sharing practices (Barrett, Cappelmann, Shoib, & Walsham, 2004; Davenport et al., 1998; DeTiene & Jackson, 2001; Janz & Prasarnphanich, 2003; Knapp & Yu, 1999; Levinthal & March, 1993; Miles, Snow, Matthews, Miles, & Coleman, 1997). In general, the studies show an important connection between the openness and collectiveness of a culture and the extent of sharing. For example, Delong and Fahey (2000) identified specific value orientations believed to facilitate or hinder knowledge sharing. They found that value orientations such as trust and collaboration led to greater willingness among firm members to share insights and expertise with each other. In contrast, value systems that emphasized individual power and competition among firm members led to knowledge-hoarding behaviors. Consequently, they argue that firms should seek to reinforce and mold those cultural values most consistent with knowledge-sharing behaviors. In another study, Jarvenpaa and Staples (2001) draws from Goffee and Jones' (1996) dimensions of culture to define the relationship between organizational culture and perception of organizational ownership of information and knowledge and the resulting relationship of this construct with knowledge-sharing practices. Their results suggest that organizations with values emphasizing the pursuit of shared objectives (solidarity) will tend to have a higher perception of organizational ownership of information and knowledge produced by its individual members. As a result, this perception should presumably lead to greater levels of organizational knowledge sharing.

Another focus of KM-culture research has been on the concept of knowledge creation. This area is less concerned with whether, and how, knowledge is shared as it is with how new knowledge emerges, particularly in groups. For example, Lee and Choi (2003) found that the culture of the extended Linux community was important in regulating the norm of open sharing, in addition to providing a quality-control mechanism. They discovered that culture acted as a social control mechanism to manage community members and to sanction those who deviated from norms. The freedom to express criticism was found to be a significant underpinning of the development process that enabled knowledge to expand. Likewise, Styhre, Roth, and Ingelgard's (2002) study of a major pharmaceutical company found that values emphasizing caring relationships facilitated a greater level of knowledge creation within the organization.

In a broader attempt to isolate specific values that affect knowledge creation, Lee and Choi (2003) examined the organizational values of collaboration, trust, and learning. They found support for their hypothesis of a positive relationship of organizational culture (defined by collaboration, trust, and learning) and knowledge-creation processes and conclude that shaping an organization's cultural factors are key to a firm's ability to manage knowledge effectively.

A third area of KM-culture research has been that of the effect of culture on the effectiveness of KM. Organizations with more open and supportive value orientations have been found to have more constructive knowledge behaviors such as firm members sharing insights with others (Gold, Malhotra, & Segars, 2001). Another study reveals that constructive-type values (achievement, self-actualization, encouragement, and affiliation) have a positive impact on certain organizational factors (role clarity, communication quality, organizational fit, creativity, and job satisfaction) believed to promote KM effectiveness (Baltahazard & Cooke, 2003).

Drawing from the literature review, some key questions emerge. We know that culture influences KM approaches, behaviors, and effectiveness, but little is known about how this influence occurs and how lasting its effect is. We are thus interested in understanding the process through which culture affects organizations' knowledge management strategy as well as the process through which culture influences the primary KM activities of knowledge creation, sharing, and use. To address these questions, we employ a case study of two organizations.

METHODOLOGY

A case-study method involving multiple (two) cases was used. The approach of the study is depicted in Figure 2. The figure, based on the work of Yin (1994), displays the replication approach to multiple-case studies. As illustrated in Figure 2, the initial step in the study involved the development of a theoretical framework on the relationship between organizational culture and organizational KM strategies. This step was then followed by the selection of the two specific cases (the data collection sites) and the design of the data collection protocol. Following the case selection and data collection steps, the individual

Figure 2. Case-study methodology (adapted from Yin, 1994)

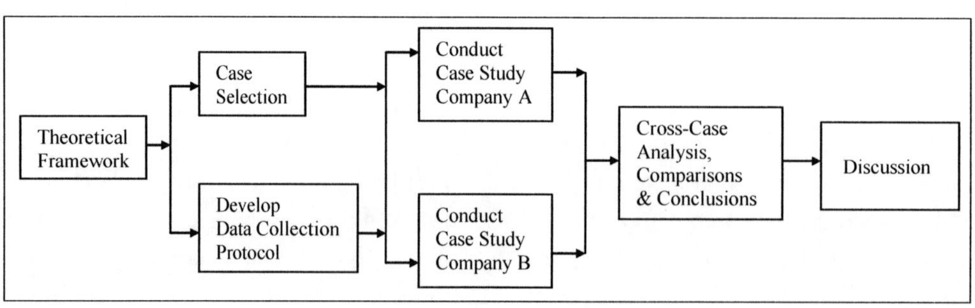

case reports were developed. A cross-case analysis of the findings was then undertaken. This analysis provided the basis for the theoretical and normative discussions and implications presented in the final section of the chapter.

The two case studies involve two very large and global corporations: Company A and Company B. Company A is a global consumer goods company with 369,000 employees worldwide. The company is headquartered in the United States and operates in four other regions: Europe, Middle East and Africa, Central and South America, and Asia. The company revenues consistently exceed $20 billion. In Company A, large-scale knowledge management projects were initiated at the North American region in 1996. Company B is a high-tech global company with multiple product lines and services. Similar to Company A, Company B is headquartered in the United States and operates globally in other regions of the world. With approximately 316,000 employees, its revenues exceed $80 billion. Large-scale knowledge management projects were initiated in Company B in 1995.

These two particular companies were selected for the purpose of this study for the following reasons. First, significant opportunities and challenges are associated with knowledge management activities in large and geographically dispersed companies. Thus, the identification of factors such as organizational culture that may influence KM outcomes in this type of organizations can potentially lead to high payoffs. Second, considering the high levels of organizational resources required for the implementation of large-scale knowledge management initiatives, these initiatives are most likely encountered in very large firms. Thus, the phenomenon of interest to these researchers could be best investigated in the context of very large firms with an established track record in KM projects. Lastly, past contacts that one of the researchers had with these two firms facilitated their recruitment as case-study sites.

Data Collection

Data for this study were collected through semistructured interviews with a small group of managers and professional employees at the two company locations in the United States. Identical approaches to data collection were used at Company A and Company B.[1] Six individuals at each of the two companies were interviewed. In each of the two companies, three of the interviewees were the current or the potential users of the KM systems. The remaining three interviewees in each company were the KM system sponsors or supporters. The interviews took between 45 to 85 minutes and were conducted between October 2001 and January 2002. All the interviews were tape-recorded and then transcribed for data analysis. The interviews all followed the same protocol. The informants were first asked to characterize their organization's culture in their own words. The three cultures described by Wallach (1983) were then portrayed and the informants were requested to identify which one best described their organization. The interviewees were next asked to describe and characterize the KM practices in their company. A set of specific questions guided the discussions of these practices. For example, each informant was asked to describe the specific KM activities that he or she engaged in and to discuss the effects of these activities on the self and/or on peers. Each informant was also asked to describe any resistance and impediments to KM that he or she might have noticed in the organization. The same interviewer, using identical data collection protocols, conducted all the interviews in Company A and Company B. The interviewer carefully read the transcripts to ensure accuracy.

Data Analysis

An author not involved in the interviews, and hence having no predisposed interpretation of the transcripts, conducted the data analysis. Based upon the transcribed interviews, 12 profiles were written, each one based upon the perspective of a single informant. These profiles described the informants' perspective of culture and their perspective of KM. The profiles of informants for Company A were compared and contrasted with each other, as were those of Company B. Cases for each company, reported in the next section, were then written based upon the within-case analysis. The cases for each company were then interpreted from the perspective of how the culture appeared to be influencing the organizational KM initiative. This is also reported in the next section. After the two cases and their within-case analysis were complete, a cross-case comparison and contrast was undertaken, leading to the formulation of the discussion section.

CASE DESCRIPTIONS AND ANALYSES

Knowledge Management at Company A

Knowledge management at Company A began as a top-down idea, courted by senior management "as a way of helping the company become more leading edge," according to one informant. A small group of eight or nine individuals at the headquarters was charged with driving knowledge management and facilitating knowledge sharing. As a result of larger issues surfacing, most notably the economic downturn that rocked U.S.-based businesses in early 2000, the top-level initiative "fell into the background" and the small dedicated group was disbanded. Thus, at the organizational level, KM was an idea that received neither funding nor action. However, at the business-unit level, successful KM initiatives have been built around intranet or around Lotus Notes team rooms.

Intranet-Based KM Projects

One initiative in the marketing area of the corporate headquarters is called MIC (Marketing Information Center). MIC serves the global marketing community of several thousand individuals around the world. It is an intranet-based library containing links to agencies, compensations, human resource (HR) information, and contracts, among other things. MIC is opportunity oriented rather than problem oriented. The members did not use the community to post problem inquiries and await responses, but rather to look for ideas performed in other parts of the company and think about adopting the ideas for their local groups.

MIC is intended to be a catalyst for collaboration and to "propel a universal worldwide marketing community." Because the chief marketing officer no longer allowed the budgeting of glossy manuals or brochures, MIC was widely accepted as the primary means of obtaining such static information. In fact, as attempts were made to include best practices in MIC, the initiative encountered resistance. Explains one informant, "we could never nudge the culture enough to have people understand and be motivated to enter their information." Another informant felt that there were challenges in overcoming "people's fear of being judged for their ideas and their indifference to yet another information site."

CM Connection (CMC) is another KM initiative within the North America marketing unit. This is a Web-based marketing repository used to disseminate information such that wholesalers that are responsible for store-level execution can have access to the most recent information on how to merchandise the latest promotions. As with MIC, the major impact of CMC has been the reduction of the number of printed catalogs, in this case by 80%. Among the challenges experienced with CMC has been convincing content providers to own the information in the sense of both providing it and keeping it up to date. Another issue has been that CMC is seen by some as distracting them from their relationships with clients. Even while CMC may reduce the amount of time spent traveling, this is not necessarily welcome in "a sales and marketing oriented relationship company because you are taking away relationship points."

The human resources unit with the corporate functions unit also has an intranet-based KM initiative, referred to as My Career. My Career is designed for managers and employees to help provide information about what tools, classes, and coaching is available for development. One of the goals of My Career has been to merge all of the training information into one place.

Many such intranet-based KM projects have been developed throughout Company A, so many that a portal project was initiated to alleviate the problem of "too much information in too many places, different IDs and passwords for each database, having to remember what is in the database to even go to get the information." However, despite some initial receptiveness to the idea from the head of the new-business ventures unit, IT budgets were frozen and the project never got under way.

The common thread running through the intranet-based KM projects at Company A is that they all are geared toward housing static information with the most major of impacts being the reduction in printed catalogs. Among the greatest resistances, according to informants, is that these KM projects appear to try to standardize work practices in a company comprised of "creative assertive people who want to do it their way and make their own individual mark."

Lotus-Notes-Based KM

Lotus Notes forms the basis of other KM initiatives within Company A. What distinguishes the Lotus-Notes-based KM projects from the intranet-based KM projects is the added focus on facilitating teamwork. The Lotus-Notes-based initiatives developed independently from the intranet-based initiatives. The North American marketing group developed a Lotus-Notes-based community of interest. The system contains examples of briefs, shared research, shared examples of different sites, and information on internal research. This micro KM initiative has 50 to 60 regular users. An important feature of the system is that whenever new information is added, community members receive an e-mail. In this way, members visit the community when new information that is relevant to them has been posted. This KM project has served as a means of sharing best practices. For example, a marketing manager from the United Kingdom posted information concerning a successful auction initiative that was then emulated by five other countries. On an individual level, the KM project has helped to increase the frequency of communication among members of the community. Similarly, human resources developed HR Source, a Lotus-Notes-based general bulletin board where meeting notes, follow-up action items, strategy documents, and work plans are placed. It is shared by the HR community on a global basis.

Lotus Notes is also the platform used to develop team rooms. The individual responsible for managing team rooms for North America has what he calls the "6-month rule": If a team room is not getting regular utilization over 6 months, it is deleted so that they can save money on the server expense. He says that he deletes about 70 to 80% of team rooms. He thinks the lack of reward is the biggest barrier toward KM system usage: "People who don't have technology in their title don't take it upon themselves and are not generally rewarded for exploiting technology." Also, content management is a barrier: "This is the responsibility of the end user but it is perceived as the responsibility of the technology group." However, a marketing manager had another opinion, attributing lack of use of the team rooms to self-preservation:

Table 2. Characteristics of culture, KM initiatives, and KM behaviors

Culture Characteristics	KM Characteristics	KM Behaviors
Dominant culture is bureaucratic Emphasis on individual: *individuals are risk averse *individuals fear being criticized for ideas *individuals are uneasy and prefer to go unnoticed *relationships externally, particularly within the marketing unit, are perceived as critical to their success	Intranet-based static repositories of information Failed top-down effort Bottom-up initiatives largely targeted creation of repositories Some use of Lotus Notes to create team rooms Team rooms have high failure rate	Individuals access information on as-needed basis Individuals reluctant to contribute information Individuals reluctant to own and maintain content Individuals uncomfortable using ideas from the systems since they do not own the ideas Individuals use a repository when rules prohibit printing brochures Individuals reluctant to use tools that would result in a loss of touch points with customers

"Even if someone took the time to put something out there, even if I knew it was there, went and got it, had the time to review it, and understand it, I am going to create this other thing by myself. I might look at that as input, but then it is the new XYZ program and I created it."

Analysis of Company A's Knowledge Management: The Impact of Culture on KM Behaviors and Outcomes

The Perceptions of Culture

While each individual interviewed gave their own perception of the culture at Company A and the perceptions naturally contain some variance, there is a marked theme running through the individuals' views. Informants describe Company A as "risk averse" and bureaucratic. They speak of an environment where people "don't want to be noticed," where direction is "unclear," and where "individual survival" trumps teamwork. Moreover, informants state that people "work in silos," "feel isolated," and "are afraid of being criticized for their ideas." The "slow, bureaucratic, hierarchical" culture at Company A has resulted in "silos of information." As a consequence, managers indicate that even though they have "great consumer and customer information," they "end up reinventing the wheel 1,000 times." However, our informants also maintained that although they characterize the culture as bureaucratic, they also sense that Company A is "striving" to become more innovative and supportive.

The Possible Impacts of Culture on KM

The statements and observations of our informants point to two largely shared perspectives: (a) that the culture emphasizes the individual and (b) that the culture is in a state of transition. In understanding the impacts of KM, one can see the influence of individuality within Company A. Table 2 lists the characteristics of the culture, KM initiatives, and KM behaviors as expressed by the informants.

At work within Company A seems to be a tension between a culture that demands individuality and the communal aspects of KM. The informants talk about a culture that is one of "individual survival" where individuals "fear being judged for their ideas," where there is individual isolation, and where individuals try to go unnoticed. The overall feeling is that of individuals trying to avoid being noticed. Such a culture does little to foster the sense of community that may be necessary to enable KM to move beyond static repositories of information into the kind of dynamic system envisioned by developers where ideas flow freely and where KM provides a catalyst for collaborative engagement. Not only are individuals reluctant to share their information for fear of being criticized for their ideas, they are also reluctant to use information posted in a KM project for lack of credit for the idea. Such behaviors can spring from a culture that emphasizes individual ideas and contribution.

The individual aspects of the culture go well beyond individuals behaving a certain way because of a reward system, but reflects an underpinning notion that to succeed in a marketing-oriented organization, one must be creative and that creativity is, perforce, of an individual nature so that to survive as an individual, one must capture ideas and only share them if they are going to be favorably judged. One must not look to others for learning or for problem solving, but might look to reuse creative ideas in some circumstances (like the auction site example from the United Kingdom) where one may tailor the idea to one's environment. It is telling that the informants speak of using outsiders (e.g., consultants) to assist with problem solving and learning instead of attempting to use any of the existing KM initiatives to post queries, and this in spite of the fact that it is recognized that the company "reinvents the wheel 1,000 times."

Another tension within Company A seems to stem from the expectations of what should occur in a bureaucratic culture and what was occurring. The top-down approach to KM, an approach that would be consistent with a bureaucratic organization, had failed at Company A. Yet, despite the failure of the top-down approach to KM and the seeming success of several bottom-up approaches such as MIC and the marketing team room for the community of 50, one informant still proffered the need for top-management leadership to be the key to success with KM. He considered the bottom-up approaches as "band-aid approaches." In his opinion, power within Company A comes "from knowledge hoarding not knowledge sharing." In order for KM to be assimilated in this environment, "behavior really has to come from the top. Leadership needs to walk the walk." In a bureaucratic culture, individuals become accustomed to clear guidance from senior management. The absence of clearly stated support from senior management may be sufficient to deter many from experimenting with the KM tools available to help them.

Summary: Company A has many KM initiatives that have largely been developed as bottom-up initiatives. The KM tools seem well designed and housed with valuable information. The informants are able to use the tools to facilitate the retrieval of information they need in the performance of their jobs. However, the tools have not, as yet, progressed to the level of fostering collaboration. While there are some successful communities from the standpoint of providing a place to share meeting notes and plans, the majority of team rooms remains unused and, if used, become as much a library of information as a communication tool. In some ways, the culture of Company A appears to foster the types of KM behaviors observed in that the individual is seen as the primary source of innovation and ideas as opposed to the community being the ultimate source of success. Thus, individuals will use the systems as needed, but are mostly occupied with their individual roles and work and do not attribute value to the collaborative features of technology.

The Case of Company B

Company B is organized into seven major units. Our interviews were concentrated within the innovations services group of the consulting wing (referred to as worldwide services group, or WSG).

Knowledge management at Company B began in 1996 with the view that KM was about "codifying and sharing information," leading to the creation of "huge repositories of procedures and process approaches." It was assumed that people would go to a central site, called the intellectual capital management system (ICM), pull information, and "would all be more knowledgeable." ICM is under the protection of Company B. There is a process one must undertake to have information submitted and approved. The process is complicated by legalities and formalities. As a result, ICM is not used as widely as it could be. What was discovered from the initial foray into knowledge management was that the information was not being refreshed and that the approach was not complimenting the way people really learned, which was through communities. Consequently, the KM initiative began to shift toward providing tools to communities that would help foster collaboration "both within teams and within locations and across the globe." Among the tools are team rooms and communities.

Team Rooms

Lotus-Notes-based team rooms are widely used at Company B to coordinate virtual teams and to share important documents. Access to the databases of a team is limited to the members because of the confidential nature of a lot of the issues. The project manager or someone delegated by the project manager takes the responsibility of sanitizing the material and posting the most relevant parts to a community system such as OC-zone (to be discussed below) and/or to ICM after the team's project has been completed.

The team rooms are valuable tools to help members keep track of occurrences as well as to help newly assigned members get quickly up to speed. Because of the itinerant nature of the Company B consultant's job, it is invaluable to have the documents he or she needs stored in an easily accessible manner that does not require sending and receiving files over a network. Team-room databases are also used for managing the consulting practices. It is important in helping new people with administrative tasks: how to order a piece of computer equipment, or how to order business cards, for example. The team rooms keep track of such metrics as utilization so that members of the team know "who's on the bench and who's not." One informant gave the example of a recent project she was put on at the last minute involving a selling project to a government department in another country. She was able to access all the documentation from the team room and was able to become a productive member of a new team very quickly: "I can go in and start getting information about a particular topic and work with colleagues almost immediately. It allows me to work more easily with colleagues across disciplines."

Although the team rooms are invaluable in organizing and coordinating project teams, there are also some potential drawbacks. Some view the team rooms as engendering "a false sense of intimacy and connectedness." This sense of intimacy can be productive for the team as long as things are going well. However, "if things go south," says an informant, "you don't have the history or skill set to really deal with difficult situations." As a result, instead of "dealing with the conflict," the group is more likely to "just take someone off the team" and replace the person with another. In this sense, problems are not solved so much as they are avoided, and team members take on an expendable quality.

Communities

Communities serve members based not upon project or organizational position, but based upon interest. By 2000, a group referred to as the organizational change (OC) group had established a successful community of 1,500 members cutting across all lines of business and was beginning to act as consultants to the other groups trying to set up communities. The OC community has gone so far as to quantify the business return of such a community in terms of cycle-time reductions and sophistication of responses to clients. The OC community is comprised of tools, events, and organization.

1. **Tools:** The technology tools at the disposal of the OC community are databases of information submitted by team rooms, including such things as white papers, projects, and deliverables, as well as client information. The databases also contain pictures of community members with personal information about the members.
2. **Events:** An important aspect of the OC community is the events that are organized for community members. These include monthly conference-call meetings, which generally get attendance of between 40 and 90 members, and replay meetings, which draw another 40 t0 70 members. In the past, the community had sponsored a face-to-face conference for members. Members often meet others for the first time, yet they already feel they know each other.
3. **Organization:** The organization of the community is managed by two community leaders. When someone requests information or has a query to post to members, he or she sends a message to one of the community leaders. The leader first tries to forward the message directly to a subject-matter expert (SME). If the leader does not know offhand of an appropriate SME, he or she will post the question to the entire group. In this event, the group members respond to the leader rather than to the community in order to avoid an inundation of messages. The leader normally receives responses within an hour. The leader then forwards the responses to the individual with the query. He or she later sends an e-mail to the person who made the inquiry, asking how the response was, and how many days-time it saved. The leader says that as many as 28 responses will be made to a particular inquiry. The leader has manually loaded a portion of what they've developed in the past 7 months: There are 114 pieces of intellectual capital that was loaded, and it is just a portion of what was received.

The community has a structure that consists of a senior global board of 30 members representative of different parts of the business. There is a subject-matter council that constantly scans the intellectual capital as well as an expert council and the health-check team.

The health-check team examines such things as how well members communicate with each other. It conducted an organizational network analysis to help better understand the communication networks. The team has a series of questions to help assess how the group is doing in terms of high-performance teaming. It uses a survey that measures perceptions from the community members about what they see is happening and does a gap analysis on what is actually happening. Finally, the team does a self-assessment of where they are compared with the community maturity model developed by the OC community leaders. There is a community mission, vision, and goals, and it is working on capturing data to support the metrics to demonstrate value to the company and community members.

The goal is to attain Level 5 maturity, when a group is considered "an adaptive organization." There are 13 areas of focus that the community leaders look at in building a sustained community. While communities are felt to be organic, there is also a community developers' kit with an assessment tool to determine at what level of maturity a community is and what steps need to be taken to move the community forward. One community leader says that the purpose of the development kit "is not to confine, but to provide a road map in which to navigate and build." For this leader, the essence of community is "continuous learning." Of the initial KM efforts focused on information repositories, the leader says, "I could see the technology coming that was going to enslave people, like an intellectual sweat shop." By contrast, the primary tools for a community are "passion and environment."

Impact of OC

Among the major impacts of the OC-zone is that having a community helps people "not feel isolated. People feel they are affiliated, that they are part of the company." Thirty percent of Company B employees do not have offices and work instead from home or from client sites. Such a work environment can easily be associated with isolation. However, the community is claimed by some to provide a clarity of purpose: "I see it as a conduit for both developing thought leadership and enabling thought leadership to get into the hearts and minds of the workers so that they all have a common vision, goals, and objectives."

Community members view the purpose of the community as a knowledge-sharing forum and as a means to creating a sense of belonging. One member went so far as to suggest that she would "not be at Company B any longer if it wasn't for this community." The reason is that most of her connections at Company B have been made through the community. Also, being in the community helps her get assigned to projects. For example, the leader of a new project will call someone in the community and say that they are looking for a person with a certain profile. She finds that she gets asked to work on projects this way.

Other members refer to the community as a "supportive family" and state that within the community is someone who has already encountered any issue they will encounter on a project, so that the community keeps them from reinventing the wheel. The norms of operation exist to help OC-zone be as effective as possible. No one is under obligation to contribute, but individuals contribute in order to help other members. One member credits the success of the community to the two leaders, whom she feels "in their hearts, care about the members of the community." She feels that the community is more than a community of people who like the topic of organizational change, but it is a community of people who support one another.

Table 3. Characteristics of Company B culture, KM initiatives, and KM behaviors

Culture Characteristics	KM Characteristics	KM Behaviors
Hierarchical, yet collaborative and innovative Individuals largely responsible for their own careers, yet competition is undertaken in a cooperative manner The team is the unit of success, more so than the individual Absence of extreme supervision of individuals' work; individuals have a sense of control	Company-wide information repository consisting of hundreds of information databases Team rooms used by project teams Communities of practice emerging; these communities include tools, events, and structures OC community used as an example of a successful community and as a consultant to other emerging communities	Team members actively coordinate via the team rooms Community members obtain a sense of belonging to the community Community members post information from completed team projects to the community out of a sense of commitment, not coercion Community members are more loyal to the company (less likely to depart) because of their belonging to the community Assignments to projects made through community references

The primary resistance to the OC community has been the practice managers. Most of the community members report to practice managers. The practice managers are used to thinking in terms of billable hours. Indeed, the performance evaluation system requires that an individual's goals support those of his or her boss. The community leaders hope that one day, participating in a community will be included as a standard part of this evaluation system.

Analysis of Company B Knowledge Management: The Impact of Culture on KM Behaviors and Outcomes

The Perceptions of Culture

All of the respondents from Company B work within the same business unit. The respondents describe the culture of Company B as a blend of hierarchical and innovative. The hierarchical aspects are evident in that little innovation is undertaken until senior management has officially supported the innovation, but once senior management does give the green light to an idea, "everybody jumps on it."

One aspect of culture that is highlighted by the informants is the importance of collaboration. Informants characterize the "street values" within Company B as "win, team, and execute." The informants recognize a duality of culture that on the one hand gives individuals control over their work and at the same time is highly supportive of the individual. The culture is autonomous in the sense of "not having someone looking over your shoulder and telling you what to do." And while there is certainly competition as everyone has objectives they are trying to meet, things "are always done in a collaborative, helpful spirit."

The other dominant aspect of culture as related by the informants is that of hierarchy. The hierarchy is as much a hierarchy of experience as of structure. Community members, for example, proffered that becoming a subject-matter expert is more about length of service to the company than one's inherent knowledge. Another aspect of the bureaucratic culture is that "there is very much a correct way to do things."

Table 3 lists the characteristics of culture, KM initiatives, and KM behaviors expressed by the Company B informants.

Company B's emphasis on collaboration seems to have enabled the progression of KM from a static information repository system to an active, vital community of interest wherein individuals feel a sense of belonging to the extent that they identify themselves first with the community, and only second, if at all, with the actual formal business unit. One informant claimed to not identify herself at all with the innovation services unit. Of course, one could ponder whether such identity transfer from the business unit to the community serves the best interest of the unit.

At the same time, the bureaucratic and innovative aspects of the culture have also helped. Having senior management show interest in KM was a catalyst to individual groups undertaking KM initiatives with great enthusiasm. In addition, rather than ad hoc communities that are entirely organic, the community model emerging at Company B is a relatively structured one.

While one can make the argument that Company B's culture influences KM development and use, one can also argue that KM at Company B is itself influencing the culture. OC members claim that without a sense of connection provided by the OC community, the company would be nothing but a "big and scary" company in which individuals "get lost." The community, though, allows and enables a culture of connection. In effect, one informant believes that the OC community attempts to shift a very technical, phone-oriented, work-product-oriented way of communicating with each other into a more personal work-in-process movement toward what Company B refers to as "thought leadership." When asked why members take the time to participate in the community when there is no formal reward for doing so, one informant said simply, "it's just how we do business." Thus, the community has infused the culture of the members.

Yet this does not suggest that an organizational utopia has been, or will be, achieved. While the culture is becoming more connected, there is another angle. One informant believes that when you have widespread access to knowledge management, you can also have a culture where people that know very little about something have access to enough information to be dangerous. People get too comfortable with having access to knowledge and then they feel free to share it. This informant remained unconvinced that the knowledge one acquires through the network is as solid a foundation as the knowledge one has acquired through experience and traditional learning. Moreover, she feels that the notion of dialogue can get redefined in a way that you lose the quality of participation that one might be looking for.

Summary: Company B has many KM databases, collectively referred to as ICM. While these databases serve an important role of housing and organizing information in a huge organization, they do not go so far as to foster collaboration. Instead, team rooms and communities of interest, largely left to the discretion of team members and community members, have proven to be vital tools to achieving collaboration, community, and belonging. And as the culture of Company B has been receptive to individual group setting and pursuing community agendas, the culture is also being subtly altered by the communities as members feel more belonging to the community than to their business units.

DISCUSSION

The two cases offer insights into the role that organizational culture plays in the inception and maturation of knowledge management. This section summarizes the key findings that help us answer the question, how does organizational culture influence KM approaches? We suggest four responses to this question.

1. Organizational culture influences knowledge management through its influence on the values organizational members attribute to individual vs. cooperative behavior.
   The two companies we examined share several similarities: Both are huge multinational organizations widely regarded by organizational members as being predominantly bureaucratic in culture, both organizations had initial KM approaches that were strongly supported by senior management, and both had initial KM approaches focused on the creation of a large centralized repository of organizational knowledge to be shared across the organization. These two large bureaucratic organizations began their KM quests with the process approach. The most striking difference between the organizational cultures of these two companies was the emphasis at Company A on the individual and the emphasis at Company B on the collectivity: the team or the community. This evinces itself even in the interpretation of innovation. While individuals at both companies spoke of the need for innovation in their organizations and of the striving of their organizations to develop an innovative culture, in the case of Company A, innovation was perceived as an individual attribute whereas at Company B, innovation was perceived as a team-level attribute.

The individualistic view of innovation at Company A seemed to militate against the requisite sharing and cooperation that makes the evolution of KM from the process approach to a community-of-practice approach possible. In both companies, microlevel experimentation of the various possibilities of KM was undertaken within teams or business units. The value placed on individualism vs. cooperation seems to have played a significant role in the nature and form of the KM approach. The microlevel experimentations by teams or business units were carried out with their own assumptions about the usefulness of repositories of knowledge and the usefulness of communities of practice. We suggest that it is not organizational culture at the organizational level or even the subunit level that has the most significant influence on KM approach, but organizational culture as embodied in the individualistic vs. cooperative tendencies of organizational members. Thus, organizational culture influences KM approaches through its influence on individualism vs. cooperation. From a theoretical view, it seems that Wallach's (1983) cultural dimensions and those of Earley (1994) were both valuable at explaining organizational-level culture. However, Earley's cultural dimensions at the organizational level seem best able to explain why a KM approach tended to become more process or more practice based.

2. Organizational culture influences the evolution of knowledge management initiatives. Our findings suggest that firms do not decide in advance to adopt a process or practice approach to KM, but that this evolves. The most natural starting point is one of process, perhaps because the benefits seem more evident and because it can more closely align with the existing organizational structure. Moreover, the practice approach may not only fail to align with existing structure, but it may engender a virtual structure and identity. It is interesting that at Company B, having a culture dominantly viewed as bureaucratic, once the initial organizational change community was established, the evolution of the community then became a highly structured process of maturation. The community leaders developed a tool kit to help other communities develop, and developed a maturation model to help them determine how mature a community was and a plan to move the community forward. What some might see as an organic process—that of establishing and developing a community of practice—became in a bureaucratic organization a structured process. Even if the idea for the community emerged from interested potential members, the evolution took on a structured form with tools, kits, assessments, and plans. The cooperative aspect of culture at the individual level made the community possible; the bureaucratic elements of culture at the organizational level enabled the community to mature. Hence, the evolution of the community was highly dependent on the individual willingness of organizational members to sustain and nurture their community. This appeared tied to the importance they placed on cooperation with their community members, most of whom they had never met.

3. Organizational culture influences the migration of knowledge. In the case of Company A, where as mentioned the informants seemed to identify the individual as the ultimate unit of responsibility in the organization, the individuals were also viewed as the owners of knowledge and had the responsibility to share their knowledge. This in fact created a major challenge since the individuals rejected this new responsibility. At Company B, where the team seemed to be the focus of responsibility, knowledge migrated from

the team to the community to the organizational-level system and back down to the team. The leader of the team would take responsibility for cleaning the team's data and submitting them to the community and to the central information repository. Thus, knowledge migrated upward from the team to the central repository. Interestingly, the most useful knowledge was claimed to be that at the team and community level. Once the knowledge had completed its migration to the central repository, it was seen primarily as an item of insurance: for use in case of need. Knowledge sharing and transfer occurred primarily at the team and community level whereas knowledge storage was the function of the central repository. The migration of knowledge is also influenced by the structural processes put in place to ensure that knowledge finds its way to the appropriate persons. Of key importance seems to be the way the queries are handled. The marketing group at Company A adopted the approach of notifying individuals when new information had been added to the KM system. However, little interference was put in place to either guide people to the appropriate knowledge or encourage people to contribute knowledge. Contrarily, believing that the community should not become a bulletin board of problems and solutions, the leaders of the organizational change community at Company B worked arduously to find the subject-matter experts so that queries would be submitted to the community leader who would serve as an intermediary between the individual with the query and the expert.
It has been widely reported that the use of knowledge directories is a primary application of KM in organizations. Our study suggests that the facilitated access to experts rather than direct access via the location of an individual through a directory or via a problem posted to a forum may lead to a more favorable community atmosphere.

4. Knowledge management can become embedded in the organizational culture. Over time, as KM evolves and begins to reflect the values of the organization, the KM itself can become a part of the organizational culture. At Company B, individuals spoke of their community involvement and their team rooms as simply the "way we work." In fact, the communities became so much part of the culture that even though they were not part of the organizational structure, they were part of individuals' implicit structure.

Table 4. Summary of organizational culture's influence on KM

Cultural Perspective	Influence of Culture on Knowledge Management
Bureaucratic (Wallach, 1983)	Favors an initial process approach to KM Creates expectation among members that senior-management vision is essential to effective KM
Innovative (Wallach, 1983)	Enables subgroups in organizations to experiment with KM and develop KM useful to their groups
Individualistic (Earley, 1994)	Inhibits sharing, ownership, and reuse of knowledge
Cooperative (Earley, 1994)	Enables the evolution of process-oriented KM to practice-oriented KM Enables the creation of virtual communities

The sense of belonging that the individuals reported feeling toward their community suggests that the community had become an essential aspect of their value system, and hence, had become part of organizational culture. That the organizational change community members at Company B identified themselves first and foremost with their community, in spite of receiving neither reward nor recognition within their formal reporting unit for participating in the community, indicates the extent to which community participation had become a value and an aspect of the individual culture.

IMPLICATIONS AND CONCLUSION

The findings of our study suggest that a dominantly bureaucratic culture seems to tend toward an initial process-based KM approach. Furthermore, a bureaucratic culture seems to create the expectation among organizational members that senior management needs to provide a vision of purpose for KM before the organizational members should embark on KM activities. As well, the members view senior management support as validating any KM activities that they undertake. Innovative cultures, even if not the dominant culture at the organizational level, seem to enable subgroups to experiment with KM or create micro KM initiatives. In essence, in organizations having dominant bureaucratic cultures with traces of innovativeness, senior management support legitimizes KM, but the innovativeness of the culture enables it to expand far beyond an organization-wide repository. Specific KM behaviors such as the ownership and maintenance of knowledge, knowledge sharing, and knowledge reuse seem largely influenced by the individualistic or cooperative nature of the culture. Individualistic cultures inhibit sharing, ownership, and reuse, while cooperative cultures enable the creation of virtual communities. Earley's (1994) work on organizational culture emphasized the individualistic and collectivistic aspects of culture. Organizations encouraging individuals to pursue and maximize individual goals and rewarding performance based on individual achievement would be considered as having an individualistic culture whereas organizations placing priority on collective goals and joint contributions and rewards for organizational accomplishments would be considered collectivist (Chatman & Barsade, 1995; Earley). This dimension of organizational culture emerged as critical in our examination of the influence of culture on KM initiatives. These findings are summarized in Table 4.

This research set out to examine the influence of organizational culture on knowledge management approaches. Using a case study approach, we have gathered the perspectives of individuals in two firms that share some cultural similarities yet differ in other aspects. The findings suggest that organizational culture influences the KM approach initially chosen by an organization, the evolution of the KM approach, and the migration of knowledge. Moreover, the findings suggest that KM can eventually become an integral aspect of the organizational culture. Much remains to be discovered about how organizational cultures evolve and what role information technology takes in this evolution. This case study is an initial effort into a potentially vast array of research on the issue of the relationship of information technology and organizational culture.

REFERENCES

Alavi, M., Kayworth, T., & Leidner, D. (2005). An empirical examination of the influence of organizational culture on knowledge management practices. *Journal of Managment Information Systems, 22*(3), 191-224.

Baltahazard, P. A., & Cooke, R. A. (2003). *Organizational culture and knowledge management success: Assessing the behavior-performance continuum.* AZ: Arizona State University West.

Barrett, M., Capplemann, S., Shoib, G., & Walsham, G. (2004). Learning in knowledge communities: Managing technology and context. *European Management Journal, 22*(1), 1-11.

Bloor, G., & Dawson, P. (1994). Understanding professional culture in organizational context. *Organization Studies, 15*(2), 275-295.

Brown, S. J., & Duguid, P. (2000). Balancing act: How to capture knowledge without killing it. *Harvard Business Review*, 73-80.

Chatman, J. A., & Barsade, S. G. (1995). Personality, organizational culture, and cooperation: Evidence from a business simulation. *Administrative Science Quarterly, 40*(3), 423-443.

Davenport, T. H., DeLong, D. W., & Beers, M. C. (1998). Successful knowledge management. *Sloan Management Review, 39*(2), 43-57.

DeLong, D. W., & Fahey, L. (2000). Diagnosing cultural barriers to knowledge management. *Academy of Management Executive, 14*(4), 113-127.

Dennis, A., & Vessey, I. (2004). Three knowledge management strategies: Knowledge hierarchies, knowledge markets, and knowledge communities. *MIS Quarterly Executive, 3*(4), 399-412.

DeTiene, K. B., & Jackson, L. A. (2001). Knowledge management: Understanding theory and developing strategy. *Competitiveness Review, 11*(1), 1-11.

Earley. (1994). Self or group? Cultural effects of training on self-efficacy and performance. *Administrative Science Quarterly, 39*, 89-117.

Enz, C. (1986). *Power and shared values in the corporate culture.* Ann Arbor, MI: The University of Michigan Press.

Goffee, R., & Jones, G. (1996). What holds the modern company together? *Harvard Business Review, 74*(6), 133-148.

Gold, A. H., Malhotra, A., & Segars, A. H. (2001). Knowledge management: An organizational capabilities perspective. *Journal of Management Information Systems, 18*(1), 185-214.

Gupta, A. K., & Govindarajan, V. (2000). Knowledge management's social dimension: Lessons from Nucor Steel. *Sloan Management Review, 42*(1), 71-80.

Hansen, M. T., Nohria, N., & Tierney, T. (1999). What's your strategy for managing knowledge? *Harvard Business Review*, 106-115.

Hargadon, A. B. (1998). Firms as knowledge brokers: Lessons in pursuing continuous innovation. *California Management Review, 40*(3), 209-227.

Hasan, H., & Gould, E. (2001). Support for the sense-making activity of managers. *Decision Support Systems, 31*(1), 71-86.

Janz, B. D., & Prasarnphanich, P. (2003). Understanding the antecedents of effective knowledge management: The importance of knowledge-centered culture. *Decision Sciences, 34*(2), 351-384.

Jarvenpaa, S. L., & Staples, S. D. (2001). Exploring perceptions of organizational ownership of information and expertise. *Journal of Management Information Systems, 18*(1), 151-183.

Johnson, G. (1992). Managing strategic change: Strategy, culture and action. *Long Range Planning, 25*(1), 28-36.

Kanungo, S., Sadavarti, S., & Srinivas, Y. (2001). Relating IT strategy and organizational culture: An empirical study of public sector units in India. *Journal of Strategic Information Systems, 10*, 29-57.

Knapp, E., & Yu, D. (1999). How culture helps or hinders the flow of knowledge. *Knowledge Management Review, 2*(1), 16-21.

KPMG Management Consulting. (1998). *Knowledge management: Research report.*

Lee, H., & Choi, B. (2003). Knowledge management enablers, processes, and organizational performance: An integrative view and empirical examination. *Journal of Management Information Systems, 20*(1), 179-228.

Levinthal, D., & March, J. (1993). The myopia of learning. *Strategic Management Journal, 14*(8), 95-112.

Miles, R., Snow, C., Matthews, J., Miles, G., & Coleman, H. (1997). Organizing in the knowledge age: Anticipating the cellular form. *Academy of Management Executive, 11*(4), 7-24.

O'Dell, C., & Grayson, C. J. (1998). If only we know what we know: Identification and transfer of best practices. *California Management Review, 40*(3), 154-174.

O'Reilly, C. A., Chatman, J., & Caldwell, D. F. (1996). Culture as social control: Corporations, cults, and commitment. *Research in Organizational Behavior, 18*, 157-200.

Quinn, R. E., & Rohrbaugh, I. (1983). A spatial model of effectiveness criteria: Towards a competing values approach to organizational analysis. *Management Science, 29*, 363-377.

Ruggles, R. (1998). The state of the notion: Knowledge management in practice. *California Management Review, 40*(3), 80-89.

Schein, E. H. (1985). *Organizational culture and leadership.* San Francisco: Jossey-Bass.

Schultze, U., & Boland, R. (2000). Knowledge management technology and the reproduction of knowledge work practices. *Journal of Strategic Information Systems, 9*(2-3), 193-213.

Styhre, A., Roth, J., & Ingelgard, A. (2002). Care of the other: Knowledge-creation through care in professional teams. *Scandinavian Journal of Management, 18*(4), 503-522.

Von Krogh, G. (1998). Care in knowledge creation. *California Management Review, 40*(3), 133-153.

Wallach, E. J. (1983). Individuals and organizations: The cultural match. *Training and Development Journal.*

Watson, S. (1998, January 26). Getting to "aha!" companies use intranets to turn information and experience into knowledge: And gain a competitive edge. *Computer World.*

Wenger, E. C., & Snyder, W. M. (2000). Communities of practice: The organizational frontier. *Harvard Business Review*, 139-145.

ENDNOTE

1 After this initial data collection, we returned to Company B a year later and conducted more widespread interviews across different business units. This data collection and analysis is discussed in Alavi, Kayworth, and Leidner (2005).

Chapter XIV
The Desire for Cohesion in Virtual Teams:
Be Careful What You Wish For

John McAvoy
University College Cork, Ireland

Tom Butler
University College Cork, Ireland

ABSTRACT

Cohesion is regarded as something to strive for in virtual teams yet difficult to attain. What happens, though, when cohesion is achieved; does cohesion, as assumed, enhance the virtual team? During a longitudinal participant observation study of a virtual software development team, a strange paradox was noted. A new software development methodology was introduced to the project, and the developers were initially committed to its use. Over time, the commitment gradually decreased to the stage where aspects of the new methodology were practically ignored. As the team was a virtual team, with group members rarely congregating as a whole for any length of time, it was hard to explain why this diminishing of commitment occurred. The remoteness and part-time participation of group members meant that the team deciding themselves to ignore aspects of the methodology was not a likely possibility. A review of existing research suggested that the concepts behind the diffusion of innovations (specifically software process innovations) may have a bearing. Although pertinent to the area of introducing new software development methodologies, diffusion theories did not provide a complete explanation for the decrease in commitment that was observed. The theory of competing commitments was applied, and it was discovered that one cause of the decreased commitment among team members was groupthink. Groupthink should not be a problem with virtual teams as there should be less cohesion: a lack of contact between members dictating the low level of cohesion. Further analysis showed that traditional peer groupthink was not the issue, but hierarchical groupthink influenced by the project manager had a large influence. These findings are in contrast to most expectations concerning cohesion and virtual teams, including the project management of virtual teams.

INTRODUCTION

Researchers have noted the differences between virtual and face-to-face teams. One of the major differences is in the area of team cohesion. While Mark (1998) found the importance of building relationships and trust for members of virtual teams, Casey and Richardson (2005) provide an example where the use of virtual teams leads to conflict: An us-vs.-them mentality arose. A team that had originally been colocated and worked well together became hostile when working in different locations. De Pillis and Furumo (2006) found cohesion to be lower in virtual teams (although this problem is not necessarily universal). Powell, Picolli, and Ives (2004), in a review of existing research, note that the debate continues as to whether virtual teams achieve cohesion levels similar to traditional teams. Whether virtual teams achieve this or not, it is noticeable from the research described by Powell et al. that cohesion is something to be strived for.

Discussions on change management and the introduction of change in virtual teams should therefore take the differences between virtual and face-to-face teams into account. The focus of this chapter is specifically an examination of why the change to a new software development process, although initially supported by a virtual development team, never materialises. Authors have referred to the escalation of commitment to a failing course of action (Beynon-Davies, 1995; Keil, Mann, & Rai, 2000; Newman & Sabherwal, 1996). This chapter describes the de-escalation of commitment to a succeeding course of action in a virtual team. It does this by concentrating on the social environment surrounding the individuals in the team: one of the six key conceptual elements of electronic collaboration as described in Koch (2005b).

Change within software development projects is an area of importance to the success of the project as projects, by their very nature, are about change. Although Cushway and Lodge (1999) emphasise the importance of managing change, their description of change management is a restrictive one. For them, the concern is in developing strategies and structures. No mention is made of the teams and individuals who will effect, and be affected by, change. The sole mention of employees is a list of expectations, or required behaviours, such as roles must be carried out in a dependable fashion, and there must be innovation in achieving organisational objectives. In the context of a virtual team, there are further considerations regarding change that need to be addressed.

This chapter describes a case study, undertaken by the authors, that examined the change involved in introducing a software development methodology. The case study is based on a software development project to develop a knowledge management system (KMS) for a European government. A longitudinal study of the development project was undertaken using participant observation as its primary method. The study concentrates solely on the software project team—a virtual team—as opposed to involving the various high-level project sponsors. One aspect of agile software development employed in the project to develop a KMS is the use of user stories. Rather than relying on complex design documents, agile methods espouse the writing of customer requirements in simple language. The stories should describe what is required of a part of the final software project. The longitudinal research into the software development project highlighted a problem with the change to this new process. The developers in the virtual team were initially committed to its use. Over time, the commitment gradually decreased to the stage where aspects of the new methodology were practically ignored. As the team was a virtual team, with group members rarely congregating as a whole for any length of time, it was hard to explain why this diminishing of commitment occurred. The remoteness and part-time participation of group members meant that the team deciding themselves to ignore aspects of the methodology was not a likely possibility.

The investigation into this dilution of commitment became a two-phase process. In Phase 1, to determine the reasons behind this reduction in commitment to the change, Kegan and Lahey's (2001a, 2001b) competing-commitments process was followed. This process aims to determine the reasons, often subconscious, why a change that was originally committed to is not successful. These reasons are known as competing commitments as they work against the original commitment to change. Analysis of these competing commitments was still insufficient in explaining the lack of success of the methodology change. Therefore, in Phase 2, the output of the competing-commitment process was then aligned with observations from the longitudinal case study and existing research literature on groupthink to determine a cause for this lack of success. This cause, the explanation for the failure to adopt user stories, is then elaborated on.

Groupthink should not be a problem with virtual teams as there should be less cohesion, a lack of contact between members dictating the low level of cohesion. Further analysis showed that traditional peer groupthink was not the issue, but hierarchical groupthink influenced by the project manager had a large influence. These findings are in contrast to most expectations regarding virtual teams and the project management of virtual teams.

RESISTANCE TO CHANGE

Virtual teams are a common element of many organisations (Beise, 2004; Cascio & Shurygailo, 2002; Paul, Seetharamann, Samarah, & Mykytyn, 2004; Potter & Balthazard, 2002). They have developed to such an extent that they are vital to the success of many organisations (Lurey & Raisinghani, 2001). Virtual teams bring their own unique attributes and characteristics to a software development project, in fact any project, so it is necessary to take these into account when investigating areas of software development. One of these areas is change: the resistance to change and the management of change, which may take a different form than in face-to-face teams.

Cushway and Lodge (1999, p. 180) emphasise the fact that "probably one of the key skills required of managers in today's organizations is the ability to manage change." LaFleur (1996) stresses the fact that change is constant in a project. Interestingly, many others regard projects as a usual method of implementing change (Alsene, 1999; Boody & Macbeth, 2000; Clarke, 1999; McElroy, 1996; Pellegrinelli, 1997; Turner & Muller, 2003).

Changes are resisted and individuals or teams can cause this resistance. What is questionable is whether a virtual team of software developers would resist change in the same manner as a face-to-face team would. Before examining resistance to change in virtual teams, a situation is described where change is welcomed.

The Relevance of Change to Agile Software Development Projects

Metzger (1981) describes how change is part of every software project. Software development projects that follow agile methods (of which user stories is one aspect) regard change as one of their core aspects. Beck (1999) describes agile techniques as embracing change. Agile manufacturing, which shares many principles with agile software development, "is the ability to thrive and prosper in an environment of constant and unpredictable change" (Maskell, 2001, p. 5). Authors such as McConnell (2001) believe that virtual teams are not appropriate for agile software development, yet others, such as Maurer (2002) and Sepulveda (2003), show that virtual teams can actually be used. The agile manifesto, the core values of agile software development, is listed below.

- Individuals and interactions over processes and tools.
- Working software over comprehensive documentation.
- Customer collaboration over contract negotiation.
- Responding to change over following a plan (Fowler & Highsmith, 2001; Lindvall et al., 2002)

The fourth point shows the importance of change in agile projects. Change is not only accepted but also encouraged. The first point shows the importance of the developers and their interactions to an agile project. This would suggest that resistance to change should not be a major concern in agile development, whether the team is virtual or not.

Resistance to Change in a Software Development Project

Whether teams are virtual or face to face, software development projects experience change and resistance to change. Humphrey (1989) describes how process change in software development will only be successful with continual commitment, not just from senior management, but from everyone associated with the development of software. In addition, Thomsett (1998) agrees that change is only effective when all impacted by the change are involved.

Nader (1993) describes the problems that arise when people resist change. People require a sense of security and this is achieved through stability. Changes imposed on an individual remove this stability. Whitehead (2001) describes how change can evoke stress and an emotional reaction against the change. Change, though, cannot be avoided in software development projects: "Everything in software changes. The requirements change. The design changes. The team changes. The business changes. The team members change. The problem isn't change, per se, because change is going to happen; the problem, rather, is the inability to cope with change when it comes" (Beck, 2000, p. 28). Zmud (1983) adds to the list of areas that change in software development projects by discussing changes to processes (such as user stories in this research). Zmud argues that trying to implement process change by changing people will lead to resistance. Rainwater (2002) describes the danger for software projects of not assessing the impact of change. Although, virtual teams have their own unique attributes, there is nothing in the literature to suggest that a virtual team would have different attitudes or behaviours with specific regard to change than would a face-to-face team. Similarly, there is nothing to suggest that a virtual team would not differ from a face-to-face team.

Resistance to change can indicate a lack of commitment to the change, but there can be other explanations (Bowe, Lahey, Kegan, & Armstrong, 2003a). Software developers are a good demonstration of this resistance. Programmers' resistance to change is shown in a description of programmers from the 1960s: "like converts to a new religion, they often display a destructive closed-mindedness bordering on zealotry" (Zachary, 1994, p. 13). Yourdon (1993) describes change as one of the major dislikes of programmers. Changing anything, ranging from procedures to hardware platforms to methodologies, will result in complaints. Resistance to change is not helped with various attitudes to change, demonstrated in others' work, such as change being the enemy of the project (Birmingham, 2002), and change being something that must be coped with and its disruptive impact minimised (Field & Keller, 1998).

Returning to the point made by Bowe et al. (2003a), there can be other explanations for resistance than a lack of commitment. Lawrence (1996) believes that this resistance to change can highlight the fact that something is being overlooked. Rather than resistance to change having negative connotations, it should be regarded as an indicator that the change itself needs further examination. With the lower levels of communication and the absence of cues, inherent in virtual teams, it would be easier to overlook a problem and simply describe the situation as resistance. Kegan and Lahey (2001a) propose that some resistance to change is easy to explain (for example, the stress of learning a new skill), but other resistance is not as easily explained. A paradox exists where people show a commitment to and support the change, yet still resist that change. Robbins and Finley (1998) state that resistance can be a subconscious act. We can all agree on an idea, but then do nothing to implement it. It was necessary to use a method that would identify and explain the causes for this paradox, where initial commitment evolved into resistance.

THE THEORY OF COMPETING COMMITMENTS

To understand the need for a new approach to determine why a change was resisted and failed, it is necessary to understand the context in which the resistance occurred.

The Case Study: An Agile Software Development Project

A longitudinal study was performed on a virtual software development team that was changing over to an agile software development process. The project's objective was the design and development of a knowledge management system for a European government. A team of seven developers and one project manager was involved in the project. One developer worked in a different city, only attending meetings when required, a common scenario in virtual teams according to Lurey and Raisinghani (2001) and Cascio and Shurygailo (2003). Another developer spent a portion of his time in a different city and a portion of his time in the development centre. Three developers and the project manager spent less than 5% of their time in the development centre, but were situated on the same campus and available if required. One developer was permanently assigned to the development centre, although he did perform other duties outside of the project. Again, this hybrid type of virtual teams is common (Cascio & Shurygailo, 2002; Zigurs, 2003). The team, although not necessarily all together, were colocated briefly for meetings. Communication outside meetings took place via e-mail and instant messaging, again a common scenario in virtual teams (Cascio & Shurygailo, 2002; Paul et al., 2004).

The project manager suggested, and the team approved, the use of agile software development for this new project. Prior to this, a more formal approach was used, although a specific methodology was not followed (it should be noted though that Fitzgerald, 1997, doubts if formal methodologies are actually completely adhered to anyway). One of the new processes was the use of user stories, which involved the creation of user stories in place of traditional requirements documentation.

During the longitudinal study, over a period of 8 months, the authors noted that the initial commitment to the use of user stories decreased. This resisting force was manifested in sighs when user stories were mentioned, an unwillingness to take ownership of the stories, and the fact that they were not applied properly. Correctly implemented, user stories are applied throughout the project as opposed to just the design phase, so this necessitated the longitudinal aspect of the study. At the start of the development project, there appeared to be full commitment to the use of user stories but, as time passed, the commitment effectively disappeared. It is not unusual for traditional teams to resist change (although this could also be better described as diminishing commitment rather than resistance). It does appear unusual that a virtual team could change their beliefs and commitments concurrently and over a period of time. According to Boody, Boonstra, and Kennedy (2005, p. 210), "Staff working at a distance may not develop the beliefs and values held by those in close, daily, informal contact." People are less restrained in virtual teams (Johnson, Suriya, Yoon, Berrett, & LaFleur, 2002), so there should be less consensus seeking. Despite this, the virtual team did seem to be reducing their commitment in unison.

The investigation of this decrease in commitment involved the use of the competing-commitment theory, which, to the best of the author's knowledge, has not been applied before in the field of virtual teams. This theory was used as existing theories provided an inadequate explanation for the failure of the change to user stories.

The Inability of Existing Theories on Change to Completely Explain the Observations

Problems seen with the introduction of a new software process or methodology can often be explained by the diffusion-of-innovation theory, described in Rogers (1962), and the variations of this theory by authors such as Fichman and Kemerer (1999), Green, Collins, and Hevner (2004), Toleman, Ally, and Darroch (2004), Gallivan (2001, 2004), and Agarwal and Prasad (2000). Although diffusion of innovations often refers to the diffusion of technologies, researchers have used the theory (and variations of the theory) to describe the introduction of software processes, methodologies, languages, and so forth.

Fichman and Kemerer (1999) describe the assimilation gap between acquisition and deployment, referring to the illusory diffusion of innovation. Fitzgerald (1997) and Lee and Truex (2000), while not discussing diffusion theories, describe how the reality of methodology use is inconsistent with its prescribed use. This maps onto what was observed in the adoption of user stories in this case study. What was originally committed to in the use of user stories and what was actually adopted could be described as an assimilation gap. Further, the assimilation gap, in diffusion of innovations, increases over time. Again, this mirrors what was seen in the case study as the decrease in commitment to user stories widened the gap over time between what was required from user stories and what was actually adopted.

Technologies used in the software development process demonstrate assimilation gaps for two primary reasons: increasing returns and knowledge barriers. Fichman and Kemerer (1999) accept that there are other potential factors, which they list as structural, managerial, political, and social. They concentrate on increasing returns and knowledge barriers as they are not attempting to explain the assimilation gap for an individual instance, but explain why some instances demonstrate this gap and why the gap differs across technologies.

It was found in this research that the other potential factors, which Fichman and Kemerer (1999) do not concentrate on (such as managerial and social), did have a bearing on the failure to adopt user stories. This may be explained by the fact that this research concentrated on explaining a specific instance, while Fichman and Kemerer were generalising the phenomenon across multiple instances. An example of this would be Fichman and Kemerer's concentration on increasing returns, which describe the difference between the performance of an average adopter and the potential performance of a larger user base. This does not apply in this research case as user stories were used only for one instance; there was no intention to adopt universally.

A further possible explanation offered by Fichman and Kemerer (1999) is the problem of knowledge barriers. The adoption of innovations can be hindered by the learning required to successfully deploy the technology (or methodology). Orlikowski (1993) demonstrates how this can occur in the adoption of CASE tools. Lundell and Lang (2004) show the complexity of CASE tools, which would explain how these knowledge barriers arise. Again, though, this did not apply in this case study as user stories were initially implemented and used. The knowledge and skills required to adopt user stories were seen to be present. Future use of user stories did not require additional knowledge or skills.

Fichman and Kemerer (1999) list potential causes of resistance that could lead to an assimilation gap (a similar list is seen in Agarwal & Prasad, 2000). Although the causes listed (risk, deskilling, competence destroying, incompatibility) were not noted during the 8 months of observations, it is the concept of resistance that points toward a problem with change, as much as or more so than a problem with the innovation. Green et al. (2004), again discussing diffusion, describe how developers can resist what they perceive as threats to their autonomy while Agarwal and Prasad note the importance of a developer's attitude toward the innovation. As the developers initially accepted and applied user stories, it is not clear that user stories presented a threat to their autonomy. What is apparent is that change, and the resistance to change, could have had an impact. In fact, Mustonen-Ollila and Lyytinen (2003) noted with surprise that resistance to change receives little attention from decision makers in attempts to diffuse process innovation.

As change is a core characteristic of agile software development methodologies, the methodology itself causing, or partly causing, a decrease in commitment appears as a possibility. Fowler (2000) and Cohn and Ford (2003) describe how agile methods are suited to unpredictable environments with constant change, which accurately describes the change environment in this project. As requirements were seen to be prone to regular change and revision, the use of user stories was deemed an appropriate technique. Although the previous methodology experience of the developers was a more formalised system, the developers agreed and committed to using an agile method. There were no observed problems, or comments in interviews, with dealing with a constantly changing environment.

The fact that the team was a virtual team is seen by some authors as problematic in applying an agile approach. A single development site is recommended (Schwaber, 1996) to minimise communication requirements and to keep the team together, thus promoting teamwork. Reichheld (2001) lists one of the benefits of small, close-knit teams as the easily understood, and visible, ownership of tasks. Kalita (2003) demonstrates that colocated teams, although not a formal principal of the dynamic systems development method (DSDM), another agile method, should be used in DSDM projects. From this we can deduce that splitting teams across buildings or countries makes such ownership ill-defined at best. Maurer (2002), Reeves and Zhu (2004), and Stotts, Williams, Nagappan, Baheti, Jen, and Jackson (2003) disagree with the necessity for colocated teams, stating that virtual teams can work with the extreme-programming agile method through the use of appropriate tools. Furthermore, Sepulveda (2003) describes the success of remote teams with agile development. Virtual teams, therefore, should not be a major obstacle to the adoption of an agile process.

This again leads us back to resistance to change and its potential causes. Burnes (2000) lists the three schools of thought in the field of change management. The Individual perspective states that change is achieved through stimuli that influence human behaviour. There are two camps within this school: behaviourists and the Gestalt field. The behaviourists believe that change is achieved through reward. The Gestalt-field psychologists believe that individuals accept change through the use of reasoning. If an individual is helped to understand the need for change, then behaviour will change.

Neither explains the resistance to change seen in this development project. Rewards were present: the financial reward of successfully completing the project on time. The agile software development methodologies are aimed at successful and timely completion of a project, and user stories are part of these methodologies. Similarly, the use of reasoning does not appear to explain the paradox as user stories were explained, and their benefits described, after which the team accepted their usefulness. The diminished commitment was not an act of reasoning; the limited interaction of a virtual team, as described in Balthazard, Waldman, Howell, and Atwater (2004), would negate the possibility of the team reasoning as a group.

The open-systems school of thought examines change through organisations and their subsets. Any change in one subset will impact the others. It is necessary, therefore, to take a holistic view of the organisation when implementing change. Boody and Macbeth (2000) add subtly to this by stating that change in one area needs to be accompanied by (as opposed to causing) appropriate changes elsewhere. Hunt and Thomas (2000) refer to this as nonorthogonal systems. Birmingham (2002) describes these small changes affecting other areas as having a ripple effect or cascading effect. A small change in one area can affect the project in a seemingly independent area. Again, this does not adequately explain the paradox seen. User stories involve changes in other areas, such as how testing of the product is performed, but these changes were initially implemented in the project.

The group-dynamics school, of importance when discussing small teams, emphasises bringing about change through groups. As individuals work in groups, changes occur through changing the group's norms and practices. Initial considerations of this do not explain the paradox. The team was committed to the use of user stories, yet it was the team that failed to implement them. If the team was initially willing to adopt the use of user stories, then it implies that group norms and practices would not present an obstacle. In fact, Boody et al. (2005) argue that it is more difficult for these norms to develop in virtual teams. The term *initial considerations* is used now, in hindsight, as the use of the competing-commitments theory did show the relevance of the group-dynamics school of thought.

Kegan and Lahey (2001a, 2001b), as organisational psychologists (or adult development psychologists), propose that resistance to change, in these paradoxical situations, does not imply opposition; rather, it implies the existence of a competing commitment. Competing commitments are hidden in a person's mind, and are observed as energy being unwittingly applied against the commitment already made. In this research, a resistance to user stories was undermining the commitment to them. This resisting force was seen, ultimately, by the failure of the use of user stories, but also by the lack of application or desire to apply them. This resisting force is caused by commitments that act against the initial commitment to the user stories. It was necessary to determine what these competing commitments were, and what effect they had on the virtual team.

Competing-Commitments Theory

Abrahamsson and Jokela (2000), Abrahamsson and Iiavara (2002), Tavares-Rixon (2003), and Newman and Sabherwal (1996) describe commitments and how they are evaluated and reevaluated. While the negative effects of commitment, such as resistance, are discussed, what is not described is how commitments can work against each other in a hidden way, leading to a resistance to change. Kegan and Lahey (2001a, 2001b) prefer the term *immunity to change* in place of *resistance to change*. Resistance implies knowingly working against something; competing commitments are not obvious, even to the individual who has them. Banerjee (2003, p. 74) describes competing commitments as "self-defeating behaviour." These behaviours, even if subconscious, act against change. Competing commitments, also known as the big-assumptions theory, proposes a process through which the competing commitments that affect change can be identified. This process was originally proposed in Kegan and Lahey (2001a), and further discussed and demonstrated in Kegan and Lahey (2001b), Sparks (2002), Nash (2002), Bowe et al. (2003a, 2003b), and Banerjee. Competing commitments have some similarity with the view of Milgram (1971), who argues that public declarations of adherence to group decisions do not imply that the individual will translate this adherence into action. Kegan and Lahey's competing commitments describe the reasons why this initial acceptance is not acted upon. It should be pointed out, though, that Milgram's experiments showed that adherence can be translated into action.

Kegan and Lahey, who originally proposed the theory, do not restrict the domain of its application. While Nash (2002) and Bowe et al. (2003a, 2003b) apply the process in the field of medicine and medical education, and Banerjee (2003) applies it at the organisational level, this chapter applies it to the context of a virtual software development team. To the best of our knowledge, this process has not been applied in this field before. The focus of this chapter is specifically an examination of why the change to a new software development process, although initially supported by a virtual development team, never materialised.

The suitability of this approach is identified in Bowe et al. (2003b, p. 723), which describes the technique being used to examine why problems arose "during implementation when unanticipated or unaddressed organizational resistance surfaces." Nash (2002, p. 592) describes the use of the competing-commitments process to go beyond "buy-in." The same problems arose during the software development project being studied in this chapter. User stories were initially supported: There was buy-in. However, they were never fully implemented. Kegan, in an interview in Sparks (2002), describes what Nash refers to as buy-in as espousing commitment. Bowe et al. (2003a) described the problem as follows: "like many new years resolutions, sincere intent to change may be short lived and followed by a return to old behaviours." Again, the analogy of New Year's resolutions applies in this project. At the beginning of the project, the virtual team felt that some of the problems with a previous project highlighted the need for more processes, or new processes, within the project. Although the previous project was successful, the developers were able to acknowledge failings that they would like to overcome in the new projects. This in itself is noteworthy as most teams and individuals find it hard to acknowledge their own faults, as described by cognitive dissonance theory (Festinger & Carlsmith, as cited in Coats & Feldman, 2001; Harmon-Jones, as cited in Coats & Feldman; Schelling, 1989; Weinberg, 1971) and self-justification theory (Keil et al., 2000). The success of the previous project may explain part of the problem, though not all. Arrow, McGrath, and Berdahl (2000) highlight the seemingly contradictory theory that successful teams are more problematic when it comes to change. A team that has failed in the past is more likely to adapt new responses to change.

THE PROCESS OF UNCOVERING COMPETING COMMITMENTS

Before examining the competing-commitments methodology used in this case, it is necessary to justify the initial use of participant observation in this study, which both highlighted the need for the use of competing commitments and subsequently provided further data to explain the competing commitments.

The Use of Participant Observation

Participant observation was chosen as the method for examining the failure to adopt user stories. Jorgensen (1989, p. 12) argues that participant observation is relevant where "the phenomenon is obscured from the view of outsiders." Asch (1952) goes further than this by arguing that observations of groups can discover areas that even group members themselves are unaware of. Competing commitments are an effect that an individual, or group of individuals, are unaware of.

The development team was observed over an 8-month period. Both researchers had roles on the team. In longitudinal research, observations lasting weeks or months allow the researcher to develop a relationship with those being studied that goes beyond a superficial short-term relationship (Fetterman, 1991; Gurney, 1991). Describing those performing research as participant observers, Burgess (1982, p. 45) argues that their activities involve sharing "in the lives and activities of those whom they study and take roles which are effective in the setting under study." For Ezey (2003), integration into the group under study is vital in participant observation. According to Spradley (1980, p. 51), "Participation allows you to experience activities directly to get a feel of what events are like, and to record your own perceptions." Gephart (1988), for example, used participant observation to examine the informal structures of an organisation.

As in this study, others have used participatory observation to investigating agile software development projects. Martin, Biddle, and Noble (2004) argue that interpretative in-depth case studies are the best method of investigating agile software development. Robinson and Sharp (2004) use a qualitative approach in their investigation of the characteristics of an agile team involving participant observation. Koch (2005a) argues that observation allows a researcher of e-collaboration access to more data than would be available using surveys or experiments (it should be noted that Koch was discussing action research rather than participant observation, yet the argument still holds true). Despite these advantages, research though observation is still the minority in e-collaboration research.

Examining a specific case provided insights that may not be visible in other research methods. Hartley (1984) and Traylen (1994) make an argument for case studies in general. A benefit of the trust gained by the researcher is the ability to examine areas that may be secret or concealed. For Schwartzman (1993, p. 4) it is that which is "taken for granted" that is worth observing, for it is the seemingly trivia of daily work that influence organisations. It is this everyday life that provides a description of reality (Jorgensen, 1989).

It should be noted, though, that the main process in this research was the competing-commitments process. Participant observation highlighted a problem and helped to further refine the outcome of the competing-commitment process, but it was a secondary source of data.

Competing Commitments Applied

Traditional interviews and questionnaires are inadequate when determining competing commitments. These competing commitments are often hidden even from those having the commitments. It was necessary to use a technique that would extract these competing commitments. The word *extract* is deliberately used here as it involves delving deeply into an individual's personal and private beliefs.

Kegan and Lahey (2001a, 2001b) developed a technique that is used to determine competing commitments. Various authors describe this technique (Bowe et al., 2003a, 2003b; Nash, 2002; Sparks, 2002). The technique comprises six steps in the form of questions, although different authors merge some steps. The examples used below are those given as examples in Kegan and Lahey (2001b).

In the examples below (examples of the technique provided by Kegan & Lahey, 2001b), the questions asked by the interviewer are listed on the left-hand side with the interviewee's responses on the right. The responses are fed into the subsequent question. (As an example, the interviewee is asked to describe a complaint about the work. The response is that the team does not provide enough information. Based on that response, the interviewee is asked what this complaint implies about him or her—what he or she is committed to.) At the end of this process, we have moved from a complaint about a team not keeping the individual in the loop to the big assumption that people will think the person is incompetent if he or she cannot do everything. The individual states a commitment to full communication, yet the competing commitment—not learning about things that cannot be controlled—effectively works against the commitment to full communication. The technique, as described in the examples below, was applied in the virtual-team case study. Obviously, the answers and subsequent questions based on these answers followed the same format as above, but differed in their content.

OBSERVATIONS AND ANALYSIS

Observation and analysis involved two subphases. Phase 1 involved discovering the competing commitments of the developers in the virtual team. This provided the cause of the failure to change, but did not explain the *why*. Phase 2 took these competing commitments and explained why they caused failure to change by using data from the longitudinal study of the virtual team's change to an agile methodology. Phase 2 also required a review of existing literature on groupthink and virtual teams to guide the analysis.

Table 1. Determining competing commitments

Step	Question	Example Response
1	What problem are you experiencing in work: a gripe or complaint?	My team does not tell me what's happening in a project.
2	The complaint identifies something about you. What commitment does it imply?	I am committed to maximising the flow of information within the project.
3	What are you doing or not doing that goes against this commitment?	Sometimes I don't go out of my way to find out what is happening.
4	What do you think would happen if you were not doing what you described in Question 3—if you did the opposite of the undermining behaviour? What would worry you about this?	I might find out things from my team that I can do nothing about, or something I can't fix.
5	What does this worry imply that you are committed to?	I am committed to not learning about things I can't control.
6	Inverting the answer from Step 5, and making it into the beginning of an assumption, complete the sentence: I assume that if I...	I assume that if I learned about things I couldn't control, people would realise that I am not able to do my job.

Phase 1: Competing Commitments Identified

At the conclusion of the field observations, which lasted 8 months (approximately 10 hours of observation per week), five developers were interviewed out of the team of seven. Each interview lasted approximately 2 hours. One problem with a virtual team is availability for interview. Although this is a small group, the argument for its use is that the proposition of this research is the effect that virtual teams, which are predominately small, have on change.

Each developer went through the interview process and discovered two competing commitments. Although the process aims to find one competing commitment, each developer felt that two areas were important. Out of the five interviewed, five had a competing commitment directly related to the team. Each felt that a good relationship with their coworkers was vital, and that they were committed to ensuring that this relationship was sustained. This is contrary to existing research on virtual teams that suggest that intrateam relations is not of prominence. Johnson et al. (2002) illustrate this with a quote from a virtual team member who stated that fellow team members were not known well enough to develop a trusting relationship with them. The team's commitments expressed are listed in Table 2.

In terms of the interviewees' second competing commitment (second in terms of display; it is not meant to imply a level of importance), two of these had relevance to relationships within the team. One was specific about avoiding conflict within the development team, while the other concerned recognition from the project manager. Each of the original problems that the developers had with user stories, and their related original commitments, were unique to the developer. For each of these to evolve through the questions asked in the competing-commitments process into the importance of team relationships does show its relevance. It should be pointed out, though, that although developers accepted that it was important to them, it was not always their ultimate commitment. This was illustrated when the developers were asked to comment on their competing commitments.

- "It plays a part but is not the major one."
- "They are valid but not complete."
- "They are not commitments; they are more like traits."

It was highlighted to the developers that these were their commitments based on their problem with user stories and not their commitments regarding other areas of their work. The developers accepted this. The fact that intrateam relationships were not the ultimate competing commitments but still important commitments is summed up by one developer who stated that even if the major problems were sorted out, influences like intrateam relationships could still cause user stories to fail.

Table 2. The developers' competing commitments

Developer	Competing Commitment 1	Competing Commitment 2
1	A good relationship with coworkers	Results are more important than process
2	To be accepted as part of the team	Do not want to be associated with a failed project
3	Recognition from team and project manager	The path of least resistance
4	Respect from the team	Avoiding conflict
5	Being part of the team	Ensuring the boss sees the work I am doing

Phase 2: Competing Commitments in Context

From the previous phase, we can see that the commitment to the team was having an effect, even if at times subconscious, on the change to the use of user stories. What was still not clear was how the social relationships within a virtual team had impacted the change. A review of the research literature, however, revealed a possible explanation.

The importance of group relationships pointed to the possibility of groupthink having an impact on the issues at hand. One of the main symptoms of groupthink (described in Cartwright, 2002; Griffin, 1997; Janis, 1972; Moorhead, Neck, & West, 1991) is a pressure to conform to the group's views. This appeared to be relevant to this study as initially there was group agreement on the benefit of changing the process to use user stories. Over time, there was group consensus that they were not beneficial. The group's opinion changed as a group. Contrary to this is the proposition that the ultimate factor in groupthink is a highly cohesive team (Griffin; Janis; Kim, 2001; Martin, 1991; Neck, 1996). Kim (p. 175) states, "Group cohesiveness refers to the degree to which members of the group desire to remain in the group." Martin refers to cohesiveness as the forces between group members that keep them as a unit, cohesiveness being generally regarded as highly beneficial. Von Bergen and Kirk (1978) use the terms *esprit de corps* and *close-knit* while describing group cohesiveness. Jones and Roelofsma (2000) simplify their definition to that of a group knowing and liking each other, while Baron, Kerr, and Miller (1999, p. 8) refer to cohesion as "the overall strength of positive relationships within the group." Even the proximity of the team can increase cohesiveness (Sunstrom, DeMeuse, & Futrell, 1990). All of these descriptions of group cohesiveness are contrary to descriptions of virtual teams. Virtual teams face problems in building and maintaining intrateam relationships, thus cohesiveness and trust are difficult to achieve (Balthazard et al., 2004b; Powell et al., 2004; Zigurs, 2003). Added to this, a lack of cues in virtual teams can ultimately create problems for members in relating to each other (Balthazard et al.; Beise, 2004; Cascio & Shurygailo, 2002; Erickson, Halveron, Kellog, Laff, & Wolf, 2002; Zigurs). This appeared to contradict the conclusion that groupthink was causing the decrease in commitment to user stories by the development team. It should be highlighted, though, that existing research on virtual teams discusses cohesion in virtual teams, but it does not discuss groupthink (although the two are linked in other literature). Furst, Blackburn, and Rosen (1999) list their proposed research agenda for research into virtual teams. Groupthink is mentioned, but it is not made clear how, why, or if it affects virtual teams. Furst et al. are simply noting it as an area for possible research. Balthazard, Potter, and Warren (2004) propose that it should be possible to improve virtual teams by differentiating low- and high-performance face-to-face teams. Although group cohesion is discussed, groupthink is not mentioned in the research. Based on this analysis, groupthink cannot be discounted as being a problem in the virtual team studied.

There was still an incomplete explanation of why the change to user stories was not effective. Groupthink appeared to be having an effect, yet it was unclear why. A study of field notes made during the 8-month observation of the project provides an explanation.

The project manager was liked and respected by the team. Unlike some project managers, the project manager in this case was regarded as part of the development team as opposed to an outsider. Several developers specifically stated this. The majority of the team had worked under the project manager on a previous successful project. The project manager suggested the change from traditional development methods to agile software development, of which user stories are a major part. At this stage, the development team agreed to the change and was committed to the use of user stories. Over time, the project manager himself allowed the use and application of user stories to be diluted. The project management approach, or style, was against the rigid interpretation of the method of writing and using user stories, and tended to apply a less rigid application. One of the reasons that the project management approach was liked and respected was the manager's willingness to let the developers "get on with it," avoiding anything that was perceived to slow them down. An example of this dilution of the new methodology was seen during the prioritisation of user stories. One of the core concepts of user stories is that each story must be prioritised and ranked numerically, the most important story being developed first, and so on. The project manager was against this and argued for a rating of high, medium, and low. Further examples of the dilution of the application of user stories occurred, such as customer involvement in the stories being minimal. Concurrent with the project manager's dilution of the application of stories, the commitment from the developers to the stories diminished. As the project manager in a virtual team occupies such a core position due to the remoteness of others, his or her influence could be postulated to be greater than in a face-to-face team.

This matches what is described as hierarchical groupthink as opposed to the commonly described peer groupthink. Maoz (1990) and Wright and Schaal (1988) argue that groupthink originates from a desire for concurrence with the group's ideas or the leader's ideas. Cartwright (2002) specifically differentiates, and names, two types of groupthink. Peer groupthink originates in a need for conformity and close integration within a team. Huczynski and Buchanan (1991) argue that synergy and loyalty, which are regarded as a team's greatest benefits, are the same factors that lead to groupthink. Hierarchical groupthink originates in a desire to please a leader, specifically the desire not to disagree with him or her. It is similar to approval-seeking behaviour as found in Lippit, Polansky, Redl, and Rosen (1968) and ingratiation through conformance with the leader's view described in Jones, Gergen, Gumpert, and Thibaut (1968) and Hurwitz, Zander, and Hymovitch (1968). Kohl (1975) presents the argument that the dominance of Kissinger in Nixon's cabinet could have lead to groupthink as his opinion tended to be sacrosanct. Neck (1996) lists leader preference for a certain option as a potential factor in groupthink.

As the project manager changed from a firm advocate of user stories to a diluted advocate of user stories, commitment among the virtual team's developers decreased. This appears to be an example of hierarchical groupthink: a desire to please the leader. Returning to the developers' competing commitments, this hierarchical groupthink can be seen in the competing commitments of two developers who specifically stated that one of their competing commitments was a desire for recognition from the project manager. The other competing commitments, involving good relationships with the team, also have relevance because, as already noted, the project manager was regarded as part of the team. Another developer's competing commitment was that results are more important than the process. This is practically verbalising the reason for the project manager's dilution of the user stories' process. Hierarchical groupthink is therefore seen to have an impact on the ineffective change to the use of user stories. The role of the project manager as a bureaucracy buster, while beneficial in many aspects of the project (especially in a virtual team), eventually assisted in the failure of the change to a new methodology.

Abrahamsson and Jokela (2000) and Tavares-Rixon (2003) refer to three forms of commitment when discussing software process improvement (SPI). Affective commitment is attachment and involvement with the entity (organisation, SPI), continuance commitment deals with the costs involved in leaving the entity (i.e., cancelling the project), and normative commitment is a sense of duty in continuing association with the entity (i.e., the SPI project). The developers in this case had expressed affective commitment: the desire to use user stories. Working against that was a competing commitment—a normative commitment. This was the desire to continue membership with the project manager (a social or organisational pressure from the project manager).

CONCLUSION AND RECOMMENDATIONS FOR FUTURE RESEARCH

Before examining the conclusions to this research, some limitations of the study need to be addressed. Although diffusion-of-innovation theories did not originally appear to have an impact on this study, it should be noted that elements of the theories may have had an impact, and can therefore not be discounted. Green et al. (2004) describe the results of their work, which suggested that the less voluntary the use of a software process is, the more likely it is to be used. Direction from management is required for the process innovation to be used. In this case, the inverse occurred: The project manager indirectly directed the team away from the use of user stories, which ultimately caused to team to drop its commitment to its use. Although this is the inverse of Green et al.'s conclusion, their work does still appear to have a bearing on this case.

It is accepted that it is difficult to generalise the findings of one case study across the entire IS field. The authors accept this, and it was not their aim to do so. This research highlights a problem in one case that has the potential to cause problems elsewhere. It should also be noted that all developers were of the same nationality, and therefore culture may have played a part. It is unclear if this problem of groupthink in virtual teams would affect cross-cultural initiatives. The impact of culture could not be determined as a similar culture between the observer and observed is recommended for participant observation (Jorgensen, 1989; Tedlock, 2000). Other researchers from different cultures would be best placed to examine if the findings of this research can be replicated in other cultures and virtual teams.

Powell et al. (2004) regard the technological support of virtual teams as being in place, and propose that research should concentrate on the issues of managing them, as virtual teams are generally recognised as less effective than traditional face-to-face teams (Johnson et al., 2002). Leadership is as vital in virtual teams as in face-to-face teams (Cascio & Shurygailo, 2002), yet little is known about leadership in virtual teams (Zigurs, 2003). To date, there has been limited research into how the idiosyncrasies of virtual teams are affected by, or affect, their team leaders (project manager in this research). Piccoli and Ives (2000) argue that the lack of a socialization process adds to the difficulties of managers of virtual teams. Zigurs and Balthazard, Waldman, et al. (2004) elaborate on this by proposing that a lack of leadership presence makes leadership more difficult. This research casts doubts on these assertions. While considerable research on virtual teams argues that, for effective leadership of virtual teams, a manager should concentrate on increasing the social interaction and developing cohesion and trust, this chapter proposes that caution is required in this objective. This study also found that traditional explanations of change, and the management of change, did not explain the phenomena observed in this particular case study.

This chapter does not claim that the role of the project manager of a virtual team is not to create cohesion in the virtual team. Nor does it imply that a project manager who is respected by the team, and central to its functioning, will cause problems in a virtual team. The longitudinal aspect of the overall research showed the benefits that an empowering and supportive style of project management brings. What is clear, though, is that striving for cohesion in a virtual team can be counterproductive. The two main findings of this chapter are the following:

- The project management style, one that the developers themselves preferred, was found to negatively impact the change to a new process. This particular style of project management, one that fosters cohesion, is often mentioned as favourable to virtual teams: "An effective leader of a virtual team needs to be more flexible and willing to let others take the lead where necessary" (Powell et al., 2004, p. 18).
- The level of cohesion found in the virtual team was so high (usually regarded as a goal for a virtual team) that it ultimately assisted in the failure to change. The cohesiveness of the team, and the groupthink that evolved from it, was a factor in the failure to change to a new methodology.

To summarise, two desirable aspects of any virtual team—a cohesive team, and a respected project manager central to the team—were factors in the failure of a change initiative. This would go against traditional views and findings in existing research into virtual teams and their project management.

It is possible that a more considered view of the adoption of user stories would have avoided the problems seen. The win-win model (described in Boehm & Egyed, 1998; Boehm, Egyed, Shah, Kwan, & Madachy, 1998; Boehm & Port, 2001) may have put a structure on the decision to adopt user stories. Using a model such as MBASE (Boehm & Port) may have provided this structure. Boehm and Egyed describe the win-win negotiation mode with its four artefact types: win conditions, issues, options, and agreements. In this case, the debate was noncontroversial with immediate agreement. Perhaps some debate, which would have given rise to issues and conditions, would have avoided the decrease in commitment that was seen in this case study. On the other hand, the fact that the team was a virtual team may have caused further problems when using an approach such as this.

Although previously not used in the study of virtual teams, the use of competing commitments as a research approach has its advantages. Problems with virtual teams have been well reported so, if we accept the human impact on these problems, competing commitments could be used as another research tool to determine why virtual teams are viewed as less productive than traditional teams.

REFERENCES

Abrahamsson, P., & Iiavara, N. (2002). Commitment in software process improvement: In search of the process. In *Proceedings of the 35th Annual Hawaii International Conference on System Sciences,* 260b.

Abrahamsson, P., & Jokela, T. (2000). Development of management commitment to software process improvement. In *Proceedings of Information Systems Research, 23,* 487-505.

Agarwal, R., & Prasad, J. (2000). A field study of the adoption of software process innovations by information systems professionals. *IEEE Transactions on Engineering Management, 47*(3), 295-308.

Alsene, E. (1999). Internal changes and project management structures within enterprises. *International Journal of Project Management, 17*(6), 367-376.

Arrow, H., McGrath, J., & Berdahl, J. (2000). *Small groups as complex systems.* Thousand Oaks, CA: Sage Publications.

Asch, S. (1952). *Social psychology.* Upper Saddle River, NJ: Prentice-Hall.

Balthazard, P., Potter, R., & Warren, J. (2004). Expertise, extraversion and group interaction styles as performance indicators in virtual teams. *The Database for Advances in Information Systems, 35*(1), 41-64.

Balthazard, P., Waldman, D., Howell, J., & Atwater, L. (2004). Shared leadership and group interaction styles in problem-solving virtual teams. In *Proceedings of the 37th Hawaii International Conference on System Sciences*, 10043a.

Banerjee, R. (2003). Organisational character: Issues, imperatives and practices. *International Journal of Human Resources Development and Management, 3*(1), 72-83.

Baron, R., Kerr, N., & Miller, N. (1999). *Group process, group decision, group action.* Buckingham, UK: Open University Press.

Beck, K. (1999). Embracing change with extreme programming. *IEEE Computer, 32*(10), 70-77.

Beck, K. (2000). *Extreme programming explained: Embrace change.* Boston: Addison Wesley.

Beise, C. (2004). IT project management and virtual teams. *Proceedings of the 2004 SIGMIS Conference on Computer Personnel Research*, 129-133.

Beynon-Davies, P. (1995). Information systems "failure": The case of the London Ambulance Service's Computer Aided Despatch Project. *European Journal of Information Systems, 4*(3), 171-184.

Birmingham, D. (2002). *Software development on a leash.* New York: Springer-Verlag.

Boehm, B., & Egyed, A. (1998). Software requirements negotiation: Some lessons learned. In *Proceedings of the 20th International Conference on Software Engineering*, 503-506.

Boehm, B., Egyed, A., Shah, A., Kwan, J., & Madachy, R. (1998). A stakeholder win-win approach to software education. *Annals of Software Engineering, 6*, 295-321.

Boehm, B., & Port, D. (2001). Balancing discipline and flexibility with the spiral model and MBASE. *Crosstalk, 14*(12).

Boody, D., Boonstra, A., & Kennedy, G. (2005). *Managing information systems: An organisational perspective.* Harlow, UK: Pearson Education.

Boody, D., & Macbeth, D. (2000). Prescriptions for managing change: A survey of their effects in projects to implement collaborative working between organizations. *International Journal of Project Management, 18*(5), 297-306.

Bowe, C., Lahey, L., Kegan, R., & Armstrong, E. (2003a). Questioning the big assumptions: Part I. Addressing personal contradictions that impede professional development. *Medical Education, 37*(8), 715-722.

Bowe, C., Lahey, L., Kegan, R., & Armstrong, E. (2003b). Questioning the big assumptions: Part II. Recognizing organizational contradictions that impede organisational change. *Medical Education, 37*(9), 723-733.

Burgess, R. (1982). Some role problems in field research. In R. Burgess (Ed.), *Field research: A sourcebook and field manual* (pp. 45-49). London: George Allen & Unwin Publishers.

Burnes, B. (2000). *Managing change: A strategic approach to organisational dynamics.* Harlow, UK: Pearson Education Ltd.

Cartwright, R. (2002). *Mastering team leadership.* Wales, UK: Palgrave Macmillan.

Cascio, W., & Shurygailo, S. (2002). E-leadership and virtual teams. *Organizational Dynamics, 31*(4), 362-376.

Casey, V., & Richardson, I. (2005). Uncovering the reality within virtual software teams. In *Proceedings of the 2006 International Workshop on Global Software Development for the Practitioner*, 66-72.

Clarke, A. (1999). A practical use of key success factors to improve the effectiveness of project management. *International Journal of Project Management, 17*(3), 139-145.

Coats, E., & Feldman, R. (2001). *Classic and contemporary readings in social psychology* (3rd ed.). Upper Saddle River, NJ: Pearson Education.

Cohn, M., & Ford, D. (2003). Introducing an agile process to an organisation. *IEEE Computer, 36*(6), 74-78.

Cushway, B., & Lodge, D. (1999). *Organization behaviour and design.* London: Kogan Page Ltd.

De Pillis, E., & Furumo, K. (2006). Virtual vs. face-to-face teams: Deadbeats, deserters, and other considerations. In *Proceedings of the 2006 ACM SIGMIS CPR Conference on Computer Personnel Research*, 318-320.

Duck, J. (1993). Managing change: The art of balancing. *Harvard Business Review, 71*(6), 109-118.

Erickson, T., Halveron, C., Kellog, W., Laff, M., & Wolf, T. (2002). Social translucence: Designing social infrastructures than make collective activity visible. *Communications of the ACM, 45*(4), 40-44.

Ezey, P. (2003). Integration and its challenges in participant observation. *Qualitative Research, 3*(2), 191-205.

Fetterman, D. (1991). A walk through the wilderness: Learning to find your way. In W. Shaffir & R. Stebbins (Eds.), *Experiencing fieldwork: An inside view of qualitative research* (pp. 87-96). Thousand Oaks, CA: Sage Publications.

Fichman, R., & Kemerer, C. (1999). The illusory diffusion of innovation: An examination of assimilation gaps. *Information Systems Research, 10*(3), 255-275.

Field, M., & Keller, L. (1998). *Project management.* London: Thomson Learning.

Fitzgerald, B. (1997). The use of systems development methodologies in practice: A field study. *Information Systems Journal, 7*(3), 210-212.

Fowler, M. (2000). Put your process on a diet. *Software Development, 8*(2), 32-36.

Fowler, M., & Highsmith, J. (2001). The agile manifesto. *Software Development, 9*(8), 28-32.

Furst, S., Blackburn, R., & Rosen, B. (1999). Virtual team effectiveness: A proposed research agenda. *Information Systems Journal, 9*(4), 249-269.

Gallivan, M. (2001). Organizational adoption and assimilation of complex technological innovations: Development and application of a new framework. *The Database for Advances in Information Systems, 32*(3), 51-85.

Gallivan, M. (2004). Examining IT professionals' adaptation to technological change: The influence of gender and personal attributes. *The Database for Advances in Information Systems, 35*(3), 28-49.

Gephart, R. (1988). Status degradation and organizational succession. In J. Van Maanen (Ed.), *Qualitative studies of organizations* (pp. 159-191). Thousand Oaks, CA: Sage Publications.

Green, G., Collins, R., & Hevner, A. (2004). Perceived control and diffusion of software process innovations. *Journal of High Technology Management, 15*(1), 123-144.

Griffin, E. (1997). *A first look at communication theory.* New York: McGraw-Hill.

Gurney, J. (1991). Female research in male-dominated settings: Implications for short-term versus long-term research. In W. Shaffir & R. Stebbins (Eds.), *Experiencing fieldwork: An inside view of qualitative research* (pp. 53-61). Thousand Oaks, CA: Sage Publications.

Hartley, J. (1984). Case studies in organizational research. In C. Cassell & G. Symon (Eds.), *Qualitative methods in organizational research* (pp. 208-229). London: Sage Publications.

Huczynski, A., & Buchanan, D. (1991). *Organizational behaviour: An introductory text* (2nd ed.). New York: Prentice Hall.

Humphrey, S. (1989). *Managing the software process.* New York: Addison-Wesley.

Hunt, A., & Thomas, D. (2000). *The pragmatic programmer.* Boston: Addison Wesley Longman.

Hurwitz, J., Zander, A., & Hymovitch, B. (1968). Some effects of power on the relations among group members. In D. Cartwright & A. Zander (Eds.), *Group dynamics* (pp. 291-297). New York: Harper & Row Publishers.

Janis, I. (1972). *Victims of groupthink.* Boston: Houghton Mifflin Company.

Johnson, S., Suriya, C., Yoon, S., Berrett, J., & LaFleur, J. (2002). Team development and group processes of virtual learning teams. *Computers & Education, 39*, 379-393.

Jones, E., Gergen, K., Gumpert, P., & Thibaut, J. (1968). Some conditions affecting the use of ingratiation to influence performance evaluation. In D. Cartwright & A. Zander (Eds.), *Group dynamics* (pp. 278-289). New York: Harper & Row Publishers.

Jones, P., & Roelofsma, P. (2000). The potential for social contextual and group biases in team decision making: Biases, conditions and psychological mechanisms. *Ergonomics, 43*(8), 1129-1152.

Jorgensen, D. (1989). *Participant observation: A methodology for human studies.* Thousand Oaks, CA: Sage Publications.

Kalita, T. (2003). DSDM in process improvement. In J. Stapleton (Ed.), *DSDM business focused development* (2nd ed., pp. 175-191). London: Pearson Education.

Kegan, R., & Lahey, L. (2001a). *How the way we talk can change the way we work.* CA: Jossey-Bass.

Kegan, R., & Lahey, L. (2001b). The real reason people won't change. *Harvard Business Review, 79*(10), 85-89.

Keil, M., Mann, J., & Rai, A. (2000). Why software projects escalate: An empirical analysis and test of four theoretical models. *MIS Quarterly, 24*(4), 631-660.

Kim, Y. (2001). A comparative study of the "Abilene paradox" and "groupthink." *Public Administration Quarterly, 25*(2), 168-187.

Koch, N. (2005a). Using action research to study e-collaboration. *International Journal of e-Collaboration, 1*(4), i-vii.

Koch, N. (2005b). What is e-collaboration? *International Journal of e-Collaboration, 1*(1), i-vii.

Kohl, W. (1975). The Nixon-Kissinger foreign policy system and US-European relations: Patterns of policy making. *World Politics, 28*(1), 1-43.

Labaree, R. (2002). The risk of "going observationalist": Negotiating the hidden dilemmas of being an insider participant observer. *Qualitative Research, 2*(1), 97-122.

LaFleur, R. (1996). Project management: Getting control and keeping control of complex projects. *American Programmer, 9*(4), 24-28.

Lawrence, B. (1996). Unresolved ambiguity: The silent source of risk in your project. *American Programmer, 9*(4), 18-22.

Lee, J., & Truex, D. (2000). Exploring the impact of formal training in ISD methods on the cognitive structure of novice information systems developers. *Information Systems Journal, 10*(4), 347-367.

Lindvall, M., Basili, V., Boehm, B., Costa, P., Dangle, K., Shull, F., et al. (2002). Empirical findings in agile methods. *Extreme Programming and Agile Methods: XP/Agile Universe Chicago 2002*, 197-207.

Lippit, R., Polansky, N., Redl, D., & Rosen, S. (1968). The dynamics of power. In D. Cartwright & A. Zander (Eds.), *Group dynamics* (pp. 236-250). New York: Harper & Row Publishers.

Lundell, B., & Lings, B. (2004). Changing perceptions of CASE technology. *The Journal of Systems and Software, 72*(2), 271-280.

Lurey, J., & Raisinghani, M. (2001). An empirical study of best practices in virtual teams. *Information & Management, 38*(8), 523-544.

Maoz, Z. (1990). Framing the national interest: The manipulation of foreign policy decisions in group settings. *World Politics, 43*(1), 77-110.

Mark, G. (1998). Building virtual teams: Perspectives on communication, flexibility and trust. *ACM SIGGROUP Bulletin, 19*(3), 38-41.

Martin, A., Biddle, R., & Noble, J. (2004). When XP met outsourcing. *Extreme Programming and Agile Processes in Software Engineering: 5th International Conference*, 51-59.

Martin, R. (1991). Working in groups. In M. Smith (Ed.), *Analysing organisational behaviour* (pp. 154-177). London: Macmillan Press.

Maskell, B. (2001). The age of agile manufacturing. *Supply Change Management, 6*(1), 5-11.

Maurer, F. (2002). Supporting distributed extreme programming. *Extreme Programming and Agile Methods: XP/Agile Universe Chicago 2002*, 13-22.

McConnell, S. (2001). Managing outsourced projects: Project management inside out. In L. Constantine (Ed.), *Beyond chaos: The expert edge in managing software development* (pp. 103-110). NJ: Addison-Wesley.

McElroy, W. (1996). Implementing strategic change through projects. *International Journal of Project Management, 14*(6), 325-329.

Metzger, P. (1981). *Managing a programming project.* New York: Prentice Hall Inc.

Milgram, S. (1971). Group pressure and action against a person. In D. Taylor (Ed.), *Small groups* (pp. 157-170). Chicago: Markham Publishing Company.

Moorhead, G., Neck, C., & West, M. (1991). The tendency towards defective decision making with self-managing teams: The relevance of groupthink. *Organizational Behaviour & Human Decision Process, 73*(2), 327-351.

Mustonen-Ollila, & Lyytinen, K. (2003). Why organizations adopt information system process innovations: A longitudinal study using diffusion of innovation theory. *Information Systems Journal, 13*(3), 275-297.

Nader, D. (1993). Concepts for the management of organisational change. In C. Mabey & B. Major-White (Eds.), *Managing change* (2nd ed., pp. 5-19). London: Paul Chapman Publishing.

Nandhakumar, J., & Jones, M. (2002). Development gain? Participant observation in interpretive management information systems research. *Qualitative Research, 2*(3), 323-341.

Nash, D. (2002). Beyond buy-in. *P&T Journal, 27*(11), 617-618.

Neck, C. (1996). Letterman or Leno: A groupthink analysis of successive decisions made by the national broadcasting company. *Journal of Managerial Psychology, 11*(8), 3-17.

Newman, M., & Sabherwal, R. (1996). Determinants of commitment to information systems development: A longitudinal investigation. *MIS Quarterly, 20*(1), 23-54.

Orlikowski, W. (1993). CASE tools as organizational change: Investigating incremental and radical changes in systems development. *MIS Quarterly, 17*(3), 309-340.

Pellegrinelli, S. (1997). Programme management: Organising project-based change. *International Journal of Project Management, 15*(3), 141-149.

Paul, S., Seetharamann, P., Samarah, I., & Mykytyn, P. (2004). Impact of heterogeneity and collaborative conflict management style on the performance of synchronous global virtual teams. *Information & Management, 41*(3), 303-321.

Piccoli, G., & Ives, B. (2000). Virtual teams: Managerial behaviour control's impact on team effectiveness. *Proceedings of the 21st International Conference on Information Systems*, 575-580.

Potter, R., & Balthazard, P. (2002). Virtual team interaction styles: Assessment and effects. *International Journal of Human-Computer Studies, 56*(4), 423-443.

Powell, A., Picolli, G., & Ives, B. (2004). Virtual teams: A review of current literature and directions for future research. *The Database for Advances in Information Systems, 35*(1), 6-36.

Rainwater, J. (2002). *Herding cats: A primer for programmers who lead programmers.* New York: Springer-Verlag.

Reeves, M., & Zhu, J. (2004). Moomba: A collaborative environment for supporting distributed extreme programming in global software development. *Extreme Programming and Agile Processes in Software Engineering: 5th International Conference*, 38-50.

Reichheld, F. (2001). Lead for loyalty. *Harvard Business Review, 79*(7), 76-84.

Robbins, H., & Finley, M. (1998). *Why change doesn't work.* London: Orion Publishing.

Robinson, H., & Sharp, H. (2004). The characteristics of XP teams. *Extreme Programming and Agile Processes in Software Engineering: 5th International Conference*, 139-147.

Rogers, E. (1962). *Diffusion of innovations.* New York: The Free Press.

Schelling, T. (1989). The mind as a consuming organ. In D. Bell, H. Raiffa, & A. Tversty (Eds.), *Decision making: Descriptive, normative, and prescriptive interactions* (pp. 343-357). Cambridge, UK: Cambridge University Press.

Schwaber, K. (1996). Controlled chaos: Living on the edge. *American Programmer, 9*(4), 11-16.

Schwartzman, H. (1993). *Ethnography in organizations.* Thousand Oaks, CA: Sage Publications.

Sepulveda, C. (2003). Agile development and remote teams: Learning to love the phone. *Proceedings of the Agile Development Conference*, 140-145.

Siegel, A., & Siegel, S. (1968). Reference groups, membership groups, and attitude change. In D. Cartwright & A. Zander (Eds.), *Group dynamics* (pp. 74-79). New York: Harper & Row Publishers.

Sparks, D. (2002). Inner conflicts, inner strengths. *Journal of Staff Development, 23*(3), 66-71.

Spradley, J. (1980). *Participant observation.* New York: Holt, Rinehart, & Winston.

Stotts, D., Williams, L., Nagappan, N., Baheti, P., Jen, D., & Jackson, A. (2003). Virtual teaming: Experiments and experiences with distributed pair programming. In F. Maurer & D. Wells (Eds.), *Extreme Programming and Agile Methods: XP/Agile Universe 2003*, 129-141.

Sunstrom, E., DeMeuse, K., & Futrell, D. (1990). Work teams: Applications and effectiveness. *American Psychologist, 45*(2), 120-133.

Tavares-Rixon, D. (2003). Exploring the nature of commitment to software quality initiatives: A political perspective. *Proceedings of the First Australian Undergraduate Students Computing Conference*, 87-92.

Tedlock, B. (2000). Ethnography and ethnographic presentation, In N. Denzin & Y. Lincoln (Eds.), *Handbook of qualitative research* (2nd ed., pp. 455-486). Thousand Oaks, CA: Sage Publications.

Thomsett, R. (1998). The care and feeding of project managers. *American Programmer, 11*(2), 5-11.

Toleman, M., Ally, M., & Darroch, F. (2004). Aligning adoption theory with agile system development methodologies. *Proceedings of the 8th Pacific Asia Conference on Information Systems*, 458-471.

Traylen, H. (1994). Confronting hidden agendas: Cooperative inquiry with health visitors. In P. Reason (Ed.), *Participation in human inquiry* (pp. 59-81). London: Sage Publications.

Turner, J., & Muller, R. (2003). On the nature of the project as a temporary organization. *International Journal of Project Management, 21*(1), 1-8.

Von Bergen, C., & Kirk, R. (1978). Groupthink: When too many heads spoil the decision. *Management Review, 67*(3), 44-49.

Weinberg, G. (1971). *The psychology of computer programming.* New York: Van Nostrand Reinhold Co.

Whitehead, R. (2001). *Leading a software development team: A developers guide to successfully leading people and projects.* London: Addison-Wesley.

Wright, J., & Schaal, D. (1988). Groupthink: The trap of consensus investing. *Journal of Financial Planning, 1*(1), 41-44.

Yourdon, E. (1993). *Decline and fall of the American programmer.* Upper Saddle River, NJ: Prentice Hall Inc.

Zachary, G. (1994). *Show-stopper! The breakneck race to create Windows NT and the next generation at Microsoft.* New York: The Free Press.

Zigurs, I. (2003). Leadership in virtual teams: Oxymoron or opportunity? *Organizational Dynamics, 31*(4), 339-351.

Zmud, R. (1983). The effectiveness of external information channels in facilitating innovation within software development groups. *MIS Quarterly, 7*(2), 43-58.

Chapter XV
Distributed Knowledge Management in Organizations

Ganesh Vaidyanathan
Indiana University South Bend, USA

ABSTRACT

A framework to capture and manage distributed knowledge can address distributed knowledge management from creation to facilitation. Knowledge generation and dissipation needs to be embedded in corporate processes. These processes need to have an underlying principle that eliminates the obstacles of collecting multiple knowledge perspectives within complex organizations. Moreover, extrinsic motivators, social-psychological forces, and organizational climate factors are believed to influence knowledge sharing. This study discusses a framework that provides a synergized view to collect, share, and manage distributed corporate knowledge using organizational knowledge models and technology knowledge models. Structural, cognitive, relational, and technological factors derived from a synthesized literature review aid to formulate this framework. Using this framework, the role of peer-to-peer networks on distributed knowledge management in organizations is examined.

INTRODUCTION

Whirlpool Corporation has set a clear expectation for everyone in its development group to devote 20% of their jobs to being consultants to those in other areas of development and deployment. To encourage this type of culture, the company has its staff focus on deliverables and further collaborating with others who may have insights and thus institutes a culture of knowledge sharing (Penzias, 2005). How can this knowledge be managed and distributed inside an organization? Distributed knowledge management (DKM) is important in today's knowledge-based economy (Desouza & Evaristo, 2004; Ezingeard, Leigh, & Chandler-Wilde, 2000; Pedersen & Larsen, 2001;

Spangler & Peters, 2001; Un & Cuervo-Cazurra, 2004). The concept of the management of intellectual capital is well established in the academic arena (Grant, 1996; Lynn, 1999; Masoulas, 1998; Nonaka, 1994; Templeton, Lewis, & Snyder, 2002). Firm-specific knowledge that is a part of the intellectual capital is difficult for competitors to imitate even when employees are hired away since that knowledge is specific to the original work environment (Hatch & Dyer, 2004). However, the firm-specific human capital of knowledge can be retained by using knowledge management systems (KMS). Knowledge management is of paramount importance to organizations and is emerging as a powerful source of competitive advantage (Hahn & Subramani, 2000). Scholars have recognized interorganizational knowledge transfer and knowledge flows and their link with competitive success (Gupta & Govindarajan, 2000; Inkpen & Tsang, 2005).

One of the reasons why multinational corporations exist is because of their ability to transfer and exploit knowledge more efficiently and effectively (Gupta & Govindarajan, 2000). Knowledge flows in such firms provide them the ability to be flexible, to respond more quickly to changing market conditions, to be more innovative, and to improve decision making and productivity (Alavi & Leidner, 1999). The goal of such organizations is to be aware of the existence and management of collective and individual knowledge (Bennet & Bennet, 2003). KM incorporates the organization of corporate knowledge according to a single, supposedly shared and objective classification. However, most of the KMS do not have this vision of knowledge (Bonifacio, Bouquet, & Traverso, 2002). Moreover, in the process of knowledge extraction and refinement, all subjective and contextual aspects of knowledge are eliminated (Bonifacio, Bouquet, Mameli, & Nori, 2003). DKM is an approach to KM based on the principle that multiplicity and heterogeneity of perspectives within complex organizations should not be viewed as obstacles to knowledge exploitation, but rather as an opportunity that can foster innovation and creativity (Bonifacio et al., 2002). DKM is a distinct and explicit process that attempts to leverage various perspectives of organizational knowledge into shared institutional capital. While DKM is a process and a strategy, KMS is an advanced information technology tool that is essential to implement the knowledge management process and strategy.

Literature on KM may be broadly classified into four dimensions that include structural, cognitive, and relational dimensions in management journals (Inkpen & Tsang, 2005; Rulke & Galaskiewicz, 2000), and the technology solutions dimension in information systems research (Hahn & Subramani, 2000; Lee & Choi, 2003). The KM enablers that include social and technical perspectives were mapped to a knowledge creation process (Nonaka, 1994) to initiate an integrated view of KM (Lee & Choi). This integrated model discusses only the knowledge creation process. However, KM is more than just knowledge creation. In general, KM is the creation, representation, storage, dissemination, transformation, application, maintenance, and facilitation of distributed organizational knowledge (Alavi & Leidner, 2001; Schultze & Leidner, 2002). In DKM, knowledge discovery, knowledge transfer, and knowledge facilitation are as important as knowledge creation (Briggs, De Vreede, Nunamaker, & Sprague, 2002; Grover & Davenport, 2001). The literature lacks a general framework that captures all these facets of DKM to represent a unified model of DKM. Such a framework is needed for academicians to further research distributed knowledge systems and for practitioners to understand the implications of DKM implementations.

A framework for DKM using structural, cognitive, relational, and technology dimensions has been proposed (Vaidyanathan, 2006). The framework is based on a balanced view of organizational managerial, social, cognitive, and technological facets of DKM. The next section defines DKM and provides the background literature for the framework. The third section derives the four dimensions of DKM supported by existing research literature. Peer-to-peer (P2P) and grid computing are discussed as technologies for organizational knowledge management using these dimensions of DKM. The fourth section defines P2P technology and grid computing in the context of organizational knowledge management and discusses how these technologies can be used. The final section discusses the contributions of the study and presents avenues for future research.

DISTRIBUTED KNOWLEDGE MANAGEMENT

The concept of DKM has received widespread attention primarily because organizations have come to realize that a large amount of corporate knowledge resides in the heads of workers and is dispersed in the processes, practices, and documents of the organization (Davenport, DeLong, & Beers, 1998). Sharing this knowledge is a critical catalyst for creativity and subsequent innovation (Peters, 1999) as it provides a means by which ideas can be conceptualized, shared, and tested. To share this disparate knowledge, a distributed knowledge strategy is needed. Without an effective distributed knowledge strategy, firm-specific, individual-owned organizational knowledge is lost when employees leave the organization (DeLong, 2004).

It is appropriate to define knowledge in an organizational context before discussing DKM approaches. Most researchers distinguish knowledge from data and information (Courtney, 2001). Davenport et al. (1998) define knowledge as information combined with experience, context, interpretation, and reflection. There are several classifications of KM in the research literature. Courtney describes several types of knowledge recognized in the literature that includes explicit vs. tacit, procedural vs. declarative, esoteric vs. exoteric, and shallow vs. deep. The distinction between explicit and tacit knowledge is of particular importance in KM. While both types of knowledge are important, organizations have focused primarily on managing explicit knowledge. The attempt to manage tacit knowledge is relatively a recent development.

Nonaka (1994) was among the first researchers to suggest the importance of managing tacit knowledge, and he proposed the spiral model of organizational knowledge creation. The spiral model is comprised of four modes of knowledge conversion that include socialization, combination, externalization, and internalization. The socialization mode refers to the conversion of tacit knowledge to new tacit knowledge through social interactions and shared experience among organizational members, for example, through mentoring. The combination mode refers to the creation of new explicit knowledge by merging, categorizing, reclassifying, and synthesizing existing knowledge, for example, the statistical analysis of sales data. Externalization and internalization are the critical steps in the spiral of knowledge. Externalization refers to converting tacit knowledge to new explicit knowledge, for example, writing requirement specifications after interviewing a potential client. Internalization refers to the creation of new tacit knowledge from explicit knowledge, for example, gaining new research ideas through reading a journal article.

Earl (2001) proposes a more extensive classification of KM strategies. His research identifies the following schools of KMS that include cartographic, engineering, commercial, organizational, spatial, and strategic. The "learning ladder" model (Ciborra & Andreu, 2001) discusses a compact way of describing the unfolding of multiple organizational knowledge creation, transformation, and transfer processes to three different organizational contexts.

For a distributed decision support system, a classification based on dimensions of system design and their attributes were developed (Ba, Stallaert, & Whinston, 2001). In this view of classification, three dimensions that include software engineering, technology acceptance, and incentive alignment are discussed. All three of these dimensions are key attributes to any generic information systems design. However, there are important factors to be considered for DKM other than the generic design attributes. Collaboration (Lee & Choi, 2003), social networks (Cross & Cummings, 2004), and the facilitation of knowledge (Kwok, Ma, & Vogel, 2002) are essential to DKM implementation.

Grover and Davenport (2001) developed a framework that focuses on the knowledge process and the context in which the process is embedded. Even though this framework considers the generation, codification, transfer, and realization of KM, it can be extended with specifics of how technology may be integrated toward the success of firms. A model for KM success takes into account key managerial influences, key resource influences, and key environmental influences (Massey, Montoya, & O'Driscoll, 2002). KM initiatives will not be successful by focusing on the knowledge transfer processes alone. The knowledge transfer processes combined with knowledge acquisition using the efficiencies of the technology will likely be more successful in managing knowledge. These studies pose a question. What technologies are to be used for successful knowledge management? Even though there is a great deal of advanced technologies, due to variations in organizational nature and different KM approaches, executives confront the challenging task of deciding the type of information technology to deploy in support of their KM initiatives (Kankanhalli, Tanudidjaj, Sutanto, & Tan, 2003).

The effective use of tools like KMS results in the successful management of knowledge and is manifested in a variety of implementations (Davenport et al., 1998). IT tools and IT infrastructures are important enablers of DKM systems (King, Marks, & McCoy, 2002). Collaborative computing tools, knowledge servers, enterprise knowledge portals, electronic document management tools, knowledge harvesting tools, search engines, and knowledge management suites are some IT tools that support DKM. These tools help firms to produce, manipulate, store, communicate, and disseminate information and knowledge. These IT tools need to harmonize with existing infrastructure as firms have invested enormous amounts of capital in their existing computer systems (King et al.). IT can play an important role in transcending organizational boundaries such

Figure 1. Four dimensions of distributed project knowledge management

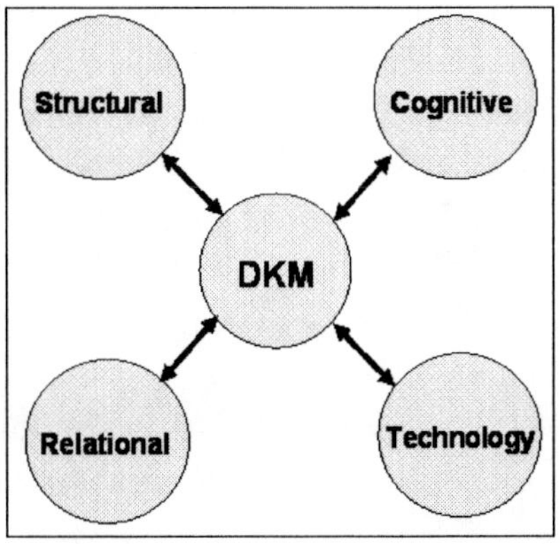

as in the use of functional knowledge repositories and geographically dispersed employees. While the most obvious role of IT is seen in the creation, capture, and maintenance of knowledge repositories, IT also plays a dominant role in the access and dissemination of distributed organizational knowledge (Hendriks, 1999).

These KM strategies and technology may be used to formulate a framework of networked knowledge management. Organizations should strive to balance the efficiencies of process capabilities and technology infrastructure capabilities (Gold, Malhotra, & Segars, 2001). The process capabilities of organizations, in the form of structural, cognitive, and relational dimensions found in management journals along with the technology solution dimension in information systems research can be integrated to form a framework as described in the next section.

FOUR-DIMENSIONAL FRAMEWORK

Firms incorporate knowledge repositories to store knowledge. The knowledge of individual employees does not transform easily into organizational knowledge even with the implementation of knowledge repositories (Bock, Zmud, Kim, & Lee, 2005). Individuals tend to hoard knowledge for various reasons. Factors such as extrinsic motivators, social-psychological forces, and organizational climate factors are believed to influence individuals' knowledge sharing intentions (Bock et al.). While these factors contribute to knowledge sharing, other attributes of social capital networks are important to knowledge transfer in firms (Inkpen & Tsang, 2005). In addition, technology to manage the creation, sharing, and transfer knowledge is much needed to realize knowledge flows (Grover & Davenport, 2001). Firms use technology tools to support knowledge transfer and management (Alavi & Leidner, 2001). All of these organizational attributes of distributed knowledge management may be organized in four dimensions. These four dimensions are illustrated as a framework in Figure 1. Each one of these dimensions is comprised of various factors that are discussed in the following sections and detailed in Table 1. The dimensions and factors have been adapted from multidisciplinary research literature.

Structural Dimension

The structural dimension (Inkpen & Tsang, 2005; Nahapiet & Ghoshal, 1998) involves the pattern of relationships between the project team members. The structural dimension factors include the ties, configuration, stability, informality, and management of project teams.

Ties

Team member ties deal with the specific way members are related. Social ties facilitate interteam member social interactions and provide channels for knowledge transfer. Strong ties between members are needed for knowledge transfer to occur. Factors that support these ties are prior team relationships and repeated transactions during projects (Inkpen & Tsang, 2005). Griffith, Sawyer, and Neale (2003) proposed location and time spent interacting with team members via various media as one of the constructs in their research. In group performances, Fjermestad (2005) illustrated that teams using constructive consensus procedures in a virtual environment have a greater willingness to work together than the teams using structured conflict procedures, especially after the first task, and have a tendency toward stronger decision acceptance.

Configuration

The configuration determines the pattern of linkages among project members. Hierarchy, density, and connectivity affect the flexibility and ease of knowledge transfer through their impact on the extent of the contact and accessibility among team members (Krackhardt, 1992). Intracorporate project teams are arranged in a hierarchical way, providing connectivity among the team members. The decentralization of authority to the team members is important so that they can determine how to make the best use of their knowledge (Inkpen & Tsang, 2005). The quality and pattern of relationships existing among group members (Argote, McEvily, & Reagans, 2003) is important as specialists and generalists perform well in decentralized teams (Rulke & Galaskiewicz, 2000).

Stability

Organizational learning depends partially on the memories of individuals and their learning abilities (Carley, 2005). Intracorporate teams are usually stable, but this stability may not help if there is a high personnel turnover. Maintaining a stable pool of personnel can help individuals develop relationships. This stability retains knowledge as well as promotes knowledge sharing by establishing rapport and friendship (Inkpen & Tsang, 2005).

Flexibility

Knowledge creation requires flexibility and less emphasis on work rules. Flexibility gives rise to new ideas. Increased flexibility in an organizational structure can result in increased creation of knowledge (Lee & Choi, 2003). This flexibility enables team members to communicate and interact with one another (Jarvenpaa & Staples, 2000).

Management

The competence of knowing what others know enhances a team's overall knowledge capacity (Grant, 1996). Maintaining this directory of the environment is needed for managing distributed projects. The ability to develop and maintain relationships with others as well as interacting and working with team members is crucial to managing others (Joseph, Ang, & Tan, 1996). Leadership and disciplined project management helps in managing project teams, enables project managers to interact and find ways of combining business processes, and establishes an appreciation of knowledge assets relative to core business processes (Bassellier & Benbasat, 2004; Massey et al., 2002). Managerial influences that establish enabling conditions for KM includes leadership, coordination, control, and measurement. Measurement is essential for evaluation of the impacts of KM initiatives and consequently leadership, coordination, and control (Massey et al.). Knowledge managers facilitate knowledge creation, transfer, sharing, and maintenance (Grover & Davenport, 2001).

Cognitive Dimension

The cognitive dimension represents the resources that provide shared meaning and understanding between project team members (Inkpen & Tsang, 2005; Nahapiet & Ghoshal, 1998). Shared goals and shared culture encouraged in organizations enable project team members to achieve tasks and outcomes (Inkpen & Tsang). Learning is part of this cognitive dimension as leaning can be defined as the degree to which it is encouraged in organizations (Hurley & Hult, 1998; Lee & Choi, 2003). Through a learning environment, organizations can help team members play active roles in knowledge creation (Lee & Choi). The cognitive dimension factors include shared goals, shared culture, and learning as resources and environments provided by organizations to share knowledge.

Shared Goals

Shared vision (Tsai & Ghoshal, 1998) involves collective goals and aspirations for project team members. Members have similar perceptions on interacting with each other. This promotes mutual understanding and exchanges of ideas and resources. A shared vision is a bonding mechanism that helps different parts of distributed project teams to integrate knowledge (Inkpen & Tsang, 2005). Lower levels of goal clarity increases frustration and dissatisfaction among team members (Schnake & Cochran, 1985).

Shared Culture

Distributed project teams may be culturally diverse. Local or national culture needs to be understood and accommodated so that when knowledge is transferred from one to another, cultural conflicts will not hinder the knowledge transfer (Inkpen & Tsang, 2005). The overall effect of cultural diversity should be beneficial to knowledge transfer (Inkpen & Tsang).

Learning

The organizational level is where we find stores of knowledge and learning in such nonhuman elements as structures, systems, procedures, routines, and strategy (Crossan & Berdrow, 2003). Analysis at the organization level shows us the importance of aligning these nonhuman resources with the competitive environment (Mintzberg, Ahlstrand, & Lampel, 1998). Successful KM requires the appropriate management style, work environ-

Table 1. Distributed project KM dimensions

Dimensions	Factors	Description	References
Structural	Ties	Strong ties between members; prior partner relationships and repeated transactions; constructive consensus	Inkpen & Tsang (2005); Griffith et al. (2003); Fjermestad (2005)
	Configuration	Decentralization of authority by headquarters; group structure	Inkpen & Tsang (2005); Rulke & Galaskiewicz (2000)
	Stability	Personnel relationships; low personnel turnover	Inkpen & Tsang (2005)
	Flexibility	Flexible work rules; lack of formal structure; synergistic group processes	Lee & Choi (2003)
	Management	Disciplined project management; leadership	Massey et al. (2002); Bassellier & Benbasat (2004); Grover & Davenport (2001)
Cognitive	Shared Goals	Shared vision; collective goals; goal clarity	Inkpen & Tsang (2005)
	Shared Culture	Cultural diversity; accommodation	Inkpen & Tsang (2005)
	Learning	Learning environment; training; mentoring	Lee & Choi (2003); Alavi & Leidner (2001)
Relational	Trust	Clear and transparent reward; incentives to reduce mistrust	Inkpen & Tsang (2005); Ba et al. (2001)
	Collaboration	Active available help within team; sharing of knowledge; expertise credibility; perceived status	Lee & Choi (2003); Herzog (2001); Thomas-Hunt, Ogden, & Neale (2003); Sussman & Siegal (2003)
Technical	Tools	IT to enable DKM; technology acceptance and adoption; compatibility; ontology; security	Bonifacio et al. (2003); Ba et al. (2001); Grover & Davenport (2001); King et al. (2002); Gregor & Benbasat (1999); Edgington, Choi, Henson, Raghu, & Vinze (2004)
	Support and Maintenance	Support from IT; knowledge maintenance; loose structure	Lee & Choi (2003); Hahn & Subramani (2000)

ment, and incentives. For successful knowledge creation, organizations should develop a deeply integrated learning culture and provide resources such as education, training, and mentoring (Alavi & Leidner, 2001; Lee & Choi, 2003).

Relational Dimension

The relational dimension focuses on the role of the direct ties between team members. This dimension represents relational as opposed to structural outcomes of project team interactions (Inkpen & Tsang, 2005). The relational dimension factors include trust and collaboration among distributed project teams.

Trust

Trust is essential for the coherence of any social system and is an essential and pervasive feature of individual and organizational relationships (Maclagan, 1997). Trust plays an important part in the willingness of team members to share knowledge. Trust in a firm focuses on the role of direct ties between actors and the relational, as opposed to structural, outcomes of interactions (Inkpen & Tsang, 2005). As trust develops over time, opportunities for knowledge transfer should increase (Inkpen & Tsang). When team members compete against one another for resources, suspicion may replace trust in their relationships and, consequently, knowledge sharing is sacrificed (Inkpen & Tsang). Hence, it is important that corporations establish clear and transparent reward criteria so that the team members concerned will not suspect any under-the-table favoritism (Inkpen & Tsang). Knowledge sharing is both constrained and difficult without proper and necessary incentives for doing so (Ba et al., 2001).

Collaboration

Collaboration is defined as the degree to which team members actively help one another in their work (Hurley & Hult, 1998). Electronic collaboration is defined as collaboration among different individuals to accomplish a common task using electronic technologies (Kock & D'Arcy, 2002). The collaborative culture fosters this type of exchange by reducing fear and increasing openness to other members. There is a positive relationship between collaboration and the knowledge creation process (Lee & Choi, 2003). Trust has been identified as a factor in successful collaboration. Successful collaborative project teams use shared processes and conditions to build trust, and enable open and honest communication. The collaborative sharing of knowledge can be a valuable and powerful tool for building collaborative team trust (Herzog, 2001). In the context of text-based messages, which is one of the media of collaboration in distributed projects, opinions change significantly in the direction advocated by the communicator when the material is attributed to a high-credibility source than when it is attributed to a low-credibility source (Sussman & Siegal, 2003). Individuals within a project team may derive status both from social interaction with other members and their knowledge or expertise (Thomas-Hunt et al., 2003). Socially isolated members participate more in discussions and express greater unique knowledge than socially connected members. Social status can promote the differential emphasis of shared knowledge, one's own knowledge, and other unique knowledge, as well as biased evaluations of member knowledge and contribution (Thomas-Hunt et al.).

Technical Dimension

Technology plays an important role in all aspects of KM (Grover & Davenport, 2001). Collaborative computing tools, knowledge servers, knowledge portals, search engines, document management tools, knowledge harvesting tools, and KM suites encompass the technologies used in distributed projects. Firms are making significant IT investments in deploying systems to manage knowledge

(Hahn & Subramani, 2000). The major challenge of managing knowledge is less its creation and more its capture and integration (Grant, 1996). To effectively manage knowledge, organizations should strive to balance the efficiencies of process capabilities and technology infrastructure capabilities (Gold et al., 2001). To achieve this balance, technology and its support functions need to be addressed as a dimension. The technology dimension includes IT tools, support, and maintenance.

Tools

System design plays an important role in KM initiatives (King et al., 2002). As with most information systems, the success of a DKM system partially depends upon the extent of use, which itself may be tied to system quality, information quality, and usefulness (Delone & McLean, 1992). One difference between traditional IT and DKM systems is the ex ante nature of the objectives, outcomes, and processes of the system to be developed (Hahn & Subramani, 2000). Traditional approaches to system development may not be appropriate for KMS. A distributed knowledge representation is needed. Knowledge schemas change over time and different schemas may be used for different parts of the organization (Hahn & Subramani). Further, the quality of the DKM system may be addressed using information systems product measurements. The quality of a product is measured with respect to whether the software functions correctly under every possible contingency, whether the system can be ported easily, whether the system has been well documented, whether standards have been adhered to, whether open architecture has been followed, and so forth (Ba et al., 2001).

Technology acceptance has been a subject in research for a number of years (Davis, Bagozzi, & Warshaw, 1989; Devaraj, Fan, & Kohli, 2002). User friendliness, user acceptance, perceived ease of use, perceived usefulness, user satisfaction, cognitive fit, and task-technology fit along with incentives make users adapt and contribute to a system (Ba et al., 2001). A KMS needs to adapt this view to be more successful. The size of the KMS, especially in the technology sense, needs to be addressed. Users of P2P networks have established that there is an optimal size in network externalities and that they are less likely to contribute resources to the system as the network size increases (Asvanund, Clay, Krishnan, & Smith, 2004). Explanations, when suitably designed, improve performance and perceptions of the system, especially when the users perceive an anomaly, when they want to learn, and when they want a specific piece of knowledge to solve a problem (Gregor & Benbasat, 1999). In a distributed project system, the support of local languages as well as company-specific functional language translations of knowledge are needed (Massey et al., 2002).

KM success is enhanced when applying ontology as an enabler to integrate KM practices and processes (Edgington et al., 2004). A centrally located index for knowledge in projects helps foster an efficient coordination mechanism as this index contains the sources of knowledge in projects (Desouza & Evaristo, 2004). Security of the stored knowledge is of utmost importance as the intellectual wealth of an organization is captured electronically. Also, the DKM infrastructure has to be compatible with the existing computer systems (King et al., 2002). Finally, as with any information systems, technical standards need to be established for DKM (King et al.). In summary, IT tools, which include data mining, learning, bulletin boards, repositories, knowledge directories, expert systems, and workflow systems, enable knowledge creation, storage, retrieval, transfer, and application.

Support and Maintenance

Maintaining contributions to distributed knowledge content is an important issue. Motivating team members to contribute depends on the structuring of content (Hahn & Subramani, 2000). When the knowledge structure is loosely structured, the response behavior of members is enhanced, thus providing a source for motivation and greater visibility, and in general, engendering a higher social status (Hahn & Subramani). IT facilitates the collection, storage, and exchange of distributed project knowledge; integrates fragmented flows of knowledge; and fosters all modes of knowledge creation (Lee & Choi, 2003). This support from information technologies is critical in all aspects of DKM. Recent research has established relevant technologies that pertain to DKM in organizations (Bonifacio et al., 2003; Desouza & Evaristo, 2004).

The framework established in this study embraces technology as one of the main dimensions of networked knowledge management. In large global firms and in hypercompetitive environments, IT will be interlaced with organizational KM strategies and processes (Alavi & Leidner, 2001). The next section illustrates how technology can be used in conjunction with the proposed DKM framework.

DISTRIBUTED KNOWLEDGE MANAGEMENT TECHNOLOGIES

P2P technologies and grid computing have great potential to grow into a strategic organizational tool for knowledge sharing. Current IT solutions for DKM are mostly focused on explicit knowledge, its capture, and its codification. P2P and grid computing provides a way to circumvent the limitations of current IT for KM for distributed project purposes. Kini (2002) suggests that the corporate version of P2P will enable users to actively share knowledge and information in innovative and flexible ways. Schoder and Fischbach (2003) state that P2P refers to technology that enables two or more peers to collaborate spontaneously in a network of peers by using appropriate information and communication systems without the necessity for central coordination.

P2P Technologies

P2P technologies include file sharing, resource sharing, and collaboration systems. The primary challenge in P2P computing is to design and implement a robust and scalable distributed system composed of inexpensive, individually unreliable computers in unrelated administrative domains (Balakrishnan, Kaashoek, Karger, Morris, & Stoica, 2003). The demand of users for transparent access to information and applications, regardless of the hosts on which they reside, is changing the computing environment from a traditional centralized model to a distributed model (Bauer & Coburn, 1994). In the long term, the partitioning of components into clients and servers will become constraining, and a computing environment based on P2P interactions will be required (Bauer & Coburn). P2P technologies permit a greater degree of freedom and independence on the part of users by making resources available in a more customized manner (Schoder & Fischbach, 2003). P2P computing mirrors face-to-face human interaction more closely than do client-server architectures (Barth, 2001). Peer-mediated DKM technology has been proposed by Bonifacio et al. (2003). A hybrid approach including centralized (client-server) technology and peer-mediated technology has been proposed by Desouza and Evaristo (2004). Moreover, P2P networking naturally supports DKM by closely adopting the conventions of face-to-face human communication (Tiwana, 2003).

For P2P technologies to be established in DKM, they must support robust security. A viable hybrid model incorporating both client-server and P2P architectures will give end users the flexibility to share documents and applications and still allow a central authority to exercise basic policies and control (Desouza & Evaristo, 2004). Fattah (2002) distinguishes between two main categories of applications for the P2P architecture: active ones in which users or systems take some action, and passive ones in which idle resources are harnessed for other uses. Table 2 illustrates this classification scheme and provides examples of applications in each category. Each of these application categories offers unique solutions to specific problems and they are mostly independent of each other.

Current collaboration systems are designed to provide the ability to easily form ad hoc groups and communities without intervention from any central authority. Groove Networks' Groove platform has received the most attention in this arena. Groove has bundled instant messaging, file sharing, online chat, and videoconferencing in a single platform. When users work off line in the Groove space, their changes to group documents are synchronized with the rest of the group as soon as each person returns to the Web. GlaxoSmith-Kline PLC uses Groove Networks P2P software for collaboration with scientists at other biotech firms and universities (DiSabatino, 2001).

HP Services selected Groove Networks software because it enables secure, cross-firewall team collaboration on and off line. HP Services uses Groove Networks software for geographically dispersed project collaboration since the software dissolves cross-domain security issues; enables ad hoc, self-organizing groups, enhancing the ability to harness knowledge; reduces coordination steps; and eliminates information bottlenecks and errors associated with phone and e-mail.

To maintain its market leadership, Steelcase International has established common processes companywide to support its increasingly global customer and dealer base. One such process includes an environment where subject-matter experts lend their skills wherever and whenever they are needed in a decentralized structure in the form of virtual work teams. Such teams are global, cross-functional, multi-organizational, and often multicultural and multilingual. Recognizing that while virtual work provides efficiencies, it can also increase coordination costs and can erode team cohesion and morale. For this reason, Steelcase International required a technology solution that would allow dispersed teams to work together and interact naturally, just as they would if they were in the same place (Groove Networks, 2006). Basic collaboration tools, such as note databases and online team rooms, were in place but were used primarily by sedentary teams working within one office with document-centric needs. Steelcase understood that these tools cannot easily be extended across offices to outside consultants and customers, nor can they support needs that are process, project, or communication centric; to help with these tasks, Steelcase selected Groove Virtual Office as its next-generation virtual team collaboration solution (Groove Networks).

Since current P2P technologies are still in their rudimentary stages, the decentralized nature of P2P architecture poses special challenges. Schoder and Fischbach (2003) list P2P concerns in an organizational context that include network control, security, interoperability, metadata, and cost sharing. Finally, P2P technologies create significant legal and ethical concerns. The ability

Table 2. Classification scheme of P2P technologies

Active Applications	Idle Utilizations
User Collaboration File Sharing, Work Groups and Communications, Gaming	**Resource Utilizations** Processing Power, Storage, Bandwidth Conversion
Application Interaction Knowledge Management Systems	**Supercomputing** High-Performance Applications

to access information from an individual's computer and share this information with a number of users raises security and privacy issues. However, organizations are using this technology to foster their distributed knowledge management environments.

Knowledge Grid

Another facet in research of DKM technologies is the knowledge grid (Cannataro & Talia, 2003). Although grid computing is becoming an important framework for enabling applications to utilize widely distributed collections of computational and data resources, current grid software is still in rudimentary stages. Grid computing is the most promising framework for future implementations of high-performance, data-intensive distributed applications. The knowledge grid is a significant step in the integration architecture for distributed data mining and knowledge discovery (Cannataro & Talia). Grid computing is a concept of computing infrastructure that can utilize distributed computational resources to solve problems easily—as easy as plugging into the electric power grid (Foster, Kesselman, Nick, & Tuecke, 2001). The knowledge grid concerns knowledge-based inference and knowledge discovery (Cannataro & Talia).

A grid is a geographically distributed computation platform composed of a set of heterogeneous machines that users can access via a single interface, thereby providing common resource access and operational services across widely distributed virtual organizations composed of institutions or individuals that share resources (Talia, 2002). Several tool kits and software environments for implementing grid applications have become available, which include Legion (http://legion.virginia.edu), Condor (http://www.cs.wisc.edu/condor), Unicore (http://www.unicore.org), and Foster and Kesselman's Globus Toolkit (http://www.globus.org/toolkit). These tool kits address security, information discovery, resource management, data management, communication, fault detection, and portability issues (Talia). Some researchers (Bilykh et al., 2003) use the grid computing paradigm to integrate medical information. When data are maintained over geographically distributed sites, the computational power of distributed and parallel systems can be exploited for knowledge discovery in databases. In this scenario, the grid can provide an effective computational support for distributed knowledge discovery on large data sets (Congiusta, Pugliese, Talia, & Trunfio, 2003).

Several knowledge grid projects have been identified by Cheung and Liu (2005). One such project is KB-Grid (Wu, Chen, & Xu 2003), a knowledge grid project in China, which suggests a paradigm for organizing, discovering, utilizing, and managing Web knowledge base resources, and has been applied to support knowledge services of traditional Chinese medicine. The Access Grid (Uram, 2006) is an ensemble of resources including multimedia large-format displays, presentation and interactive environments, and interfaces to grid middleware and visualization environments. These resources are used to support group-to-group interactions that include large-scale distributed meetings, collaborative work sessions, seminars, lectures, tutorials, and training using high-end audio and visual technology.

The San Diego Supercomputer Center has announced the formation of the Pacific Rim Applications and Grid Middleware Assembly (PRAGMA), an international collaboration to advance the use of the computational grid among investigators at the leading research institutions around the Pacific Rim (Shread, 2002). Based on existing collaborations, PRAGMA partners will work to diversify the number and focus of grid-enabled applications and to conduct applications and infrastructure research of common interest. PRAGMA will enhance connections among individual investigators by promoting visiting scholars' and engineers' programs, building new collaborations, formalizing resource-sharing agreements, and continuing trans-Pacific network deployment (Shread).

CONCLUSION

The knowledge network in an organization must be supported by its technology network as well as structural, cognitive, and relational networks. KM involves the distinct but interdependent processes of knowledge creation, knowledge storage and retrieval, knowledge transfer, and knowledge application (Alavi & Leidner, 2001). The integrated four dimensions that form a unified model of DKM discussed in this chapter form a strategy to integrate these interdependent processes. The extent to which P2P and grid technologies are implemented for organizational purposes will depend on technological, economic, and legal factors. While the discussion of the "purity" of P2P and grid computing efforts is of academic interest, organizations are only interested in the practical consequences of P2P computing. The short-term benefits of P2P and grid technologies include cost savings, efficiency, and a better human interface. In the long term, P2P and grid computing technologies have the potential to change how organizations do business using KM in distributed projects.

The primary objective of this study is to characterize an integrative view of distributed knowledge management and provide a strategic guideline to practitioners and academicians. Most studies in this area have examined some of the dimensions in isolation (Inkpen & Tsai, 2005; Lee & Choi, 2003; Becerra-Fernandez & Sabherwal, 2001; Szulanski, 1996; Zander & Kogut, 1995). Researchers and practitioners have not tried a fully integrative model of DKM. The framework discussed in this chapter integrates the voluminous scholarly work in the area of DKM and is ready to be examined by researchers for further studies. One challenge is to decipher the relationships among the four dimensions proposed in this study. The results of this study need to be formulated and tested in a theoretical model. The framework may be used as a stepping-stone for further empirical research and can help formulate robust strategies that involve trade-offs between these four dimensions. The relationships among the four dimensions of DKM can provide insights to practitioners of knowledge management and IT managers, as well as academicians. The ideas, discussion, and the integration of research views in this chapter will hopefully stimulate further interest in this topic.

REFERENCES

Alavi, M., & Leidner, D. (1999). *Knowledge management systems: Emerging views and practices from the field.* Proceedings of 32nd Annual HICSS, HI.

Alavi, M., & Leidner, D. E. (2001). Review. Knowledge management and knowledge management systems: Conceptual foundations and research issues. *MIS Quarterly, 25*(1), 107-132.

Argote, L., McEvily, B., & Reagans, R. (2003). Introduction to the special issue on managing knowledge in organizations: Creating, retaining, and transferring knowledge. *Management Science, 49*(4), 571-582.

Asvanund, A., Clay, K., Krishnan, R., & Smith, M. D. (2004). An empirical analysis of network externalities in peer-to-peer music-sharing networks. *Information Systems Research, 15*(2), 155-174.

Ba, S., Stallaert, J., & Whinston, A. (2001). Research commentary: Introducing a third dimension in information systems design. The case for incentive alignment. *Information Systems Research, 12*(3), 225-239.

Balakrishnan, H., Kaashoek, M. F., Karger, D., Morris, R., & Stoica, I. (2003). Looking up data in P2P systems. *Communications of the ACM, 46*(2), 43-50.

Barth, S. (2001). The promise of peerless platforms. *Knowledge Management*, 66-68.

Bassellier, G., & Benbasat, I. (2004). Business competence of information technology professionals: Conceptual development and influence on IT business partnerships. *MIS Quarterly, 28*(4), 673-694.

Bauer, M. A., & Coburn, N. (1994). A distributed system architecture for a distributed application environment. *IBM Systems Journal, 33*(3), 399-26.

Becerra-Fernandez, I., & Sabherwal, R. (2001). Organizational knowledge management: A contingency perspective. *Journal of Management Information Systems, 18*(1), 23-55.

Bennet, A., & Bennet, D. (2003). The partnership between organizational learning and knowledge management. In C. W. Holsapple (Ed.), *Handbook on knowledge management* (Vol. 1, pp. 439-460). New York: Springer-Verlag.

Bilykh, I., Bychkov, Y., Dahlem, D., Jahnke, J. H., McCallum, G., Obry, C., et al. (2003). Can GRID services provide answers to the challenges of national health information sharing? *Proceedings of the 2003 Conference of the Centre for Advanced Studies on Collaborative Research,* 39-53.

Bock, G., Zmud, R. W., Kim, Y., & Lee, J. (2005). Behavioral intention formation in knowledge sharing: Examining the roles of extrinsic motivators, social-psychological forces, and organizational climate. *MIS Quarterly, 29*(1), 87-111.

Bonifacio, M., Bouquet, P., & Cuel, R. (2002). Knowledge nodes: The building blocks of a distributed approach to knowledge management. *Journal of Universal Computer Science, 8*(6), 652-661.

Bonifacio, M., Bouquet, P., Mameli, G., & Nori, M. (2003). *Peer-mediated distributed knowledge management* (Tech. Rep. No. DIT-03-032). University of Trento.

Bonifacio, M., Bouquet, P., & Traverso, P. (2002). Enabling distributed knowledge management: Managerial and technological implications. *Novatica and Informatik/Informatique, 3*(1).

Briggs, R. O., De Vreede, G., Nunamaker, J. F., & Sprague, R. H. (2002). Special issue: Decision-making and a hierarchy of understanding. *Journal of Management Information Systems, 18*(4), 5-9.

Cannataro, M., & Talia, D. (2003). The knowledge grid. *Communications of the ACM, 46*(1), 89-93.

Carley, K. (1992). Organizational learning and personnel turnover. *Organization Science, 33*, 353-365.

Cheung, W. K., & Liu, J. (2005). On knowledge grid and grid intelligence: A survey. *Computational Intelligence, 21*(2), 111-129.

Ciborra, C. U., & Andreu, R. (2001). Sharing knowledge across boundaries. *Journal of Information Technology, 16*(2), 73-82.

Congiusta, A., Pugliese, A., Talia, D., & Trunfio, P. (2003). Designing grid services for distributed knowledge discovery. *Web Intelligence & Agent Systems, 1*(2), 91-104.

Courtney, J. F. (2001). Decision making and knowledge management in inquiring organizations: Toward a new decision-making paradigm for DSS. *Decision Support Systems, 31*(1), 17-39.

Cross, R., & Cummings, J. N. (2004). Tie and network correlates of individual performance in knowledge-intensive work. *Academy of Management Journal, 47*(6), 928-937.

Crossan, M. M., & Berdrow, I. (2003). Organizational learning and strategic renewal. *Strategic Management Journal, 24*(11), 1087-1105.

Davenport, T. H., DeLong, D. W., & Beers, M. C. (1998). Successful knowledge management projects. *Sloan Management Review, 39*(2), 43-58.

Davenport, T. H., & Prusak, L. (1998). *Working knowledge: How organizations manage what they know.* Boston: Harvard Business School Press.

Davis, F. D., Bagozzi, R. P., & Warshaw, P. R. (1989). User acceptance of computer technology: A comparison of two theoretical models. *Management Science, 35*(8), 982-1003.

Delone, W. H., & McLean, E. R. (1992). Information systems success: The quest for the dependent variable. *Information Systems Research, 3*(1), 60-95.

DeLong, D. W. (2004). *Lost knowledge: Confronting the threat of an aging workforce.* New York: Oxford University Press.

Devaraj, S., Fan, M., & Kohli, R. (2002). Antecedents of B2C channel satisfaction and preference: Validating e-commerce metrics. *Information Systems Research, 13*(3), 316-333.

Desouza, K., & Evaristo, J. R. (2004). Managing knowledge in distributed projects. *Communications of the ACM, 47*(4), 87-91.

DiSabatino, J. (2001, April 16). Collaborative peer-to-peer technology finds Groove at GlaxoSmithKline, Raytheon. *Computer World.*

Earl, M. (2001). Knowledge management strategies: Toward a taxonomy. *Journal of Management Information Systems, 18*(1), 215-233.

Edgington, T., Choi, B., Henson, K., Raghu, T. S., & Vinze, A. (2004). Adopting ontology to facilitate knowledge sharing. *Communications of the ACM, 47*(11), 85-90.

Ezingeard, J., Leigh, S., & Chandler-Wilde, R. (2000). Knowledge management at Ernst & Young UK: Getting value through knowledge flows. *Proceedings of the Twenty First International Conference on Information Systems*, 807-822.

Fattah, H. M. (2002). *How peer-to-peer technology is revolutionizing the way we do business.* Chicago: Dearborn Trade Publishing.

Fjermestad, J. (2005). Virtual group strategic decision making using structured conflict and consensus approaches. *Internal Journal of e-Collaboration, 1*(1), 43-61.

Foster, I., Kesselman, C., Nick, J. M., & Tuecke, S. (2002). Grid services for distributed system integration. *IEEE Computer, 35*(6), 37-46.

Gold, A., Malhotra, A., & Segars, A. H. (2001). Knowledge management: An organizational capabilities perspective. *Journal of Management Information Systems, 18*(1), 185-214.

Grant, R. M. (1996). Prospering in dynamically-competitive environments: Organizational capability as knowledge integration. *Organization Science, 7*(4), 375-387.

Gregor, S., & Benbasat, I. (1999). Explanations from intelligent systems: Theoretical foundations and implications for practice. *MIS Quarterly, 23*(4), 497-530.

Griffith, T., Sawyer, J. E., & Neale, M. A. (2003). Virtualness and knowledge in teams: Managing the love triangle of organizations, individuals and information technology. *MIS Quarterly, 27*(2), 265-287.

Groove Networks. (2006). *Customers: Case studies. Steelcase Inc.* Retrieved from http://www.groove.net/index.cfm?pagename=CaseStudy_Steelcase

Grover, V., & Davenport, T. H. (2001). General perspectives on knowledge management: Fostering a research agenda. *Journal of Management Information Systems, 18*(1), 5-22.

Gupta, A. K., & Govindarajan, V. (2000). Knowledge flows within multinational corporations. *Strategic Management Journal, 21*(4), 473-496.

Hahn, J., & Subramani, M. (2000). A framework of knowledge management systems: Issues and challenges for theory and practice. *Proceedings of the Twenty First International Conference on Information Systems*, 302-312.

Hansen, M. T., & Nohria, N. (1999). What's your strategy for managing knowledge? *Harvard Business Review, 77*(2), 106-117.

Hatch, N. W., & Dyer, J. H. (2004). Human capital and learning as a source of sustainable competitive advantage. *Strategic Management Journal, 25*(12), 1155-1178.

Hendriks, P. H. J. (1999). Why share knowledge? The influence of ICT on the motivation for knowledge sharing. *Knowledge and Process Management, 6*(2), 91-100.

Herzog, V. L. (2001). Trust building on corporate collaborative project teams. *Project Management Journal, 32*(1), 28-37.

Hurley, R., & Hult, T. (1998). Innovation, market orientation, and organizational learning: An integration and empirical examination. *Journal of Marketing, 62*(3), 42-54.

Inkpen, A. C., & Tsang, E. W. K. (2005). Social capital, networks, and knowledge transfer. *Academy of Management Review, 30*(1), 146-166.

Jarvenpaa, S. L., & Staples, D. S. (2000). The use of collaborative electronic media for information sharing: An exploratory study of determinants. *Strategic Information Systems, 9*(2-3), 129-154.

Joseph, D., Ang, S., & Tan, H. T. (1996). *The nature of soft skills of information systems professionals.* Proceedings of the Americas Conference on Information Systems, Phoenix, AZ.

Kankanhalli, A., Tanudidjaj, F., Sutanto, J., & Tan, B. C. Y. (2003). The role of IT in successful knowledge management initiatives. *Communications of the ACM, 46*(9), 69-73.

King, W. R., Marks, P. V., & McCoy, S. (2002). The most important issues in knowledge management. *Communications of the ACM, 45*(9), 93-97.

Kini, R. B. (2002). Peer-to-peer technology: A technology reborn. *Information Systems Management, 19*(3), 74-85.

Kock, N., & D'Arcy, J. (2002). Resolving the e-collaboration paradox: The competing influences of media naturalness and compensatory adaption. *Information Management and Consulting, 17*(4), 72-78.

Krackhardt, D. (1992). The strength of strong ties. In N. Nohria & R. G. Eccles (Eds.), *Networks and organizations: Structure, form and action* (pp. 216-239). Boston: Harvard Business School Press.

Kwok, R. C., Ma, J., & Vogel, D. R. (2002). Effects of group support systems and content facilitation on knowledge acquisition. *Journal of Management Information Systems, 19*(3), 185-229.

Lee, H., & Choi, B. (2003). Knowledge management enablers, processes, and organizational performance: An integrative view and empirical examination. *Journal of Management Information Systems, 20*(1), 179-228.

Lynn, B. E. (1999). Culture and intellectual capital management: A key factor in successful ICM implementation. *International Journal of Technology Management, 18*(5-8), 590-603.

Maclagan, P. (1997). Corporate social responsibility as a participative process. *Business Ethics: A European Review, 8*(1), 43-49.

Masoulas, V. (1998). Organizational requirements definition for intellectual capital management.

International Journal of Technology Management, 16(1-3), 126-143.

Massey, A., Montoya, W., & O'Driscoll, T. M. (2002). Knowledge management in pursuit of performance: Insights from Nortel networks. *MIS Quarterly, 26*(3), 269-289.

Mintzberg, H., Ahlstrand, B., & Lampel, J. (1998). *Strategy safari: A guided tour through the wilds of strategic management.* New York: Free Press.

Nahapiet, J., & Ghoshal, S. (1998). Social capital, intellectual capital, and the organizational advantage. *Academy of Management Review, 23*(2), 242-266.

Nonaka, I. (1994). A dynamic theory of organizational knowledge creation. *Organization Science, 5*(1), 14-38.

Pedersen, M. K., & Larsen, M. H. (2001). Distributed knowledge management based on product state models: The case of decision support in health care administration. *Decision Support Systems, 31*(1), 139-158.

Penzias, B. (2005). Whirlpool's IT practices for continuous innovation. *SAP NetWeaver Magazine*, 20-35.

Peters, T. (1999). *The circle of innovation: You can't shrink your way to greatness.* New York: Vintage Books.

Rulke, D. L., & Galaskiewicz, J. (2000). Distribution of knowledge, group network structure, and group performance. *Management Science, 46*(5), 612-625.

Schnake, M. E., & Cochran, D. S. (1985). Effect of two goal-setting dimensions on perceived intraorganizational conflict. *Group & Organizational Studies, 10*, 168-183.

Schoder, D., & Fischbach, K. (2003). Peer-to-peer prospects. *Communications of the ACM, 46*(2), 27-30.

Schultze, U., & Leidner, D. E. (2002). Studying knowledge management in information systems research: Discourses and theoretical assumptions. *MIS Quarterly, 26*(3), 213-242.

Shread, P. (2002, April 2). Pacific Rim grid collaboration formed. *Grid Computing Planet.*

Spangler, W. E., & Peters, J. M. (2001). A model of distributed knowledge and action in complex systems. *Decision Support Systems, 31*(1), 103-125.

Sussman, S., & Siegal, W. (2003). Informational influence in organizations: An integrated approach to knowledge adoption. *Information Systems Research, 14*(1), 47-65.

Szulanski, G. (1996). Exploring internal stickiness: Impediments to the transfer of best practice within the firm. *Strategic Management Journal, 17*(10), 27-43.

Talia, D. (2002). The open grid services architecture: Where the grid meets the Web. *IEEE Internet Computing, 6*(6), 2-6.

Templeton, G. F., Lewis, B. R., & Snyder, C. A. (2002). Development of a measure for the organizational learning construct. *Journal of Management Information Systems, 19*(2), 175-218.

Thomas-Hunt, M., Ogden, T. Y., & Neale, M. A. (2003). Who's really sharing? Effects of social and expert status on knowledge exchange within groups. *Management Science, 49*(4), 464-473.

Tiwana, A. (2003). Affinity to infinity in peer-to-peer knowledge platforms. *Communications of the ACM, 46*(5), 76-80.

Tsai, W., & Ghoshal, S. (1998). Social capital and value creation: The role of intrafirm networks. *Academy of Management Journal, 41*(4), 464-478.

Un, C. A., & Cuervo-Cazurra, A. (2004). Strategies for knowledge creation in firms. *British Journal of Management, 15*(1), 27-41.

Uram, T. (2006). *Access grid.* Retrieved from http://www.accessgrid.org/

Vaidyanathan, G. (2006). Networked knowledge management dimensions in distributed projects. *International Journal of e-Collaboration, 2*(4), 35-52.

Wu, Z., Chen, H., & Xu, J. (2003). Knowledge base grid: A generic grid architecture for semantic Web. *Journal of Computer and Technology, 18*(4), 462-473.

Zander, D., & Kogut, B. (1995). Knowledge and the speed of the transfer and imitation of organizational capabilities: An empirical test. *Organization Science, 6*(1), 76-92.

Compilation of References

Abdel-Hamid, T. K. (1999). The impact of goals on software project management: An experimental investigation. *MIS Quarterly, 23*(4), 531.

Abrahamsson, P., & Iiavara, N. (2002). Commitment in software process improvement: In search of the process. In *Proceedings of the 35th Annual Hawaii International Conference on System Sciences,* 260b.

Abrahamsson, P., & Jokela, T. (2000). Development of management commitment to software process improvement. In *Proceedings of Information Systems Research, 23,* 487-505.

Abuhamdieh, A. H., Kendall, J. E., & Kendall, K. E. (2000). Beyond e-commerce: Evaluating the strategic impact of Web presence for nonprofit organizations. In M. Swink (Ed.), *Proceedings of the Decision Sciences Institute* (pp. 567-569). Atlanta, GA: Decision Sciences Institute.

Abuhamdieh, A. H., Kendall, J. E., & Kendall, K. E. (2002). An evaluation of the Web presence of a nonprofit organization: Using the balanced scorecard approach in ecommerce. In R. Traunmuller (Ed.), *Information systems: The e-business challenge* (pp. 209-222). Boston: Kluwer Publishing.

Abuhamdieh, A. H., Kendall, J. E., & Kendall, K. E. (2007). E-commerce opportunities in the nonprofit sector: The case of New York Theatre Group. *International Journal of Cases on Electronic Commerce, 3*(1), 29-48.

Adams, J. (2002). Virtual defense. *Foreign Affairs, 80*(3), 98-112.

Adams-Price, C. E., & Chandler, S. (2000). The star fleet ladies auxiliary: Evolution of an online women's mailing list. *CyberPsychology and Behavior, 3*(5), 811-816.

Agarwal, R., & Prasad, J. (2000). A field study of the adoption of software process innovations by information systems professionals. *IEEE Transactions on Engineering Management, 47*(3), 295-308.

Agile Alliance. (2001). *Agile Alliance manifesto.* Retrieved from http://www.aanpo.org

Akella, R., & Dossani, R. (2001). *IT outsourcing and the software value chain: The Indian supplier during the downturn.* Proceedings of the IT Conference, Bangalore, India.

Alavi, M., & Leidner, D. (1999). *Knowledge management systems: Emerging views and practices from the field.* Proceedings of 32nd Annual HICSS, HI.

Alavi, M., & Leidner, D. E. (2001). Review. Knowledge management and knowledge management systems: Conceptual foundations and research issues. *MIS Quarterly, 25*(1), 107-132.

Alavi, M., Kayworth, T., & Leidner, D. (2005). An empirical examination of the influence of organizational culture on knowledge management practices. *Journal of Managment Information Systems, 22*(3), 191-224.

Alexander, C. (1965). *Notes on the synthesis of form.* Cambridge, MA: Harvard University Press.

Alexander, C. (1996, October). The origins of pattern theory, the future of the theory, and the generation of a living world. *ACM Conference on Object-Oriented Programs, Systems, Languages and Applications (OOPSLA)*. Retrieved from http://www.patternlanguage.com/archive/ieee/ieeetext.htm

Alexander, C., Ishikawa, S., Silverstein, M., Jacobson, M., Fiksdahl-King, I., & Angel, S. (1977). *A pattern language: Towns, buildings, construction*. Oxford University Press.

Alger, J. (1996). Introduction. In W. Schwartau (Ed.), *Information warfare. Cyberterrorism: Protecting your personal security in the electronic age* (2nd ed., pp. 8-14). New York: Thunder's Mouth Press.

Alonzo, M., & Aiken, M. (2004). Flaming in electronic communication. *Decision Support Systems, 36*(3), 205-214.

Alsene, E. (1999). Internal changes and project management structures within enterprises. *International Journal of Project Management, 17*(6), 367-376.

Anderson, E. W., & Fornell, C. (2000). Foundations of the American Customer Satisfaction Index. *Total Quality Management & Business Excellence, 11*(7), 869-882.

Ang, S., & Straub D. W. (1998). Production and transaction economies and IS outsourcing: A study of the U.S. Banking Industry. *MIS Quarterly, 22*(4), 535.

Anonymous. (2001). *Authority warns of hacker war*. Retrieved February 17, 2004, from http://fpeng.peopledaily.com.cn/200105/04/eng20010504_69242.html

Ante, S. (2004). Shifting work offshore? Outsourcer beware; quality and security woes can eat expected savings. *Business Week*, No. 3865, p. 36.

Apte, U. M., Sobal, M. G., Hanaoka, S., Shimada, T., Saarinen, T., & Versalainen, A. P. J. (1997). IS outsourcing practices in the USA, Japan, & Finland: A comparative study. *Journal of Information Technology, 12*, 289.

Argote, L., McEvily, B., & Reagans, R. (2003). Introduction to the special issue on managing knowledge in organizations: Creating, retaining, and transferring knowledge. *Management Science, 49*(4), 571-582.

Armistead, L. (Ed.). (2004). *Information operations: Warfare and the hard reality of soft power*. Dulles, VA: Brassey's.

Arrow, H., McGrath, J., & Berdahl, J. (2000). *Small groups as complex systems*. Thousand Oaks, CA: Sage Publications.

Asch, S. (1952). *Social psychology*. Upper Saddle River, NJ: Prentice-Hall.

Asvanund, A., Clay, K., Krishnan, R., & Smith, M. D. (2004). An empirical analysis of network externalities in peer-to-peer music-sharing networks. *Information Systems Research, 15*(2), 155-174.

Avolio, B. J. (1999). *Full leadership development: Building the vital forces in organizations*. Thousand Oaks, CA: Sage Publications.

Ba, S., Stallaert, J., & Whinston, A. (2001). Research commentary: Introducing a third dimension in information systems design. The case for incentive alignment. *Information Systems Research, 12*(3), 225-239.

Baar, J. E., & Jacobson, S. M. (2004). The keys to forecasting — *Work breakdown structure. Cost Engineering, 46*(3), 12.

Babakus, E., Bienstock, C. C., & Scotter, J. R. V. (2004). Linking perceived quality and customer satisfaction to store traffic and revenue growth. *Decision Sciences, 35*(4), 713-737.

Bafoutsou, G., & Mentzas, G. (2002). Review and functional classification of collaborative systems. *International Journal of Information Management, 22*(4), 281-305.

Bagozzi, R. P. (1980). *Causal models in marketing*. New York: John Wiley & Sons.

Baker, D., Friedman, R., & Miller, D. (1983). *Intelligence warfare.* New York: Crescent Books.

Balakrishnan, H., Kaashoek, M. F., Karger, D., Morris, R., & Stoica, I. (2003). Looking up data in P2P systems. *Communications of the ACM, 46*(2), 43-50.

Baltahazard, P. A., & Cooke, R. A. (2003). *Organizational culture and knowledge management success: Assessing the behavior-performance continuum.* AZ: Arizona State University West.

Balthazard, P., Potter, R., & Warren, J. (2004). Expertise, extraversion and group interaction styles as performance indicators in virtual teams. *The Database for Advances in Information Systems, 35*(1), 41-64.

Balthazard, P., Waldman, D., Howell, J., & Atwater, L. (2004). Shared leadership and group interaction styles in problem-solving virtual teams. In *Proceedings of the 37th Hawaii International Conference on System Sciences*, 10043a.

Banerjee, R. (2003). Organisational character: Issues, imperatives and practices. *International Journal of Human Resources Development and Management, 3*(1), 72-83.

Bansal, H. S., Irving, P. G., & Taylor, S. F. (2004). A three-component model of customer commitment to service providers. *Journal of the Academy of Marketing Science, 32*(3), 234-250.

Baptista, R. (2000). Do innovations diffuse faster within geographical clusters? *International Journal of Industrial Organization, 18*(3), 515.

Baron, R. M., & Kenny, D. A. (1986). The moderator-mediator variable distinction in social psychological research: Conceptual, strategic and statistical considerations. *Journal of Personality and Social Psychology, 51*, 1173-1182.

Baron, R., Kerr, N., & Miller, N. (1999). *Group process, group decision, group action.* Buckingham, UK: Open University Press.

Barrett, M., Capplemann, S., Shoib, G., & Walsham, G. (2004). Learning in knowledge communities: Managing technology and context. *European Management Journal, 22*(1), 1-11.

Barth, S. (2001). The promise of peerless platforms. *Knowledge Management*, 66-68.

Bartlett, F. (1932). *Remembering: A study in experimental and social psychology.* Cambridge, MA: Cambridge University Press.

Bartlett, F. (1958). *Thinking: An experimental and social study.* New York: Basic Books.

Baskerville, R. (1996). A taxonomy for analyzing hazards to information systems. In S. Katsikas & D. Gritzalis (Eds.), *Information systems security: Facing the information society of the 21st century* (pp. 167-176.). London: Chapman-Hall.

Bass, B. M. (1985). *Leadership and performance beyond expectations.* New York: Free Press.

Bass, B. M., & Avolio, B. J. (1994). *Improving organizational effectiveness through transformational leadership.* Thousand Oaks, CA: Sage Publications.

Bass, B. M., & Avolio, B. J. (1996). *Multifactor leadership questionnaire manual.* Palo Alto, CA: Mindgarden.

Bass, B. M., & Steidlmeier, P. (1999). Ethics, character, and authentic transformational leadership behavior. *The Leadership Quarterly, 10*(2), 181-217.

Bassellier, G., & Benbasat, I. (2004). Business competence of information technology professionals: Conceptual development and influence on IT business partnerships. *MIS Quarterly, 28*(4), 673-694.

Bauer, M. A., & Coburn, N. (1994). A distributed system architecture for a distributed application environment. *IBM Systems Journal, 33*(3), 399-26.

Becerra-Fernandez, I., & Sabherwal, R. (2001). Organizational knowledge management: A contingency perspective. *Journal of Management Information Systems, 18*(1), 23-55.

Becerra-Fernandez, I., Del Alto, M., & Stewart, H. (2006). Postdoc: A model for Web-based collaborative decision support at NASA. *International Journal of e-Collaboration, 2*(3), 49-63.

Becerra-Fernandez, I., Gonzalez, A., & Sabherwal, R. (2004). *Knowledge management: Challenges, solutions, and technologies.* Upper Saddle River, NJ: Prentice Hall.

Beck, K. (1999). Embracing change with extreme programming. *IEEE Computer, 32*(10), 70-77.

Beck, K. (1999). *Extreme programming explained: Embrace change.* Boston: Addison-Wesley.

Behravesh, N., & Klein, L. (2004). *The comprehensive impact of offshore IT software and services outsourcing on the U.S. economy and the IT industry.* Report prepared by Global Insight of the Information Technology Association of America. Arlington, VA.

Beise, C. (2004). IT project management and virtual teams. *Proceedings of the 2004 SIGMIS Conference on Computer Personnel Research,* 129-133.

Bennet, A., & Bennet, D. (2003). The partnership between organizational learning and knowledge management. In C. W. Holsapple (Ed.), *Handbook on knowledge management* (Vol. 1, pp. 439-460). New York: Springer-Verlag.

Berkowitz, B. (2003). *The new face of war: How war will be fought in the 21st century.* New York: The Free Press.

Bernstein, P. (1997). *Challenge to God: The risk history.* Rio de Janeiro, Brazil: Campus.

Bess, K. D., Fisher, A. T., Sonn, C. C., & Bishop, B. J. (2002). Psychological sense of community: Theory, research, and application. In A. T. Fisher, C. C. Sonn, & B. J. Bishop (Eds.), *Psychological sense of community: Research, applications and implications.* New York: Kluwer Academic/Plenun Publishers.

Beulen, E., van Fenema, P., & Currie, W. (2005). From application outsourcing to infrastructure management: Extending the offshore outsourcing service provider. *European Management Journal, 23*(2), 133.

Beynon-Davies, P. (1995). Information systems "failure": The case of the London Ambulance Service's Computer Aided Despatch Project. *European Journal of Information Systems, 4*(3), 171-184.

Bharadwaj, S. S. (2004). Offshore outsourcing: Path to new efficiencies in IT and business processes. *Journal of Information Technology Cases and Applications, 6*(3), 55.

Bilykh, I., Bychkov, Y., Dahlem, D., Jahnke, J. H., McCallum, G., Obry, C., et al. (2003). Can GRID services provide answers to the challenges of national health information sharing? *Proceedings of the 2003 Conference of the Centre for Advanced Studies on Collaborative Research,* 39-53.

Birmingham, D. (2002). *Software development on a leash.* New York: Springer-Verlag.

Bishop, M. W. (2006). China's cyberwarriors. *Foreign Policy, 156*, 93.

Blanchard, A., & Markus, M. L. (2003). The experienced "sense" of a virtual community: Characteristics and processes. *The Data Base for Advances in Information Systems.*

Bloor, G., & Dawson, P. (1994). Understanding professional culture in organizational context. *Organization Studies, 15*(2), 275-295.

Boaz, N. T., & Almquist, A. J. (1997). *Biological anthropology: A synthetic approach to human evolution.* Upper Saddle River, NJ: Prentice Hall.

Bock, G., Zmud, R. W., Kim, Y., & Lee, J. (2005). Behavioral intention formation in knowledge sharing: Examining the roles of extrinsic motivators, social-psychological forces, and organizational climate. *MIS Quarterly, 29*(1), 87-111.

Boehm, B. (1991). Software risk management: Principles and practices. *IEEE Software, 8*(1), 32-41.

Boehm, B. (2006). A view of 20th and 21st century software engineering. In *Proceedings of the 28th International Conference on Software Engineering,* 12-29.

Boehm, B., & Egyed, A. (1998). Software requirements negotiation: Some lessons learned. In *Proceedings of the 20th International Conference on Software Engineering*, 503-506.

Boehm, B., & Port, D. (2001). Balancing discipline and flexibility with the spiral model and MBASE. *Crosstalk, 14*(12).

Boehm, B., & Turner, T. (2004). *Balancing agility with discipline.* Boston: Addison-Wesley.

Boehm, B., Egyed, A., Shah, A., Kwan, J., & Madachy, R. (1998). A stakeholder win-win approach to software education. *Annals of Software Engineering, 6*, 295-321.

Bolton, R. N. (1998). A dynamic model of the duration of the customer's relationship with a continuous service provider: The role of satisfaction. *Marketing Science, 17*(1), 45-65.

Bonifacio, M., Bouquet, P., & Cuel, R. (2002). Knowledge nodes: The building blocks of a distributed approach to knowledge management. *Journal of Universal Computer Science, 8*(6), 652-661.

Bonifacio, M., Bouquet, P., & Traverso, P. (2002). Enabling distributed knowledge management: Managerial and technological implications. *Novatica and Informatik/Informatique, 3*(1).

Bonifacio, M., Bouquet, P., Mameli, G., & Nori, M. (2003). *Peer-mediated distributed knowledge management* (Tech. Rep. No. DIT-03-032). University of Trento.

Boody, D., & Macbeth, D. (2000). Prescriptions for managing change: A survey of their effects in projects to implement collaborative working between organizations. *International Journal of Project Management, 18*(5), 297-306.

Boody, D., Boonstra, A., & Kennedy, G. (2005). *Managing information systems: An organisational perspective.* Harlow, UK: Pearson Education.

Briggs, B. (2005). Offshore outsourcing poses risks. *Health Data Management, 13*(2), 68.

Briggs, R. O., De Vreede, G., Nunamaker, J. F., & Sprague, R. H. (2002). Special issue: Decision-making and a hierarchy of understanding. *Journal of Management Information Systems, 18*(4), 5-9.

Brothers, L., Hollan, J., Nielsen, J., Stornetta, S., Abney, S., Furnas, G., et al. (1992). Supporting informal communication via ephemeral interest groups. *Proceedings of CSCW'92 Conference*, 84-90.

Brown, M., & Trevino, L. (2006). Socialized charismatic leadership, values congruence, and deviance in work groups. *Journal of Applied Psychology, 91*(4), 954-962.

Brown, S. J., & Duguid, P. (2000). Balancing act: How to capture knowledge without killing it. *Harvard Business Review*, 73-80.

Brownsword, L., & Clements, P. (1996). *A case study in successful product line development* (Tech. Rep. No. CMU/SEI-96-TR-035). Software Engineering Institute.

Burgess, R. (1982). Some role problems in field research. In R. Burgess (Ed.), *Field research: A sourcebook and field manual* (pp. 45-49). London: George Allen & Unwin Publishers.

Burke, K., Aytes, K., & Chidambaram, L. (2001). Media effects on the development of cohesion and process satisfaction in computer-supported workgroups: An analysis of results from two longitudinal studies. *Information Technology & People, 14*(2), 122-141.

Burnes, B. (2000). *Managing change: A strategic approach to organisational dynamics.* Harlow, UK: Pearson Education Ltd.

Burns, J. M. (1978). *Leadership.* New York: Harper & Row.

Burroughs, S. M., & Eby, L. T. (1998). Psychological sense of community at work: A measurement system and explanatory framework. *Journal of Community Psychology, 26*, 509-532.

Campbell, B. C. (1992). *Humankind emerging.* New York: HarperCollins.

Cannataro, M., & Talia, D. (2003). The knowledge grid. *Communications of the ACM, 46*(1), 89-93.

Cantor, M. R. (1998). *Object-oriented project management with UML.* Wiley Computer Publishing.

Carley, K. (1992). Organizational learning and personnel turnover. *Organization Science, 33*, 353-365.

Carlson, J. R., & Zmud, R. W. (1999). Channel expansion theory and the experiential nature of media richness perceptions. *Academy of Management Journal, 42*(2), 153-170.

Carmel, E. (1999). *Global software teams: Collaborating across borders and time-zones.* Prentice Hall.

Carmel, E., & Agarwal, R. (2001). Tactical approaches for alleviating distance in global software development. *IEEE Software*, 22-29.

Carmel, E., & Agarwal, R. (2002). The maturation of offshore outsourcing of information technology work. *MIS Quarterly Executive, 1*(2), 65-77.

Carmel, E., & Tija, P. (2005). *Offshoring information technology: Sourcing and outsourcing to a global workforce.* UK: Cambridge.

Caron, J. R., Jarvenpaa, S. L., & Stoddard, D. B. (1994). Business reengineering at CIGNA Corporation: Experiences and lessons learned from the first five years. *MIS Quarterly, 18*(3), 233-250.

Cartwright, R. (2002). *Mastering team leadership.* Wales, UK: Palgrave Macmillan.

Cascio, W., & Shurygailo, S. (2002). E-leadership and virtual teams. *Organizational Dynamics, 31*(4), 362-376.

Casey, V., & Richardson, I. (2005). Uncovering the reality within virtual software teams. In *Proceedings of the 2006 International Workshop on Global Software Development for the Practitioner*, 66-72.

CeBASE. (2004). *eWorkshop on software inspections and pair programming.* Retrieved from http://www.cebase.org

Chang, C. K., & Christensen, M. (1999). A net practice for software project management. *IEEE Software, 16*(6), 80.

Chatman, J. A., & Barsade, S. G. (1995). Personality, organizational culture, and cooperation: Evidence from a business simulation. *Administrative Science Quarterly, 40*(3), 423-443.

Chavis, D. M., & Pretty, G. H. (1999). Sense of community: Advances in measurement and application. *Journal of Community Psychology, 27*(6), 635-642.

Cheung, W. K., & Liu, J. (2005). On knowledge grid and grid intelligence: A survey. *Computational Intelligence, 21*(2), 111-129.

Chidambaram, L., & Jones, B. (1993).Impact of communication medium and computer support on group perceptions and performance: A comparison of face-to-face and dispersed meetings. *MIS Quarterly Executive, 17*(4), 465.

Chidambaram, L., & Kautz, J. A. (1993). Defining common ground: Managing diversity through electronic meeting systems. *Proceedings of the 14th International Conference on Information Systems*, 1-11.

Chin, W. W. (2001). *PLS-Graph user's guide, version 3.0.* Soft Modeling Inc.

Chinowsky, P. S., & Rojas, E. M. (2003). Virtual teams: Guide to successful implementation. *Journal of Management in Engineering, 19*(3), 98-106.

Chipuer, H. M., & Pretty, G. H. (1999). A review of the Sense of Community Index: Current uses, factor structure, reliability and further development. *Journal of Community Psychology, 27*, 643-658.

Choudhury, V., & Sabherwal, R. (2003). Portfolios of control in outsourced software development projects. *Information Systems Research, 14*(3), 291-314.

Christensen, M. J., & Thayer, R. H. (2001). *The project manager's guide to software engineering's best practices.* Los Alamitos, CA: IEEE Computer Society.

Ciborra, C. U., & Andreu, R. (2001). Sharing knowledge across boundaries. *Journal of Information Technology, 16*(2), 73-82.

Clancy, J. J. (1989). *The invisible powers: The language of business.* Lexington, MA: Lexington Books.

Clark, S. C. (2002). Employees' sense of community, sense of control, and work/family conflict in Native American organizations. *Journal of Vocational Behavior, 61*, 92-108.

Clarke, A. (1999). A practical use of key success factors to improve the effectiveness of project management. *International Journal of Project Management, 17*(3), 139-145.

Cleland, D. I. (2004). The evolution of project management. *IEEE Transactions on Engineering Management, 51*(4), 396.

Clement, A. (1994). Computing at work: Empowering action by low-level users. *Communications of ACM, 37*(1), 53-63.

Coats, E., & Feldman, R. (2001). *Classic and contemporary readings in social psychology* (3rd ed.). Upper Saddle River, NJ: Pearson Education.

Cocheo, S. (2004, May). Regulators keep a risk-based eye on offshore outsourcing. *ABA Banking Journal*, p. 64.

Cohen, E., & Cornwell, L. (1989). College students believe piracy is acceptable. *CIS Educator's Forum, 1*(3), 2-5.

Cohn, M., & Ford, D. (2003). Introducing an agile process to an organisation. *IEEE Computer, 36*(6), 74-78.

Congiusta, A., Pugliese, A., Talia, D., & Trunfio, P. (2003). Designing grid services for distributed knowledge discovery. *Web Intelligence & Agent Systems, 1*(2), 91-104.

Connaughton, S. L., & Daly, J. A. (2004). Identification with leader: A comparison of perceptions of identification among geographically dispersed and co-located teams. *Corporate Communications, 9*(2), 89-103.

Conradi, R., Nguyen, M. N., Wang, A. I., Liu, C. (2000). Planning support to software process evolution. *International Journal of SEKE, 10*(1), 31.

Cossete, P., & Audet, M. (1992). Mapping of an idiosyncratic schema. *Journal of Management Studies, 29*(3), 325-348.

Courtney, J. F. (2001). Decision making and knowledge management in inquiring organizations: Toward a new decision-making paradigm for DSS. *Decision Support Systems, 31*(1), 17-39.

Cramton, C. D. (2001). The mutual knowledge problem and its consequences for dispersed collaboration. *Organization Science, 12*(3), 346-371.

Craumer, M. (2002). How to think strategically about outsourcing. *Harvard Management Update, 7*(5).

Crilley, K. (2001). Information warfare: New battle fields, terrorists, propaganda and the Internet. *Aslib Proceedings, 53*(7), 250-264.

Cronan, T., Foltz, C., & Jones, T. (2006). Piracy, computer crime, and IS misuse at the university. *Communications of the ACM, 49*(6), 85-90.

Cronin, B., & Crawford, H. (1999). Information warfare: Its application in military and civilian contexts. *The Information Society, 15*, 257-263.

Cross, R., & Cummings, J. N. (2004). Tie and network correlates of individual performance in knowledge-intensive work. *Academy of Management Journal, 47*(6), 928-937.

Crossan, M. M., & Berdrow, I. (2003). Organizational learning and strategic renewal. *Strategic Management Journal, 24*(11), 1087-1105.

Crowston, K. (1991). A coordination theory approach to organizational process design. *Organization Science, 8*(2), 157-175.

Cuilla, J. (2004). Ethics and leadership effectiveness. In J. Antonakis, A. T. Cianciolo, & R. J. Sternberg (Eds.), *The nature of leadership* (pp. 302-328). Thousand Oaks, CA: Sage.

Cushway, B., & Lodge, D. (1999). *Organization behaviour and design.* London: Kogan Page Ltd.

Daft, R. L., & Lengel, R. H. (1986). Organizational information requirements, media richness and structural design. *Management Science, 32*(5), 554-571.

Daft, R. L., & Lengel, R. H. (1986). Organizational information requirements, media richness and structural design. *Management Science, 32*(5), 554-571.

Daft, R. L., Lengel, R. H., & Trevino, L. K. (1987). Message equivocality, media selection, and manager performance: Implications for information systems. *MIS Quarterly, 11*(3), 355-366.

Dallavalle, T., Esposito, A., & Lang, S. (1992). Groupware: One experience. *Proceedings of Groupware'92 Conference*, 470-477.

Damian, D., & Moitra, D. (2006). Guest editors' introduction: Global software development: How far have we come? *IEEE Software, 23*(5), 17-19.

Darnton, G. (2003). Information warfare and the laws of war. *2nd European Conference on Information Warfare and Security*, 107-116.

Darwin, C. (1966). *On the origin of species.* Cambridge, MA: Harvard University Press. (Original work published 1859)

Davenport, T. H., & Prusak, L. (1998). *Working knowledge: How organizations manage what they know.* Boston: Harvard Business School Press.

Davenport, T. H., DeLong, D. W., & Beers, M. C. (1998). Successful knowledge management. *Sloan Management Review, 39*(2), 43-57.

Davenport, T., & Prusak, L. (1998). *Working knowledge: How organizations manage what they know.* Boston: Harvard Business School Press.

Davidson, E. J., & Tay, A. S. M. (2003). Studying teamwork in global IT support. *Proceedings of the 36th Hawaii International Conference on Systems Sciences.*

Davis, F. D., Bagozzi, R. P., & Warshaw, P. R. (1989). User acceptance of computer technology: A comparison of two theoretical models. *Management Science, 35*(8), 982-1003.

Davis, J. A. (1985). *The logic of causal order.* London: Sage.

Dawkins, R. (1986). *The blind watchmaker.* New York: W.W. Norton & Company.

Dawkins, R. (1990). *The selfish gene.* Oxford, UK: Oxford University Press.

De Pillis, E., & Furumo, K. (2006). Virtual vs. face-to-face teams: Deadbeats, deserters, and other considerations. In *Proceedings of the 2006 ACM SIGMIS CPR Conference on Computer Personnel Research*, 318-320.

Delibasis, D. (2002). The right of states to use force in cyberspace: Defining the rules of engagement. *Information & Communication Technology Law, 11*(3), 255-268.

Delone, W. H., & McLean, E. R. (1992). Information systems success: The quest for the dependent variable. *Information Systems Research, 3*(1), 60-95.

DeLong, D. W. (2004). *Lost knowledge: Confronting the threat of an aging workforce.* New York: Oxford University Press.

DeLong, D. W., & Fahey, L. (2000). Diagnosing cultural barriers to knowledge management. *Academy of Management Executive, 14*(4), 113-127.

DeMarco, T., & Lister, T. (2003). *Waltzing with bears: Managing risks on software projects.* Dorset House Publishing.

Deming, W. E. (1986). *Out of the crisis.* Cambridge, MA: MIT, Center for Advanced Engineering Study.

Denning, D. E. (1999). *Information warfare and security.* Reading, MA: Addison-Wesley.

Dennis, A. R., Carte, T. A., & Kelly, G. G. (2003). Breaking the rules: Success and failure in groupware-supported business process reengineering. *Decision Support Systems, 36*(1), 31-47.

Dennis, A. R., Daniels, R. M., Jr., Hayes, G., & Nunamaker, J. F., Jr. (1993). Methodology-driven use of automated support in business process re-engineering. *Journal of Management Information Systems, 10*(3), 117-138.

Dennis, A. R., Hayes, G. S., & Daniels, R. M., Jr. (1994). Re-engineering business process modeling. *27th Hawaii International Conference on System Sciences*, 244-253.

Dennis, A., & Vessey, I. (2004). Three knowledge management strategies: Knowledge hierarchies, knowledge markets, and knowledge communities. *MIS Quarterly Executive, 3*(4), 399-412.

Dennis, A., George, J., Jessup, L., Nunamaker, J., & Vogel, D. (1988). Information technology to support electronic meetings. *MIS Quarterly, 12*(4), 591-624.

DeSanctis, G., & Poole, M. S. (1994). Capturing the complexity in advanced technology use: Adaptive structuration theory. *Organization Science, 5*(2), 121-147.

DeSanctis, G., Poole, M. S., Dickson, G. W., & Jackson, B. M. (1993). Interpretive analysis of team use of group technologies. *Journal of Organizational Computing, 3*(1), 1-29.

Desouza, K., & Evaristo, J. R. (2004). Managing knowledge in distributed projects. *Communications of the ACM, 47*(4), 87-91.

DeTiene, K. B., & Jackson, L. A. (2001). Knowledge management: Understanding theory and developing strategy. *Competitiveness Review, 11*(1), 1-11.

Devaraj, S., Fan, M., & Kohli, R. (2002). Antecedents of B2C channel satisfaction and preference: Validating e-commerce metrics. *Information Systems Research, 13*(3), 316-333.

Dingle, M. E. (1994). *Business process reengineering: A New Zealand perspective.* Palmerston North, New Zealand: Massey University, Department of Executive Education.

DiRomualdo, A., & Gurbaxani, V. (1998). Strategic intent for IT outsourcing. *Sloan Management Review, 39*(4), 67.

DiSabatino, J. (2001, April 16). Collaborative peer-to-peer technology finds Groove at GlaxoSmithKline, Raytheon. *Computer World.*

Dodson, J. (2000, September 18). It's time to chat with your customers. *Internetweek.* Retrieved from http://www.internetwk.com/ebizapps/col091800.htm

Dossani, R., & Kenney, M. (2003). *Went for cost, stayed for quality? Moving the back office to India.* Stanford University, Asia-Pacific Research Center. Retrieved from http://APARC.stanford.edu

Dozier, R. W., Jr. (1992). *Codes of evolution.* New York: Crown Publishers.

Dube, L., & Pare, G. (2001). Global virtual teams. *Communications of the ACM, 44*(12), 71.

Dubé, L., & Paré, G. (2004). The multi-faceted nature of virtual teams. In D. J. Pauleen (Ed.), *Virtual teams: Projects, protocols, and processes* (pp. 1-39). Hershey, PA: Idea Group Publishing.

Duck, J. (1993). Managing change: The art of balancing. *Harvard Business Review, 71*(6), 109-118.

Duncan, H. D. (1968). *Symbols in society.* New York: Oxford University.

Earl, M. (2001). Knowledge management strategies: Toward a taxonomy. *Journal of Management Information Systems, 18*(1), 215-233.

Earley. (1994). Self or group? Cultural effects of training on self-efficacy and performance. *Administrative Science Quarterly, 39*, 89-117.

Easterbrook, S. M., Beck, E. E., Goodlet, J. S., Plowman, L., Sharples, M., & Wood, C. C. (1993). A survey of empirical studies of conflict. In S. Easterbrook (Ed.), *CSCW: Cooperation or conflict* (pp. 1-68)? New York: Springer-Verlag.

Edgington, T., Choi, B., Henson, K., Raghu, T. S., & Vinze, A. (2004). Adopting ontology to facilitate knowledge sharing. *Communications of the ACM, 47*(11), 85-90.

Edwards, H. K., & Sridhar, V. (2005). Analysis of software requirements engineering exercises in a global virtual team setup. *Journal of Global Information Management, 13*(2), 21.

effectiveness. *Proceedings of the 21st International Conference on Information Systems*, 575-580.

Elden, M., & Chisholm, R. F. (1993). Emerging varieties of action research. *Human Relations, 46*(2), 121-141.

Enz, C. (1986). *Power and shared values in the corporate culture.* Ann Arbor, MI: The University of Michigan Press.

Erickson, J., & Evaristo, J. R. (2005). *Risk factors in distributed projects.* Chicago: University of Illinois, College of Business Administration.

Erickson, T., Halveron, C., Kellog, W., Laff, M., & Wolf, T. (2002). Social translucence: Designing social infrastructures than make collective activity visible. *Communications of the ACM, 45*(4), 40-44.

Espinosa, J. A., Cummings, J. N., Wilson, J. M., & Pearce, B. M. (2003). Team boundary issues across multiple global firms. *Journal of Management Information Systems, 19*(4), 157-190.

Evans, G. (2005). Agile RUP: Taming the Rational™ unified process. In L. Liu & B. Roussev (Eds.), *Management of object-oriented software development.* Hershey, PA: Idea Group.

Evaristo, R., & Munkvold, B. E. (2002). Collaborative infrastructure formation in virtual projects. *Journal of Global Information Technology Management, 5*(2), 29-47.

Evaristo, R., Scudder, R., & Desouza, K. (2004). A dimensional analysis of geographically distributed project teams: A case study. *Journal of Engineering and Technology Management, 21*(3), 175-189.

Ezey, P. (2003). Integration and its challenges in participant observation. *Qualitative Research, 3*(2), 191-205.

Ezingeard, J., Leigh, S., & Chandler-Wilde, R. (2000). Knowledge management at Ernst & Young UK: Getting value through knowledge flows. *Proceedings of the Twenty First International Conference on Information Systems*, 807-822.

Fattah, H. M. (2002). *How peer-to-peer technology is revolutionizing the way we do business.* Chicago: Dearborn Trade Publishing.

Fernandez, A., Holmer, T., Rubart, J., & Schuemmer, T. (2002). Three groupware patterns from the activity awareness family. *EuroPLoP 2002: Seventh European Conference on Pattern Languages of Programs.*

Fetterman, D. (1991). A walk through the wilderness: Learning to find your way. In W. Shaffir & R. Stebbins (Eds.), *Experiencing fieldwork: An inside view of qualitative research* (pp. 87-96). Thousand Oaks, CA: Sage Publications.

Fichman, R., & Kemerer, C. (1999). The illusory diffusion of innovation: An examination of assimilation gaps. *Information Systems Research, 10*(3), 255-275.

Field, M., & Keller, L. (1998). *Project management.* London: Thomson Learning.

Financial Times Limited. (2005). Special reports: Outsourcing — Making offshoring happen. *Accountancy, 135*(1340), 55.

Fiol, C. M., & O'Connor, E. J. (2005). Identification in face-to-face, hybrid, and pure virtual teams: Untangling the contradictions. *Organization Science, 16*(1), 19-32.

Fitzgerald, B. (1997). The use of systems development methodologies in practice: A field study. *Information Systems Journal, 7*(3), 210-212.

Fitzgerald, G., & Willcocks, L. (1993). *Information technology outsourcing practice: A U.K. survey.* London: Business Intelligence.

Fjermestad, J. (2005). Virtual group strategic decision making using structured conflict and consensus approaches. *Internal Journal of e-Collaboration, 1*(1), 43-61.

Fornell, C., & Larcker, D. F. (1981). Evaluating structural equation models with unobservable variables and measurement error. *Journal of Marketing Research, 18*(1), 39-50.

Fornell, C., Johnson, M. D., Anderson, E. W., Cha, J., & Bryant, B. E. (1996). The American Customer Satisfaction Index: Nature, purpose, and findings. *Journal of Marketing, 60*(7), 7-18.

Foster, I., Kesselman, C., Nick, J. M., & Tuecke, S. (2002). Grid services for distributed system integration. *IEEE Computer, 35*(6), 37-46.

Foster, P. M. (2004). Psychological sense of community in groups on the Internet. *Behaviour Change, 21*(2), 141-146.

Fowler, M. (2000). Put your process on a diet. *Software Development, 8*(2), 32-36.

Fowler, M., & Highsmith, J. (2001). The agile manifesto. *Software Development, 9*(8), 28-32.

French, W. (2006). Business ethics training: Face-to-face and at a distance. *Journal of Business Ethics, 66*(1), 117-126

French, W. L., & Bell, Jr. (1990). *Organization development.* Englewood Cliffs, NJ: Prentice Hall.

Fulk, J., Schmitz, J., & Steinfield, C. W. (1990). A social influence model of technology use. In J. Fulk & C. Steinfield (Eds.), *Organizations and communication technology* (pp. 117-140). Newbury Park, CA: Sage.

Furnell, S. M., & Warren, M. J. (1999). Computer hacking and cyber terrorism: The real threats in the new millennium? *Computers & Security, 18*(1), 28-34.

Furst, S., Blackburn, R., & Rosen, B. (1999). Virtual team effectiveness: A proposed research agenda. *Information Systems Journal, 9*(4), 249-269.

Gallivan, M. (2001). Organizational adoption and assimilation of complex technological innovations: Development and application of a new framework. *The Database for Advances in Information Systems, 32*(3), 51-85.

Gallivan, M. (2004). Examining IT professionals' adaptation to technological change: The influence of gender and personal attributes. *The Database for Advances in Information Systems, 35*(3), 28-49.

Gamma, E., Helm, R., Johnson, R., & Vlissides, J. (1994). *Design patterns: Elements of reusable object-oriented software.* Boston: Addison-Wesley.

García, I., Giuliani, F., & Wiesenfeld, E. (1999). Community and sense of community: The case of an urban barrio in Caracas. *Journal of Community Psychology, 27,* 727-740.

Gassmann, O., & Von Zedtwitz, M. (2003). Trends and determinants of managing virtual R and D teams. *R & D Management, 33*(3), 243-262.

Gentile, C. J. (2000). Hacker war rages in Holy Land. *Wired News.* Retrieved February 7, 2004, from http://www.wired.com/news/print/0,1294,40030,00.html

Gephart, R. (1988). Status degradation and organizational succession. In J. Van Maanen (Ed.), *Qualitative studies of organizations* (pp. 159-191). Thousand Oaks, CA: Sage Publications.

Gerpott, T. J., Rams, W., & Schindler, A. (2001). Customer retention, loyalty, and satisfaction in the German mobile cellular telecommunications market. *Telecommunications Policy, 25*(4), 249-269.

Giacalone, R., & Thompson, K. (2006). From the guest co-editors: Special issue on ethics and social responsibility. *Academy of Management Learning & Education, 5*(3), 261-265.

Gilb, T., & Graham, G. (1993). *Software inspection.* Boston: Addison-Wesley.

Gillard, S., & Joharisen, J. (2004). Project management communication: A systems approach. *Journal of Information Science, 30*(1), 23.

Glaser, B. G., & Strauss, A. L. (1971). *The discovery of grounded theory: Strategies for qualitative research.* Chicago: Aldine Publishing Company. (Original work published 1967)

Glass, R. S., & Wood, W. A. (1996). Situational determinants of software piracy: An equity theory perspective. *Journal of Business Ethics, 15*(11), 1189-1198.

Goffee, R., & Jones, G. (1996). What holds the modern company together? *Harvard Business Review, 74*(6), 133-148.

Gold, A. H., Malhotra, A., & Segars, A. H. (2001). Knowledge management: An organizational capabilities perspective. *Journal of Management Information Systems, 18*(1), 185-214.

Goodbody, J. (2005). Critical success factors for global virtual teams. *Strategic Communication Management, 9*(2), 18-21.

Gopal, R. D., Walter, Z., & Tripathi, A. K. (2001). Admediation: New horizons in effective email advertising. *Communications of the ACM, 44*(12), 91-96.

Gould, S. J. (1977). *Ever since Darwin: Reflections in natural history.* New York: W.W. Norton & Company.

Graber, D. A. (1976). *Verbal behavior and politics.* Chicago: University of Illinois Press.

Graetz, K. A., Boyle, E. S., Kimble, C. E., Thompson, P., & Garloch, J. L. (1998). Information sharing in face-to-face, teleconferencing, and electronic chat groups. *Small Group Research, 29*(6), 714-743.

Grant, R. M. (1996). Prospering in dynamically-competitive environments: Organizational capability as knowledge integration. *Organization Science, 7*(4), 375-387.

Grant, R. M. (1996). Toward a knowledge-based theory of the firm. *Strategic Management Journal, 17,* 109-122.

Green, G., Collins, R., & Hevner, A. (2004). Perceived control and diffusion of software process innovations. *Journal of High Technology Management, 15*(1), 123-144.

Gregor, S., & Benbasat, I. (1999). Explanations from intelligent systems: Theoretical foundations and implications for practice. *MIS Quarterly, 23*(4), 497-530.

Griffin, E. (1997). *A first look at communication theory.* New York: McGraw-Hill.

Griffith, T., Sawyer, J. E., & Neale, M. A. (2003). Virtualness and knowledge in teams: Managing the love triangle of organizations, individuals and information technology. *MIS Quarterly, 27*(2), 265-287.

Grojean, M. W., Resick, C. J., Dickson, M. W., & Smith, D. B. (2004). Leaders, values, and organizational climate: Examining leadership strategies for establishing an organizational climate regarding ethics. *Journal of Business Ethics, 55*(3), 223-241.

Groove Networks. (2006). *Customers: Case studies. Steelcase Inc.* Retrieved from http://www.groove.net/index.cfm?pagename=CaseStudy_Steelcase

Groth, L. (1999). *Future organizational design.* West Sussex, UK: John Wiley & Sons.

Grover, V., & Davenport, T. H. (2001). General perspectives on knowledge management: Fostering a research agenda. *Journal of Management Information Systems, 18*(1), 5-22.

Grover, V., & Segars, A. H. (1996). IT: The next 1100102 years. *Database, 27*(4), 45-57.

Grover, V., Teng, J., & Cheon, M. J. (1998). Towards a theoretically-based model of information systems outsourcing. In L. P. Willcocks, & M. Lacity (Eds.),

Strategic source of information systems: Perspective and practices. New York: Wiley.

Grudin, J. (1994). Eight challenges for developers. *Communications of the ACM, 37*(1), 93-105.

Grupe, F. H., Urwiler, R., Ramarapu, N. K., & Owrang, M. (1998). The application of case-based reasoning to the software development process. *Information and Software Technology, 40*(9), 493.

Guha, S., Kettinger, W. J., & Teng, J. T. (1993). Business process reengineering: Building a comprehensive methodology. *Information Systems Management*, 13-22.

Gupta, A. K., & Govindarajan, V. (2000). Knowledge flows within multinational corporations. *Strategic Management Journal, 21*(4), 473-496.

Gupta, A. K., & Govindarajan, V. (2000). Knowledge management's social dimension: Lessons from Nucor Steel. *Sloan Management Review, 42*(1), 71-80.

Gurney, J. (1991). Female research in male-dominated settings: Implications for short-term versus long-term research. In W. Shaffir & R. Stebbins (Eds.), *Experiencing fieldwork: An inside view of qualitative research* (pp. 53-61). Thousand Oaks, CA: Sage Publications.

Hahn, J., & Subramani, M. (2000). A framework of knowledge management systems: Issues and challenges for theory and practice. *Proceedings of the Twenty First International Conference on Information Systems*, 302-312.

Hair, J. F., Jr., Anderson, R. E., Tatham, R. L., & Black, W. C. (1995). *Multivariate data analysis with readings* (4th ed.). Englewood Cliffs, NJ: Prentice Hall.

Hale, K., Souza, R., Lo, T., Adachi, Y., & Babaie, E. (2005). *Forecast: IT services, worldwide, 2005-2009.* Gartner. Retrieved June 2005 from http://www.gartner.com

Hall, G., Rosenthal, J., & Wade, J. (1993). How to make reengineering really work. *Harvard Business Review*, 119-131.

Hall, W. M. (2003). *Stray voltage: War in the information age.* Annapolis, MD: Naval Institute Press.

Hammer, M. (1990). Reengineering work: Don't automate, obliterate. *Harvard Business Review, 68*(4), 104-114.

Hammer, M., & Champy, J. (1993). *Reengineering the corporation.* New York: Harper Business.

Hansen, M. T., & Nohria, N. (1999). What's your strategy for managing knowledge? *Harvard Business Review, 77*(2), 106-117.

Hansen, M. T., Nohria, N., & Tierney, T. (1999). What's your strategy for managing knowledge? *Harvard Business Review*, 106-115.

Hargadon, A. B. (1998). Firms as knowledge brokers: Lessons in pursuing continuous innovation. *California Management Review, 40*(3), 209-227.

Harrington, H. J. (1991). *Business process improvement.* New York: McGraw-Hill.

Harris, J. (1999). The idea of community in the study of writing. In L. Ede (Ed.), *On writing research: The Braddock essays 1975-1998* (pp. 260-271). Boston: Bedford.

Hartley, J. (1984). Case studies in organizational research. In C. Cassell & G. Symon (Eds.), *Qualitative methods in organizational research* (pp. 208-229). London: Sage Publications.

Hasan, H., & Gould, E. (2001). Support for the sense-making activity of managers. *Decision Support Systems, 31*(1), 71-86.

Hatch, N. W., & Dyer, J. H. (2004). Human capital and learning as a source of sustainable competitive advantage. *Strategic Management Journal, 25*(12), 1155-1178.

Hempel, J. (2005, January 31). A corporate peace corps catches on. *Business Week*, p. 14.

Henderson, J. C., & Lee, S. (1992). Managing IS design teams: A control theories perspective. *Management Science, 38*(6), 757-777.

Hendriks, P. H. J. (1999). Why share knowledge? The influence of ICT on the motivation for knowledge sharing. *Knowledge and Process Management, 6*(2), 91-100.

Henninger, D. (2005, April 1). Wonder land: Can Justice Scalia solve the riddles of the Internet? *Wall Street Journal*, p. A10.

Herbsleb, J. D., & Grinter, R. (1999). Splitting the organization and integrating the code: Conway's law revisited. In *Proceedings of the International Conference on Software Engineering*, 85-96.

Herbsleb, J. D., & Moitra, D. (2001). Guest editors' introduction: Global software development. *IEEE Software, 18*(2), 16-20.

Herzog, V. L. (2001). Trust building on corporate collaborative project teams. *Project Management Journal, 32*(1), 28-37.

Hinde, S. (2001). Incalculable potential for damage by cyber-terrorism. *Computers & Security, 20*, 568-572.

Hinde, S. (2002). Spam, scams, chains, hoaxes and other junk mail. *Computers & Security, 21*(7), 592-606.

Hinduja, S. (2003). Trends and patterns among online software pirates. *Ethics and Information Technology, 5*(1), 49-61.

Hollingshead, A. B. & McGrath, J. E. (1993). Group task performance and communication technology. *Small Group Research, 24*(3), 307.

Homsky, O. (2003). More patterns for group leadership. *EuroPLoP 2003: Seventh European Conference on Pattern Languages of Programs.*

Huber, N. (2005, May 3). Outsourcing "backlash" highlights need for IT leaders to sharpen management skills. *Computer Weekly*, pp. 24-27.

Huck, S., Muller, W., & Vriend, N. J. (2002). The East End, the West End, and King's Cross: On clustering in the four-player hotelling game. *Economic Inquiry, 40*(2), 231-241.

Huczynski, A., & Buchanan, D. (1991). *Organizational behaviour: An introductory text* (2nd ed.). New York: Prentice Hall.

Huff, C. W., Sproull, L., & Kiesler, S. (1989). Electronic communication and organizational commitment: Tracing the relationship in a city government. *Journal of applied social psychology, 19*, 1371-1391.

Humphrey, S. (1989). *Managing the software process.* New York: Addison-Wesley.

Hunt, A., & Thomas, D. (2000). *The pragmatic programmer.* Boston: Addison Wesley Longman.

Hurley, M., & Rundell, J. (1997). *The IT outsourcing imperative: Intelligently managing the inevitable decision* (KPMG Report). Australia.

Hurley, R., & Hult, T. (1998). Innovation, market orientation, and organizational learning: An integration and empirical examination. *Journal of Marketing, 62*(3), 42-54.

Hurwitz, J., Zander, A., & Hymovitch, B. (1968). Some effects of power on the relations among group members. In D. Cartwright & A. Zander (Eds.), *Group dynamics* (pp. 291-297). New York: Harper & Row Publishers.

IEEE. (2004). *IEEE Standard 1540-2004. IEEE standard for software life cycle processes: Risk management.* IEEE Press.

Im, G., & Baskerville, R. (2005). A longitudinal study of information system threat categories: The enduring problem of human error. *The Database for Advances in Information Systems, 36*(4), 68-79.

Industry Canada. (2004). The digital economy in Canada: Unsolicited commercial electronic mail (SPAM). Retrieved July 30, 2004, from http://e-com.ic.gc.ca/epic/internet/inecic-ceac.nsf/en/h_gv00170e.html#stats

Inkpen, A. C., & Tsang, E. W. K. (2005). Social capital, networks, and knowledge transfer. *Academy of Management Review, 30*(1), 146-166.

Insinga, R. C., &. Werle, M. J. (2000). Linking outsourcing to business strategy. *Academy of Management Executive, 14*(4), 59.

Isaac, G. L. (1993). Aspects of human evolution. In R. L. Ciochon & J. G. Fleagle (Eds.), *The human evolution source book* (pp. 263-273). Englewood Cliffs, NJ: Prentice Hall.

Jackson, M. H., & Purcell, D. (1997). Politics and media richness in World Wide Web representations of the former Yugoslavia. *Geographical Review, 87*(2), 219-239.

Jacobson, I. (1987). Object-oriented development in an industrial environment. *ACM SIGPLAN Notices, 22*(12), 183-191.

Jacobson, I., Booch, G., & Rumbaugh, J. (1999). *The unified software development process.* Boston: Addison-Wesley.

Jacoby, C. (2000). E-commerce hot spots. *Corporate Location, 3*, 18.

Janis, I. (1972). *Victims of groupthink.* Boston: Houghton Mifflin Company.

Janis, I. L. (1972). *Victims of groupthink: A psychological study of foreign policy decisions and fiascos.* New York: Houghton Mifflin.

Janz, B. D., & Prasarnphanich, P. (2003). Understanding the antecedents of effective knowledge management: The importance of knowledge-centered culture. *Decision Sciences, 34*(2), 351-384.

Jarvenpaa, S. L., & Staples, D. S. (2000). The use of collaborative electronic media for information sharing: An exploratory study of determinants. *Strategic Information Systems, 9*(2-3), 129-154.

Jarvenpaa, S. L., & Staples, S. D. (2001). Exploring perceptions of organizational ownership of information and expertise. *Journal of Management Information Systems, 18*(1), 151-183.

Jarvenpaa, S., & Leidner, D. (1999). Communication and trust in global virtual teams. *Organization Science, 10*(6), 791-815.

Jessup, L. M., Connolly, T., & Galegher, J. (1990). The effects of anonymity on GDSS group process with an idea-generating task. *MIS Quarterly, 14*(3), 313-321.

Jessup, L., & Tansik, D. (1991). Decision making in an automated environment: The effects of anonymity and proximity on group process and outcome with a GDSS. *Decision Sciences, 22*(2), 266-279.

Jiang, J., Klein, G., & Discenza, R. (2002). Pre-project partnering impact on information systems projects, project team, and project manager. *European Journal of Information Systems, 11*(2), 86-97.

Johnson, G. (1992). Managing strategic change: Strategy, culture and action. *Long Range Planning, 25*(1), 28-36.

Johnson, S., Suriya, C., Yoon, S., Berrett, J., & LaFleur, J. (2002). Team development and group processes of virtual learning teams. *Computers & Education, 39*, 379-393.

Jones, A., Kovacich, G. L., & Luzwick, P. G. (2002). Everything you wanted to know about information warfare but were afraid to ask: Part 1. *Information Systems Security, 11*(4), 9-20.

Jones, E., Gergen, K., Gumpert, P., & Thibaut, J. (1968). Some conditions affecting the use of ingratiation to influence performance evaluation. In D. Cartwright & A. Zander (Eds.), *Group dynamics* (pp. 278-289). New York: Harper & Row Publishers.

Jones, P., & Roelofsma, P. (2000). The potential for social contextual and group biases in team decision making: Biases, conditions and psychological mechanisms. *Ergonomics, 43*(8), 1129-1152.

Jones, T. M. (1991). Ethical decision making by individuals in organizations: An issue-contingent model. *The Academy of Management Review, 16*(2), 366-395.

Jonsonn, S. (1991). Action research. In H. Nissen, H. K. Klein, & R. Hirschheim (Eds.), *Information systems research: Contemporary approaches and emergent traditions* (pp. 371-396). New York: North-Holland.

Jorgensen, D. (1989). *Participant observation: A methodology for human studies.* Thousand Oaks, CA: Sage Publications.

Joseph, D., Ang, S., & Tan, H. T. (1996). *The nature of soft skills of information systems professionals.* Proceedings of the Americas Conference on Information Systems, Phoenix, AZ.

Judge, T. A., & Piccolo, R. F. (2004). Transformational and transactional leadership: A meta-analytic test of their relative validity. *Journal of Applied Psychology, 89*(5), 755-768.

Jung, D. I., & Avolio, B. J. (1999). Effects of leadership style and followers' cultural orientation on performance in group and individual task conditions. *Academy of Management Journal, 42*(2), 208-218.

Kahai, S., Avolio, B., & Sosik, J. (1998). Effects of source and participant anonymity and difference in initial opinions in an EMS context. *Decision Sciences, 29*(2), 427-460.

Kahai, S., Sosik, J., & Avolio, B. (1997). Effects of leadership style and problem structure on work group process and outcomes in an electronic meeting system environment. *Personnel Psychology, 50*(1), 121-146.

Kahai, S., Sosik, J., & Avolio, B. (2004). Effects of participative and directive leadership in electronic groups. *Groups & Organization Management, 29*(1), 67-105.

Kalita, T. (2003). DSDM in process improvement. In J. Stapleton (Ed.), *DSDM business focused development* (2nd ed., pp. 175-191). London: Pearson Education.

Kankanhalli, A., Tanudidjaj, F., Sutanto, J., & Tan, B. C. Y. (2003). The role of IT in successful knowledge management initiatives. *Communications of the ACM, 46*(9), 69-73.

Kanungo, S., Sadavarti, S., & Srinivas, Y. (2001). Relating IT strategy and organizational culture: An empirical study of public sector units in India. *Journal of Strategic Information Systems, 10*, 29-57.

Karolak, D. W. (1998). *Global software development: Managing virtual teams and environments.* Los Alamitos: IEEE Computer Society Press.

Katzy, B., Evaristo, R., & Zigurs, I. (2000). Knowledge management in virtual projects: A research agenda. *Proceedings of the 36th Hawaii International Conference on Systems Sciences.*

Kedia, B., Lahiri, S., & Lovvorn, A. (2005). Seeking competitive advantage on distant shores. *European Business Forum, 21*, 37.

Kegan, R., & Lahey, L. (2001a). *How the way we talk can change the way we work.* CA: Jossey-Bass.

Kegan, R., & Lahey, L. (2001b). The real reason people won't change. *Harvard Business Review, 79*(10), 85-89.

Keil, M. (1995). Pulling the plug: Software project management and the problem of project escalation. *MIS Quarterly, 19*(4), 421.

Keil, M., & Montealegre, R. (2000, Spring).Cutting your losses: Extricating your organization when big project goes awry. *Sloan Management Review, 396-311*, 55.

Keil, M., & Robey D (2001). Blowing the whistle on troubled software projects. *Communications of the ACM, 44*(4), 87.

Keil, M., Mann, J., & Rai, A. (2000). Why software projects escalate: An empirical analysis and test of four theoretical models. *MIS Quarterly, 24*(4), 631-660.

Kendall, J. E., & Kendall, K. E. (1993). Metaphors and methodologies: Living beyond the systems machine. *MIS Quarterly, 17*(2), 149-171.

Kendall, J. E., & Kendall, K. E. (1994). Metaphors and their meaning for information systems development. *European Journal of Information Systems, 3*(1), 37-47.

Kendall, K. E. (2000). Ecommerce aesthetics: Ecommerce for ecommerce's sake. *Information Resources Management Journal, 13*(3).

Kendall, K. E., & Kendall, J. E. (2005). *Systems analysis and design* (6th ed.). Upper Saddle River, NJ: Prentice Hall.

Kern, T., & Willcocks, L. P. (2002). Service provision and the net: Risky application sourcing? In R. Hirschheim, A. Heinzl, & J. Dibbern (Eds.). *Information systems outsourcing: Enduring themes, emergent patterns, and future directions*. Heidelberg: Springer

Kerzner, H. (2000). *Project management: A systems approach to planning, scheduling, and controlling.* John Wiley & Sons.

Khan N., & Fitzgerald (2004). Dimensions of offshore outsourcing business models. *Journal of Information Technology Cases and Applications, 6*(3), 35.

Khan, N., Currie, W. L., Weerakkody, V., & Desai, B., (2002). Evaluating offshore IT outsourcing in India: Supplier and customer scenarios. In *IEEE Proceedings of the 36th Hawaii International Conference on System Sciences.*

Khazanchi, D., & Zigurs, I. (2005). *Patterns of effective management of virtual projects: An exploratory study.* Newtown Square, PA: Project Management Institute.

Khazanchi, D., & Zigurs, I. (2007). An assessment framework for discovering and using patterns in virtual project management. *Proceedings of the 36th Hawaii International Conference on Systems Sciences.*

Kiel, L. (2003). Experiences in distributed development: A case study. In *Proceedings of the International Workshop on Global Software Development at ICSE*, 44-47.

Kim, M. K., Park, M. C., & Jeong, D. H. (2004). The effects of customer satisfaction and switching barrier on customer loyalty in Korean mobile telecommunication services. *Telecommunication Policy, 28*(2), 145-159.

Kim, Y. (2001). A comparative study of the "Abilene paradox" and "groupthink." *Public Administration Quarterly, 25*(2), 168-187.

King, N. (1990). Innovation at work: The research literature. In M. A. West & J. L. Farr (Eds.), *Innovation and creativity at work* (pp. 15-59). New York: John Wiley & Sons.

King, W. R. (2005). Outsourcing becomes more complex. *Information Systems Management, 22*(2), 89-90.

King, W. R., Marks, P. V., & McCoy, S. (2002). The most important issues in knowledge management. *Communications of the ACM, 45*(9), 93-97.

Kini, R. B. (2002). Peer-to-peer technology: A technology reborn. *Information Systems Management, 19*(3), 74-85.

Kirsch, L. J. (1996). The management of complex tasks in organizations: Controlling the systems development process. *Organization Science, 7*(1), 1-21.

Kirsch, L. J., Sambamurthy, V., Ko, D., & Purvis, R. L. (2002). Controlling information systems development projects: The view from the client. *Management Science, 48*(4), 484-498.

Kliem, R. (2004). Managing the risks of offshore IT development projects. *Information Systems Management, 21*(3), 6.

Knapp, E., & Yu, D. (1999). How culture helps or hinders the flow of knowledge. *Knowledge Management Review, 2*(1), 16-21.

Koch, N. (2005a). Using action research to study e-collaboration. *International Journal of e-Collaboration, 1*(4), i-vii.

Koch, N. (2005b). What is e-collaboration? *International Journal of e-Collaboration, 1*(1), i-vii.

Kock, N. (1995). *Process reengineering, PROI: A practical methodology.* Sao Paulo, Brazil: Editora Vozes.

Kock, N. (1998). Can communication medium limitations foster better group outcomes? An action research study. *Information & Management, 34*(5), 295-305.

Kock, N. (1999). *Process improvement and organizational learning: The role of collaboration technologies.* Hershey, PA: Idea Group Publishing.

Kock, N. (2001a). Asynchronous and distributed process improvement: The role of collaborative technologies. *Information Systems Journal, 11*(2), 87-110.

Kock, N. (2001b). Compensatory adaptation to a lean medium: An action research investigation of electronic communication in process improvement groups. *IEEE Transactions on Professional Communication, 44*(4), 267-285.

Kock, N. (2003). Communication-focused business process redesign: Assessing a communication flow optimization model through an action research study at a defense contractor. *IEEE Transactions on Professional Communication, 46*(1), 35-54.

Kock, N. (2005). *Business process improvement through e-collaboration: Knowledge sharing through the use of virtual groups.* Hershey, PA: Idea Group Publishing.

Kock, N. (2005). *Business process improvement through e-collaboration: Knowledge sharing through the use of virtual groups.* Hershey, PA: Idea Group Publishing.

Kock, N. (2005). What is e-collaboration. *International Journal of e-Collaboration, 1*(1), i-vii.

Kock, N. (2006). *Systems analysis and design fundamentals: A business process redesign approach.* Thousand Oaks, CA: Sage Publications.

Kock, N. (Ed.). (2007). *Information systems action research: An applied view of emerging concepts and methods.* New York: Springer.

Kock, N., & D'Arcy, J. (2002). Resolving the e-collaboration paradox: The competing influences of media naturalness and compensatory adaption. *Information Management and Consulting, 17*(4), 72-78.

Kock, N., & Davison, R. (2003). Can lean media support knowledge sharing? Investigating a hidden advantage of process improvement. *IEEE Transactions on Engineering Management, 50*(2), 151-163.

Kock, N., & McQueen, R. J. (1995). Integrating groupware technology into a business process improvement framework. *Information Technology & People, 8*(4), 19-34.

Kock, N., & Murphy, F. (2001). *Redesigning acquisition processes: A new methodology based on the flow of knowledge and information.* Fort Belvoir, VA: Defense Acquisition University Press.

Kock, N., Lynn, G. S., Dow, K. E., & Akgün, A. E. (2006). Team adaptation to electronic communication media: Evidence of compensatory adaptation in new product development teams. *European Journal of Information Systems, 15*(3), 331-341.

Koh, J., & Kim, Y.-G. (2003). Sense of virtual community: A conceptual framework and empirical validation. *International Journal of Electronic Commerce, 8*(2), 75.

Kohl, W. (1975). The Nixon-Kissinger foreign policy system and US-European relations: Patterns of policy making. *World Politics, 28*(1), 1-43.

Kolka, J. (2002). Silicon mountains/silicon molehills: Geographic concentration and convergence of Internet industries in the US. *Information Economics and Policy, 14*(2), 211.

Kovacich, G. L., Jones, A., & Luzwick, P. G. (2002). Global information warfare: How businesses, governments, and others achieve objectives and attain competitive advantages. *Security Management Practices, 11*(5), 15-23.

KPMG Management Consulting. (1998). *Knowledge management: Research report.*

Krackhardt, D. (1992). The strength of strong ties. In N. Nohria & R. G. Eccles (Eds.), *Networks and organizations: Structure, form and action* (pp. 216-239). Boston: Harvard Business School Press.

Krebs, B. (2001). *Experts yawn over US-China hacker war.* Retrieved February 17, 2004, from http://www.findarticles.com/cf_dls/m0NEW/2001_May_1/74015540/p1/article.jhtml

Krell, E. (2004). The total cost of IT outsourcing. *Business Finance Magazine, 10*(3), 48.

Krim, J. (2003, March 13). Spam's cost to business escalates: Bulk e-mail threatens communication arteries. *The Washington Post*, p. A01.

Krippendorff, K. (2004). *Content analysis: An introduction to its methodology.* CA: Sage Publications.

Kruchten, P. (2000). *The rational unified process: An introduction.* Boston: Addison-Wesley.

Kruchten, P. (2004). Scaling down large projects to meet the agile "sweet spot." *The Rational Edge.* Retrieved from http://www.therationaledge.com

Kumar, K., & van Dissel, H. G. (1996). Sustainable collaboration: Managing conflict and cooperation in inter-organizational systems. *MIS Quarterly, 20*(3), 279-290.

Kumar, R. L. (2002). Managing risks in IT projects: An option perspective. *Information & Management, 40*, 63-74.

Kwak, Y. H., & Stoddard, J. (2004). Project risk management: Lessons learned from software development. *Technovation, 24*(2004), 915-920.

Kwok, R. C., Ma, J., & Vogel, D. R. (2002). Effects of group support systems and content facilitation on knowledge acquisition. *Journal of Management Information Systems, 19*(3), 185-229.

Labaree, R. (2002). The risk of "going observationalist": Negotiating the hidden dilemmas of being an insider participant observer. *Qualitative Research, 2*(1), 97-122.

Lacity, M. C., & Willcocks, L.P. (1998). An empirical investigation of information technology sourcing practices: Lessons from experience. *MIS Quarterly, 22*, 363.

Lacity, M. C., Willcocks, L. P., & Feeny, D. F. (1996). The value of selective IT sourcing. *Sloan Management Review, 37*(3), 13.

Lacity, M., & Hirschheim R. (1993). *Information systems outsourcing myths, metaphors, and realities.* New York: John Wiley.

Lacity, M., & Hirschheim, R. (1995). *Beyond the information systems outsourcing bandwagon: The insourcing response.* Chichester: John Wiley.

LaFleur, R. (1996). Project management: Getting control and keeping control of complex projects. *American Programmer, 9*(4), 24-28.

Laitman, J. (1993). The anatomy of human speech. In R. L. Ciochon & J. G. Fleagle (Eds.), *The human evolution source book* (pp. 56-60). Englewood Cliffs, NJ: Prentice Hall.

Lakoff, G., & Johnson, M. (1980). *Metaphors we live by.* Chicago: The University of Chicago Press.

Lanubile, F., Damian, D., & Oppenheimer, H. L. (2003). Global software development: Technical, organizational, and social challenges. *ACM SIGSOFT Software Engineering Notes, 28*(6).

Laplante, P. A., & Neill, C. J. (2005). "The demise of the waterfall model is imminent" and other urban myths. *ACM Queue*, pp. 10-15.

Laplante, P. A., Costello, T., Pawan, S., Sudi, B., & Landon, M. (2004). The who, what, why, where, and when of IT outsourcing. *IT Professional, 6*(1), 19.

Larkowski, K. (2003). *Latest Standish Group CHAOS report shows that projects success rates have improved by 50%.* West Yarmouth, MA: Standish Group International.

Lawrence, B. (1996). Unresolved ambiguity: The silent source of risk in your project. *American Programmer, 9*(4), 18-22.

Lee, A. S. (1994). Electronic mail as a medium for rich communication: An empirical investigation using hermeneutic interpretation. *MIS Quarterly, 18*(2), 143-157.

Lee, H., & Choi, B. (2003). Knowledge management enablers, processes, and organizational performance: An integrative view and empirical examination. *Journal of Management Information Systems, 20*(1), 179-228.

Lee, J., & Truex, D. (2000). Exploring the impact of formal training in ISD methods on the cognitive structure of novice information systems developers. *Information Systems Journal, 10*(4), 347-367.

Lemos, R. (2001). *Defacements rise in China hacker war.* Retrieved January 17, 2004, from http://news.com.com/2100-1001-256732.html?legacy=cnet

Lengel, R. H., & Daft, R. L. (1988). The selection of communication media as an executive skill. *Academy of Management Executive, 2*(3), 225-232.

Leonard, D., & Swap, W. (2004). Deep smarts. *Harvard Business Review, 82*(9), 88-97.

Leonard, D., & Swap, W. (2005). *Deep smarts: How to cultivate and transfer enduring business wisdom.* Boston: Harvard Business School.

Levina, N., & Ross, J. W. (2003). From the vendor's perspective: Exploring the value proposition in information technology outsourcing. *MIS Quarterly, 27*(3), 361-364.

Levinthal, D., & March, J. (1993). The myopia of learning. *Strategic Management Journal, 14*(8), 95-112.

Liao Troth, M. A., & Griffith, T. L. (2002). Software, shareware and freeware: Multiplex commitment to an electronic social exchange system. *Journal of Organizational Behavior, 23*(5), 635-653.

Lieberman, H., Rosenzweig, E., & Singh, P. (2001). Aria: An agent for annotating and retrieving images. *IEEE Computer, 34*(7), 57-62.

Lieberman, P. (1998). *Eve spoke: Human language and human evolution.* New York: W.W. Norton & Company.

Lindvall, M., Basili, V., Boehm, B., Costa, P., Dangle, K., Shull, F., et al. (2002). Empirical findings in agile methods. *Extreme Programming and Agile Methods: XP/Agile Universe Chicago 2002*, 197-207.

Lippit, R., Polansky, N., Redl, D., & Rosen, S. (1968). The dynamics of power. In D. Cartwright & A. Zander (Eds.), *Group dynamics* (pp. 236-250). New York: Harper & Row Publishers.

LiveBroadway.com. (2005). Retrieved January 2005 from http://www.liveBroadway.com.

Lokachari, P. S., & Mohanarangan, M. (2001). Outsourcing of information technology services: A decision-making framework. In *Proceedings of the Portland International Conference on Management of Engineering and Technology (PICMET)* (Vol. 1, p. 411).

Long, D. A., & Perkins, D. D. (2003). Confirmatory factory analysis of the Sense of Community Index and development of a brief SCI. *Journal of Community Psychology, 31*(3), 279-296.

Lord, R. G., & Foti, R. J. (1986). Schema theories, information processing and organizational behaviour. In H. P. Sims & D. A. Gioia (Eds.), *The thinking organization* (pp. 20-48). San Francisco: Jossey-Bass.

Lorenz, K. (1983). *The waning of humaneness.* Boston: Little, Brown & Co.

Lundell, B., & Lings, B. (2004). Changing perceptions of CASE technology. *The Journal of Systems and Software, 72*(2), 271-280.

Lurey, J., & Raisinghani, M. (2001). An empirical study of best practices in virtual teams. *Information & Management, 38*(8), 523-544.

Luthans, F., & Avolio, B. J. (2003). Authentic leadership: A positive developmental approach. In K. S. Cameron, J. E. Dutton, & R. E. Quinn (Eds.), *Positive organizational scholarship* (pp. 241-258). San Francisco: Berrett-Koehler.

Lynn, B. E. (1999). Culture and intellectual capital management: A key factor in successful ICM implementation. *International Journal of Technology Management, 18*(5-8), 590-603.

Lyytinen, K., & Hirschheim, R. (1987). Information systems failures — A survey and classification of the empirical literature. In *Oxford Surveys in Information Technology.* Oxford University Press.

Macdonald, J. (1995). Quality and the financial service sector. *Managing Service Quality, 5*(1), 43-46.

Maclagan, P. (1997). Corporate social responsibility as a participative process. *Business Ethics: A European Review, 8*(1), 43-49.

Madsen, C. (1992). Approaching group communication by means of an office building metaphor. In D. Marca & G. Bock (Eds.), *Groupware: Software for computer-supported cooperative work* (pp. 316-328). Los Alamitos, CA: IEEE Computer Society Press.

Majchrzak, A., Cooper, L., & Neece, O. (2004). Knowledge reuse for Innovation. *Management Science, 50*(2), 174-188.

Malhotra, A., & Majchrzak, A. (2004). Enabling knowledge creation in far-flung teams: Best practices for IT support and knowledge sharing. *Journal of Knowledge Management, 8*(4), 75-86.

Malhotra, A., & Majchrzak, A. (2005). Virtual workspace technologies. *MIT Sloan Management Review, 46*(2), 11-14.

Malone, T. W, Crowston, K., Lee, J., & Pentland, B. (1999). Tools for inventing organizations: Towards a handbook of organizational processes. *Management Science, 45*(11), 65-78.

Malone, T., & Crowston, K. (1994). The interdisciplinary study of coordination. *ACM Computing Surveys, 26*(1), 87-119.

Maoz, Z. (1990). Framing the national interest: The manipulation of foreign policy decisions in group settings. *World Politics, 43*(1), 77-110.

Mariani, J. (2001). The implications of your location. *Restaurant Hospitality, 85*(8), 20.

Mark, G. (1998). Building virtual teams: Perspectives on communication, flexibility and trust. *ACM SIGGROUP Bulletin, 19*(3), 38-41.

Markus, M. L. (1990). Toward a critical mass theory of interactive media. In J. Fulk & C. Steinfield (Eds.), *Organizations and communication technology* (pp. 194-218). Newbury Park, CA: Sage.

Markus, M. L. (1994). Electronic mail as the medium of managerial choice. *Organization Science, 5*(4), 502-527.

Martin, A., Biddle, R., & Noble, J. (2004). When XP met outsourcing. *Extreme Programming and Agile Processes in Software Engineering: 5th International Conference*, 51-59.

Martin, R. (1991). Working in groups. In M. Smith (Ed.), *Analysing organisational behaviour* (pp. 154-177). London: Macmillan Press.

Martins, L. L., Gilson, L. L., & Maynard, M. T. (2004). Virtual teams: What do we know and where do we go from here? *Journal of Management, 30*(6), 805-835.

Maskell, B. (2001). The age of agile manufacturing. *Supply Change Management, 6*(1), 5-11.

Masoulas, V. (1998). Organizational requirements definition for intellectual capital management. *International Journal of Technology Management, 16*(1-3), 126-143.

Massey, A. P., Hung, Y. C., Montoya-Weiss, M., & Ramesh, V. (2001). When culture and style aren't about clothes: Perceptions of task-technology "fit" in global virtual teams. *Proceedings of the 2001 International ACM SIGGROUP Conference on Supporting Group Work*, 207-213.

Massey, A., Montoya, W., & O'Driscoll, T. M. (2002). Knowledge management in pursuit of performance: Insights from Nortel networks. *MIS Quarterly, 26*(3), 269-289.

Matloff, N. (2004). Globalization and the American IT worker. *Communications of the ACM, 47*(11), 27.

Maurer, F. (2002). Supporting distributed extreme programming. *Extreme Programming and Agile Methods: XP/Agile Universe Chicago 2002*, 13-22.

Maznevski, M. L., & Chudoba, K. M. (2000). Bridging space over time: Global virtual team dynamics and effectiveness. *Organization Science, 11*(5), 473-492.

McConnell, M. (2004). Cyberterrorism and espionage is real and growing. *Bulletin of Northwest Public Power Association, 58*(4), 13-14.

McConnell, S. (2001). Common sense. *IEEE Software, 18*(4), 5.

McConnell, S. (2001). Managing outsourced projects: Project management inside out. In L. Constantine (Ed.), *Beyond chaos: The expert edge in managing software development* (pp. 103-110). NJ: Addison-Wesley.

McCrone, J. (1991). *The ape that spoke: Language and the evolution of the human mind.* New York: William Morrow and Company.

McDonald, G., & Pak, P. C. (1996). It's all fair in love, war, and business: Cognitive philosophies in ethical decision making. *Journal of Business Ethics, 15*(9), 973-996.

McDonald, J. (2001). Why is software project management difficult? And what that implies for teaching software project management.*Computer Science Education, 11*(1), 55.

McElroy, W. (1996). Implementing strategic change through projects. *International Journal of Project Management, 14*(6), 325-329.

McGrath, J. E., & Hollingshead, A. B. (1994). *Groups interacting with technology: Ideas, evidence, issues, and an agenda.* Thousand Oaks, CA: Sage Publications.

McManus, J. (2004). A stakeholder perspective in software project management. *Management Services, 48*(5), 8.

McMillan, D. W., & Chavis, D. M. (1986). Sense of community: A definition and theory. *Journal of Community Psychology, 14*, 6-23.

Meier, A. (2000). Offering social support via the Internet: A case study of an online support group for social workers. *Journal of Technology in Human Services, 17*(2-3), 237-266.

Mellor, S. J., Kendall, S., Uhl, A., & Weise, D. (2004). *MDA distilled.* Boston: Addison-Wesley.

Meredith, J. R., & Mantel, S. J. (2000). *Project management: A managerial approach* (4th ed.). New York: John Wiley & Sons.

Mesch, G. S., & Levanon, Y. (2004). Community networking and locally-based social ties in two suburban localities. *City & Community, 2*(4), 335-351.

META Group, Inc. (2003). Retrieved from http://www.metagroup.com

Metzger, P. (1981). *Managing a programming project.* New York: Prentice Hall Inc.

Mezak, S. (2005). *The seven deadly dangers of outsourced software development.* Retrieved April 2005 from http://www.accelerance.com

Miles, R., Snow, C., Matthews, J., Miles, G., & Coleman, H. (1997). Organizing in the knowledge age: Anticipating the cellular form. *Academy of Management Executive, 11*(4), 7-24.

Milgram, S. (1971). Group pressure and action against a person. In D. Taylor (Ed.), *Small groups* (pp. 157-170). Chicago: Markham Publishing Company.

Mintzberg, H., Ahlstrand, B., & Lampel, J. (1998). *Strategy safari: A guided tour through the wilds of strategic management.* New York: Free Press.

Misra, R. B. (2004). Global IT outsourcing: Metrics for success of all parties. *Journal of Information Technology Cases and Applications, 6*(3), 21.

Moores, T., & Chang, J. (2006). Ethical decision making in software piracy: Initial development and test of a four-component model. *MIS Quarterly, 30*(1), 167-180.

Moorhead, G., Neck, C., & West, M. (1991). The tendency towards defective decision making with self-managing teams: The relevance of groupthink. *Organizational Behaviour & Human Decision Process, 73*(2), 327-351.

Morgan, G. (1986). *Images of organization.* Beverly Hills, CA: Sage Publications.

Morstead, S., & Blount, G. (2003). *Offshore ready: Strategies to plan & profit from offshore IT enabled services.* ISANI Press.

Mustonen-Ollila, & Lyytinen, K. (2003). Why organizations adopt information system process innovations: A longitudinal study using diffusion of innovation theory. *Information Systems Journal, 13*(3), 275-297.

Nader, D. (1993). Concepts for the management of organisational change. In C. Mabey & B. Major-White (Eds.), *Managing change* (2nd ed., pp. 5-19). London: Paul Chapman Publishing.

Nah, S., Veenstra, A. S., & Shah, D. V. (2006). The Internet and anti-war activism: A case study of information, expression, and action. *Journal of Computer Mediated Communication, 12*(1).

Nahapiet, J., & Ghoshal, S. (1998). Social capital, intellectual capital, and the organizational advantage. *Academy of Management Review, 23*(2), 242-266.

Nair, K. G. K., & Prasad, P. N. (2004, Summer). Offshore outsourcing: A SWOT analysis of a state in India. *Information Systems Journal,* p. 34.

Nandhakumar, J., & Jones, M. (2002). Development gain? Participant observation in interpretive management information systems research. *Qualitative Research, 2*(3), 323-341.

NASA workforce analysis report. (2003). Retrieved August 14, 2003, from http://nasapeople.nasa.gov/workforce/

Nash, D. (2002). Beyond buy-in. *P&T Journal, 27*(11), 617-618.

Natovich, J. (2003). Vendor related risks in IT development: A chronology of outsourced project failure. *Technology Analysis & Strategic Managements, 15*(4), 409-419.

Neck, C. (1996). Letterman or Leno: A groupthink analysis of successive decisions made by the national broadcasting company. *Journal of Managerial Psychology, 11*(8), 3-17.

Netz, J. S., & Taylor, B. A. (2002). Maximum or minimum differentiation? Location patterns of retail outlets. *The Review of Economics and Statistics, 84*(1), 162.

Newman, M., & Sabherwal, R. (1996). Determinants of commitment to information systems development: A longitudinal investigation. *MIS Quarterly, 20*(1), 23-54.

Nicholls, C., & Ellement, G. (1997). *Ellen Moore (A): Living and working in Korea* (Richard Ivey School of Business Case #9A97G029) (p. 1-20). Richard Ivey School of Business.

Nidumolu, S. R., & Subramani, M. R. (2003). The matrix of control: Combining process and structure approaches to managing software development. *Journal of Management Information Systems, 20,* 159.

Nonaka, I. (1994). A dynamic theory of organizational knowledge creation. *Organization Science, 5*(1), 14-38.

Nonaka, I., & Takeuchi, H. (1995). *The knowledge creating company.* New York: Oxford University Press.

Nunamaker, J. F., Dennis, A. R., Valacich, J. S., Vogel, D. R., & George, J. F. (1991). Electronic meeting systems to support group work. *Communications of ACM, 34*(7), 40-61.

O'Dell, C., & Grayson, C. J. (1998). If only we know what we know: Identification and transfer of best practices. *California Management Review, 40*(3), 154-174.

O'Loughlin, C., & Coenders, G. (2004). Estimation of the European customer satisfaction index: Maximum likelihood versus partial least squares. Application to postal services. *Total Quality Management & Business Excellence, 15*(9-10), 1231-1255.

O'Reilly, C. A., Chatman, J., & Caldwell, D. F. (1996). Culture as social control: Corporations, cults, and commitment. *Research in Organizational Behavior, 18*, 157-200.

Oates, B. J. (2006). *Researching information systems and computing.* Sage Publications.

Obst, P. L., & White, K. M. (2004). Revisiting the Sense of Community Index: A confirmatory factor analysis. *Journal of Community Psychology, 32*(6), 691-705.

Obst, P. L., Smith, S., & Zinkieqicz, L. (2002). An exploration of sense of community: Part 3. Dimensions and predictors of psychological sense of community in geographical communities. *Journal of Community Psychology, 30*(1), 119-133.

Obst, P., & White, K. M. (2005). An exploration of the interplay between psychological sense of community, social identification and salience. *Journal of Community & Applied Social Psychology, 15*, 127-135.

Oliver, R. L. (1997). *Satisfaction: A behavioural perspective on consumer.* Boston: Irwin McGraw-Hill.

orative learning: An empirical evaluation. *MIS Quarterly, 18*(2), 159-174.

Orlikowski, W. (1993). CASE tools as organizational change: Investigating incremental and radical changes in systems development. *MIS Quarterly, 17*(3), 309-340.

Orlikowski, W. J., Yates, J., Okamura, K., & Fujimoto, M. (1995). Shaping electronic communication: The metastructuring of technology in the context of use. *Organization Science, 6*(4), 423-444.

Ott, J. S. (1989). *Organizational culture perspective.* Chicago: The Dorsey Press.

Outsourcing Institute. (2005). Retrieved April 2005 from http://www.outsourcing.com

Overby, S. (2003). Bringing I.T. back home.*CIO.*

Painter-Morland, M., Fontrodona, J., Hoffman, W. M., & Rowe, M. (2003). Conversations across continents: Teaching business ethics online. *Journal of Business Ethics, 48*(1), 75-88.

Paul, S., Seetharamann, P., Samarah, I., & Mykytyn, P. (2004). Impact of heterogeneity and collaborative conflict management style on the performance of synchronous global virtual teams. *Information & Management, 41*(3), 303-321.

Paulk, M. C., Curtis, B., Chrissis, M. B., & Weber, C. V. (1993). Capability maturity model, version 1.1. *IEEE Software, 10*(4), 18-27.

Pedersen, M. K., & Larsen, M. H. (2001). Distributed knowledge management based on product state models: The case of decision support in health care administration. *Decision Support Systems, 31*(1), 139-158.

Pellegrinelli, S. (1997). Programme management: Organising project-based change. *International Journal of Project Management, 15*(3), 141-149.

Penzias, B. (2005). Whirlpool's IT practices for continuous innovation. *SAP NetWeaver Magazine*, 20-35.

Peters, M., & Robinson, V. (1984). The origins and status of action research. *The Journal of Applied Behavioral Science, 20*(2), 113-124.

Peters, T. (1999). *The circle of innovation: You can't shrink your way to greatness.* New York: Vintage Books.

Petty, R. E., & Cacioppo, J. T. (1981). *Attitudes and persuasion: Classic and contemporary approaches.* Dubuque, IA: Brown.

Piccoli, G., & Ives, B. (2000). Virtual teams: Managerial behaviour control's impact on team

Pietro, C. (1992). Groupware meetings that work. *Proceedings of Groupware'92 Conference*, 50-58.

Pinsonneault, A., & Caya, O. (2005). Virtual teams: What we know, what we don't know. *International Journal of e-Collaboration, 1*(3), 1-16.

Pinsonneault, A., Barki, H., Gallupe, R. B., & Hoppen, N. (1999). The illusion of electronic brainstorming productivity: Theoretical and empirical issues. *Information Systems Research, 10*(4), 378-380.

Pollice, G. (2001a). RUP and XP: Part I. Finding common ground. *The Rational Edge.* Retrieved from http://www.therationaledge.com

Pollice, G. (2001b). RUP and XP: Part II. Valuing differences. *The Rational Edge.* Retrieved from http://www.therationaledge.com

Pollice, G. (2004). RUP and extreme programming: Complementing processes. In L. Liu & B. Roussev (Eds.), *Management of object-oriented software development.* Hershey, PA: Idea Group.

Poole, M. S., & DeSanctis, G. (1990). Understanding the use of group decision support systems: The theory of adaptive structuration. In J. Fulk & C. Steinfield (Eds.), *Organizations and communication technology* (pp. 173-193). Newbury Park, CA: Sage.

Postmes, T., Spears, R., & Lea, M. (1998). Breaching or building social boundaries? Side-effects of computer-mediated communications. *Communication Research, 25*(6), 689-715.

Potter, R., & Balthazard, P. (2002). Virtual team interaction styles: Assessment and effects. *International Journal of Human-Computer Studies, 56*(4), 423-443.

Powell, A., Picolli, G., & Ives, B. (2004). Virtual teams: A review of current literature and directions for future research. *The Database for Advances in Information Systems, 35*(1), 6-36.

Prahalad, C. K., & Hamel, G. (1990). The core competence of the corporation. *Harvard Business Review, 68*(3), 79.

Pressman, R. S. (2004). *Software engineering: A practitioner's approach* (6th ed.). McGraw Hill.

Pries-Heje, J., Baskerville, R., & Hansen, G. I., (2005). Strategy models for enabling offshore outsourcing: Russian short-cycle-time software development. *Information Technology for Development, 11*(1), 5.

Prikladnicki, R., & Audy, J. L. N. (2006). A reference model for global software development: Findings from a case study. In *Proceedings of the First IEEE Conference on Global Software Engineering.*

Prikladnicki, R., Audy, J. L. N., & Evaristo, J. R. (2003). Global software development in practice: Lessons learned. *Journal of Software Process: Improvement and Practice, 8*(4), 267-281.

Prikladnicki, R., Yamaguti, M. H., & Antunes, D. (2004). Risk management in distributed software development: A process integration proposal. *In Proceedings of the 5th IFIP Working Conference on Virtual Enterprises*, 517-526.

Project Management Institute (PMI) Standards Committee. (1996). *A guide to the project management body of knowledge.* Newtown Square, PA: Project Management Institute.

Project Management Institute (PMI). (2004). *A guide to the project management body of knowledge* (3rd ed.). Newtown Square, PA.

Project Management Institute. (2004). *A guide to the project management body of knowledge* (3rd ed.). PA.

Qu, Z., & Brocklehurst, M. (2003). What will it take for China to become a competitive force in offshore outsourcing? An analysis of the role of transaction costs in supplier selection. *Journal of Information Technology, 18*, 53.

Quin, J. B., & Paquette, P. C. (1990). Technology in services: Creating organizational revolutions. *Sloan Management Review*, 67-77.

Quinn, R. E., & Rohrbaugh, I. (1983). A spatial model of effectiveness criteria: Towards a competing values approach to organizational analysis. *Management Science, 29*, 363-377.

Radcliff, D. (2001). We have met the enemy and he is us. *Computerworld.* Retrieved February 17, 2004, from http://www.computerworld.com/industrytopics/retail/story/0,10801,60466,00.html

Radicati Group Inc. (2004). *Market numbers summary update.* Palo Alto, CA.

Rainwater, J. (2002). *Herding cats: A primer for programmers who lead programmers.* New York: Springer-Verlag.

Rajkumar, T. M., & Dawley, D. L. (1997). Problems and issues in offshore development of software. In *Strategic sourcing of information systems: Perspectives and practices* (pp. 369-386). John Wiley & Sons.

Rajkumar, T. M., & Mani, R. V. S. (2001). Offshore software development: The view from Indian suppliers. *Information Systems Management*, 63-73

Ramarapu, N., & Parzinger, M. J. (1997). Issues in foreign outsourcing. *Information Systems Management, 14*(2), 27.

Ready, K. J., Hostager, T. J., Lester, S., & Bergmann, M. (2004). Beyond the silo approach: Using group support systems in organizational behavior classes to facilitate student understanding of individual and group behavior in electronic meetings. *Journal of Management Education, 28*(6), 770.

Reeves, M., & Zhu, J. (2004). Moomba: A collaborative environment for supporting distributed extreme programming in global software development. *Extreme Programming and Agile Processes in Software Engineering: 5th International Conference*, 38-50.

Reichheld, F. (2001). Lead for loyalty. *Harvard Business Review, 79*(7), 76-84.

Reichheld, F. F. (2003). The one number you need to grow. *Harvard Business Review*, 46-54.

Rice, R. E. (1992). Task analyzability, use of new media, and effectiveness: A multi-site exploration of media richness. *Organization Science, 3*(4), 475-500.

Ritter, B. (2006). Can business ethics be trained? A study of the ethical decision-making process in business students. *Journal of Business Ethics, 68*(2), 153-164.

Robbins, H., & Finley, M. (1998). *Why change doesn't work.* London: Orion Publishing.

Roberts, L. D., Smith, L. M., & Pollock, C. M. (2002). MOOing till the cows come home: The sense of community in virtual environments. In C. C. Sonn (Ed.), *Psychological sense of community: Research, applications, implications.* New York: Kluwer Academic/Plenum.

Robinson, H., & Sharp, H. (2004). The characteristics of XP teams. *Extreme Programming and Agile Processes in Software Engineering: 5th International Conference*, 139-147.

Robinson, M., & Kalakota, R. (2004). *Offshore outsourcing: Business models, ROI and best practices.* Mivar Press.

Rodgers, S., & Chen, Q. (2005). Internet community group participation: Psychosocial benefits for women with breast cancer. *Journal of Computer Mediated Communication, 10*(4).

Rogers, E. (1962). *Diffusion of innovations.* New York: The Free Press.

Rost, J. (2004). Political reasons for failed software projects. *IEEE Software, 21*(6), 104.

Roussev, B., & Akella, R. (n.d.). Agile outsourcing projects: Structure and management. *International Journal of e-Collaboration, 2*(4), 37-52.

Roussev, B., & Rousseva, Y. (2006). *Unreliable interlocutors in information systems development.* Proceedings of San Diego International Systems Conference, San Diego, CA.

Royal, M. A., & Rossi, R. J. (1999). Predictors of within-school differences in teachers' sense of community. *Journal of Educational Research, 92*, 259-267.

Royce, W. (1998). *Software project management: A unified framework.* Reading, MA: Addison-Wesley, Inc.

Royce, W. (2002). The case for results-based software management. *The Rational Edge.* Retrieved from http://www.therationaledge.com

Ruggles, R. (1998). The state of the notion: Knowledge management in practice. *California Management Review, 40*(3), 80-89.

Rulke, D. L., & Galaskiewicz, J. (2000). Distribution of knowledge, group network structure, and group performance. *Management Science, 46*(5), 612-625.

Russel, B., & Chattargee, S. (2003). Relationship quality: The undervalued dimension of software quality. *Communications of the ACM, 46*(8), 85-89.

Rutkowski, A. F., Vogel, D. R., Genuchten, M. v., Bemelmans, T. M. A., & Favier, M. (2002). E-collaboration: The reality of virtuality. *IEEE Transactions on Professional Communication, 45*(4), 219-230.

Rutkowski, A.-F., Vogel, D., Bemelmans, T. M. A., & Van Genuchten, M. (2002). Group support systems and virtual collaboration: The HKNet Project. *Group Decision and Negotiation, 11*(2), 101-125.

Sahay, S., Nicholson, B., & Krishna, S. (2003). *Global IT outsourcing: Software development across borders.* UK: Cambridge University Press.

Sanyal, R. N. (2000). An experiential approach to teaching ethics in international business. *Teaching Business Ethics, 4*(2), 137-189.

Saran, C. (2004). Badly-managed offshore software development costs firms millions. *Computer Weekly*, p. 5.

Sarason, S. B. (1974). *The psychological sense of community: Prospects for a Community Psychology.* San Francisco: Jossey-Bass.

Sarker, S., & Sahay, S. (2001). Information systems development by US-Norwegian virtual teams: Implications of time and space. *Proceedings of the 36th Hawaii International Conference on Systems Sciences.*

Saunders, C. S., Van Slyke, C., & Vogel, D. (2004). My time or yours? Managing time visions in global virtual teams. *Academy of Management Executive, 18*(1), 19-31.

Schary, P. B., & Coakley, J. (1991). Logistics organization and the information system. *International Journal of Logistics Management, 2*(2), 22.

Schein, E. H. (1985). *Organizational culture and leadership.* San Francisco: Jossey-Bass.

Schelling, T. (1989). The mind as a consuming organ. In D. Bell, H. Raiffa, & A. Tversty (Eds.), *Decision making: Descriptive, normative, and prescriptive interactions* (pp. 343-357). Cambridge, UK: Cambridge University Press.

Schmidt, R., Lyytinen, K., Keil, M., & Cule, P. (2001). Identifying software project risks: An international Delphi study. *Journal of Management Information Systems, 17*(4), 5-36.

Schnake, M. E., & Cochran, D. S. (1985). Effect of two goal-setting dimensions on perceived intraorganizational conflict. *Group & Organizational Studies, 10*, 168-183.

Schoder, D., & Fischbach, K. (2003). Peer-to-peer prospects. *Communications of the ACM, 46*(2), 27-30.

Schubert, P., Leimstoll, U., & Romano, N. C. (2003). Internet groupware systems for project management: Experiences from a longitudinal study. *Proceedings of the 16th Bled eCommerce Conference: eTransformation*, 611-631.

Schuemmer, T. (2003). Evolving a groupware pattern language. *ECSCW2003: Fifth International Workshop on Collaborative Editing.*

Schuh, P. (2005). *Integrating agile development in the real world.* Hingham, MA: Charles River Media, Inc.

Schultze, U., & Boland, R. (2000). Knowledge management technology and the reproduction of knowledge work practices. *Journal of Strategic Information Systems, 9*(2-3), 193-213.

Schultze, U., & Leidner, D. E. (2002). Studying knowledge management in information systems research: Discourses and theoretical assumptions. *MIS Quarterly, 26*(3), 213-242.

Schuster, E. (1998). A community bound by words: Reflections on a nursing home writing group. *Journal of Aging Studies, 12*(2), 137-148.

Schwaber, K. (1996). Controlled chaos: Living on the edge. *American Programmer, 9*(4), 11-16.

Schwalbe, K. (2004). *Information technology project management.* Boston: Thomson Learning.

Schwartzman, H. (1993). *Ethnography in organizations.* Thousand Oaks, CA: Sage Publications.

Senge, P. M. (1990). *The fifth discipline: The art and practice of the learning organization.* New York: Currency Doubleday.

Sengupta, B., Chandra, S., & Sinha, V. (2006). A research agenda for distributed software development. In *Proceedings of the 28th International Conference on Software Engineering*, 731-740.

Sepulveda, C. (2003). Agile development and remote teams: Learning to love the phone. *Proceedings of the Agile Development Conference*, 140-145.

Setiadi, R., & Kom, S. (2005, February 16). Implementing IT outsourcing in Indonesia: A smart decision or a big mistake. In *Proceedings of Issues and Challenges in Global Development Conference*, The University of Melbourne, Victoria, Australia.

Shamir, B., House, R., & Arthur, M. (1993). The motivational effects of charismatic leadership: A self-concept based theory. *Organization Science, 4*(4), 1-17.

Shani, A. B., Sena, J. A., & Stebbins, M. W. (2000). Knowledge work teams and groupware technology: Learning from Seagate's experience. *Journal of Knowledge Management, 4*(2), 111-124.

Shaw, M. (1981). *Group dynamics: The psychology of small group behavior.* New York: McGraw-Hill.

Sheffield, J., & Gallupe, B. (1993). Using electronic meeting technology to support economic development in New Zealand: Short term results. *Journal of Management Information Systems, 10*(3), 97-116.

Shenhar, A. J. (1998). From theory to practice: Toward a typology of project management styles. *IEEE Transactions on Engineering Management, 45*(1), 33-48.

Sherwood, K. D. (2002). *Overcome email overload with Microsoft Outlook 2000 and Outlook 2002: Get through your electronic mail faster.* Palo Alto, CA: World Wide Webfoot Press.

Shread, P. (2002, April 2). Pacific Rim grid collaboration formed. *Grid Computing Planet.*

Siegel, A., & Siegel, S. (1968). Reference groups, membership groups, and attitude change. In D. Cartwright & A. Zander (Eds.), *Group dynamics* (pp. 74-79). New York: Harper & Row Publishers.

Smith, J. (2003, July 13). *Mistakes of NASA touted up.* Washington Post.

Smith, M. A., Mitra, S., & Narasimhan, S. (1996). Offshore outsourcing of software development and maintenance. *Information and Management, 31*(3), 165.

Soles, S. (1994). Work reengineering and workflows: Comparative methods. In E. White & L. Fischer (Eds.), *The workflow paradigm* (pp. 70-104). Alameda, CA: Future Strategies.

Sosik, J. J., Kahai, S. S., & Avolio, B. J. (1998). Transformational leadership and dimensions of creativity: Motivating idea generation in computer-mediated groups. *Creativity Research Journal, 11*(2), 111-121.

Sosik, J., Avolio, B., & Kahai, S. (1997). Effects of leadership style and anonymity on group potency and effectiveness in a GDSS environment. *The Journal of Applied Psychology, 82*(1), 89-103.

Spangler, W. E., & Peters, J. M. (2001). A model of distributed knowledge and action in complex systems. *Decision Support Systems, 31*(1), 103-125.

Sparks, D. (2002). Inner conflicts, inner strengths. *Journal of Staff Development, 23*(3), 66-71.

Spiro, L. N., & Baig, E. C. (1999, February 22). Who needs a broker? *Business Week, 3617*, 113.

Spradley, J. (1980). *Participant observation.* New York: Holt, Rinehart, & Winston.

Sproull, L., & Kiesler, S. (1991). Computers, networks and work. *Scientific American*, 84-91.

Sproull, L., & Kiesler, S. (1991). *Connections: New ways of working in the networked organization.* Cambridge, MA: MIT Press.

Sproull, L., & Kiesler, S. B. (1991). *Connections: New ways of working in the networked organization.* Cambridge, MA: The MIT Press.

Sridhar, N., Mahapatra, R., & Mangalaraj, G. (2005). Challenges of migrating to agile methodologies. *Communications of the ACM, 48*(5).

Standish Group. (2003). *CHAOS research report.* Retrieved from http://www.standishgroup.com

Stead, B., & Gilbert, J. (2001). Ethical issues in electronic commerce. *Journal of Business Ethics, 34*(2), 75-85.

Stephens, M., & Rosenberg, D. (2003). *Extreme programming refactored: The case against XP.* San Francisco: Apress.

Stotts, D., Williams, L., Nagappan, N., Baheti, P., Jen, D., & Jackson, A. (2003). Virtual teaming: Experiments and experiences with distributed pair programming. In F. Maurer & D. Wells (Eds.), *Extreme Programming and Agile Methods: XP/Agile Universe 2003*, 129-141.

Straub, D., Weil, P., & Schwaig, K. S. (2004). *Strategy and the control of the IT resource: A test of the strategic control model.* Working Paper Series, Georgia State University.

Street, M., Douglas, S., Geiger, S., & Martinko, M. (2001). The impact of cognitive expenditure on the ethical decision-making process: The cognitive elaboration model. *Organizational Behavior and Human Decision Processes, 86*(2), 256-277.

Sturges, P., Katjihingua, M., & Mchombu, K. (2005). Information in the national liberation struggle: Modelling the case of Namibia (1966-1990). *Journal of Documentation, 61*(6), 735-750.

Styhre, A., Roth, J., & Ingelgard, A. (2002). Care of the other: Knowledge-creation through care in professional teams. *Scandinavian Journal of Management, 18*(4), 503-522.

Subramanyam, M. (2004). The impact of global IT outsourcing on IT providers. *Communications of the Association for Information Systems, 14*(25), 543-557.

Sun Tzu. (1995). *The art of war.* London: Hodder & Stoughton.

Sunstrom, E., DeMeuse, K., & Futrell, D. (1990). Work teams: Applications and effectiveness. *American Psychologist, 45*(2), 120-133.

Susman, G. I., & Evered, R. D. (1978). An assessment of the scientific merits of action research. *Administrative Science Quarterly, 23*, 582-603.

Sussman, S., & Siegal, W. (2003). Informational influence in organizations: An integrated approach to knowledge adoption. *Information Systems Research, 14*(1), 47-65.

Szulanski, G. (1996). Exploring internal stickiness: Impediments to the transfer of best practice within the firm. *Strategic Management Journal, 17*(10), 27-43.

Tabachnik, B. G., & Fidell, L. S. (2001). *Using multivariate statistics* (4th ed.). Needham Heights, MA: Allyn & Bacon.

Talia, D. (2002). The open grid services architecture: Where the grid meets the Web. *IEEE Internet Computing, 6*(6), 2-6.

Tapscott, D., & Carston, A. (1993). *Paradigm shift.* New York: McGraw-Hill.

Tavares-Rixon, D. (2003). Exploring the nature of commitment to software quality initiatives: A political perspective. *Proceedings of the First Australian Undergraduate Students Computing Conference*, 87-92.

Taylor, H. (2006). Critical risks in outsourced projects: The intractable and the unforeseen. *Communications of the ACM, 49*(11).

Te'eni, D., & Kendall, J. E. (2004). Internet commerce and fundraising. In D. Young (Ed.), *Effective economic decision-making by nonprofit organizations* (pp. 167-189). New York: The Foundation Center.

Technology Partners International (TPI). (2005). *Technology Partners International Inc. quarterly index.* Retrieved from http://www.tpi.net

Tedlock, B. (2000). Ethnography and ethnographic presentation, In N. Denzin & Y. Lincoln (Eds.), *Handbook of qualitative research* (2nd ed., pp. 455-486). Thousand Oaks, CA: Sage Publications.

Templeton, G. F., Lewis, B. R., & Snyder, C. A. (2002). Development of a measure for the organizational learning construct. *Journal of Management Information Systems, 19*(2), 175-218.

The Economist. (2005). Special report: The Bangalore paradox — Outsourcing and IT in India: India's IT and outsourcing industries. *The Economist, 375*(8423), 82.

The President's Commission on Critical Infrastructure Protection. (1997). *Critical foundations protecting America's infrastructures.* Washington, DC.

Thomas-Hunt, M., Ogden, T. Y., & Neale, M. A. (2003). Who's really sharing? Effects of social and expert status on knowledge exchange within groups. *Management Science, 49*(4), 464-473.

Thomsett, R. (1998). The care and feeding of project managers. *American Programmer, 11*(2), 5-11.

Tiwana, A. (2003). Affinity to infinity in peer-to-peer knowledge platforms. *Communications of the ACM, 46*(5), 76-80.

Tiwana, A., & Keil, M. (2004). The one-minute risk assessment tool. *Communications of the ACM, 47*(11), 73-78.

Tiwana, A., & Keil, M. (2006). Functionality risk in software development. *IEEE Transactions on Engineering Management, 53*(3), 412-425.

Toffler, A. (1970). *Future shock.* New York: Bantam Books.

Toleman, M., Ally, M., & Darroch, F. (2004). Aligning adoption theory with agile system development methodologies. *Proceedings of the 8th Pacific Asia Conference on Information Systems*, 458-471.

Toth, K., Kruchten, P., & Paine, T. (1993). *Modernizing air traffic control through modern software methods.* Proceedings of the 38th Annual Air Traffic Control Association Conference, Nashville, TN.

Traylen, H. (1994). Confronting hidden agendas: Cooperative inquiry with health visitors. In P. Reason (Ed.), *Participation in human inquiry* (pp. 59-81). London: Sage Publications.

Trevino, L. K., Brown, M., & Hartman, L. P. (2003). A qualitative investigation of perceived executive ethical leadership: Perceptions from inside and outside the executive suite. *Human Relations, 56*(1), 5-37.

Trevino, L. K., Webster, J., & Stein, E. W. (2000). Making connections: Complementary influences on communication media choices, attitudes, and use. *Organization Science, 11*(2), 163-182.

Tsai, W., & Ghoshal, S. (1998). Social capital and value creation: The role of intrafirm networks. *Academy of Management Journal, 41*(4), 464-478.

Turner, J., & Muller, R. (2003). On the nature of the project as a temporary organization. *International Journal of Project Management, 21*(1), 1-8.

Turner, N., Barling, J., Epitropaki, O., Butcher, V., & Milner, C. (2002). Transformational leadership and moral reasoning. *Journal of Applied Psychology, 87*(2), 304-311.

Turvey, M. T. (1990). Coordination. *American Psychologist, 45*(8), 938-953.

Tyler, K. (2005). Do the right thing. *HRMagazine, 50*(2), 99.

U.S. Census Bureau. (2003). Retrieved from http://eire.census.gov/popest/data/counties/tables/CO-EST2001-08/CO-EST2001-08-34.php

Un, C. A., & Cuervo-Cazurra, A. (2004). Strategies for knowledge creation in firms. *British Journal of Management, 15*(1), 27-41.

University of Waikato. (1994). *Human Research Ethics Committee: Procedures.* Hamilton, New Zealand: University of Waikato, Human Research Ethics Committee.

Uram, T. (2006). *Access grid.* Retrieved from http://www.accessgrid.org/

Vaidyanathan, G. (2006). Networked knowledge management dimensions in distributed projects. *International Journal of e-Collaboration, 2*(4), 35-52.

Valacich, J. S., Dennis, A. R., & Connolly, T. (1994). Idea generation in computer-based groups: A new ending to an old story. *Organizational Behavior and Human Decision Processes, 57*(3), 448-467.

Van Ryssen, S., & Godar, S. H. (2000). Going international without going international. *Journal of International Management, 6*, 49-60.

Vatis, M. A. (2001). *Cyber attacks during the war on terrorism: A predictive analysis.* Retrieved December 6, 2004, from http://www.ists.dartmouth.edu/library/analysis/cyber_a1.pdf

Vatis, M. A. (2006). The next battlefield. *Harvard International Review, 28*(3), 56-61.

Venkatraman, N. (1994). IT-enabled business transformation: From automation to business scope redefinition. *Sloan Management Review, 35*(2), 73-87.

Verschoor, C. C. (2006). Surveys show ethics problems persist. *Strategic Finance, 88*(4), 19-21.

Vinokur, A., & Burnstein, E. (1974). Effects of partially shared persuasive arguments on group-induced shifts: A group-problem-solving approach. *Journal of Personality and Social Psychology, 29*(3), 305-315.

Vogel, D. R., van Genuchten, M., Lou, D., Verveen, S., van Eekout, M., & Adams, A. (2001). Exploratory research on the role of national and professional cultures in a distributed learning project. *IEEE Transactions on Professional Communication, 44*(2), 114-125.

Völter, M. (2002). Hope, belief and wizardry: Three different perspectives on project management. *EuroPLoP 2002: Seventh European Conference on Pattern Languages of Programs.*

Von Bergen, C., & Kirk, R. (1978). Groupthink: When too many heads spoil the decision. *Management Review, 67*(3), 44-49.

von Branconi, C., & Loch, C. H. (2004). Contracting for major projects: Eight business levers for top management. *International Journal of Project Management, 22,* 119.

Von Krogh, G. (1998). Care in knowledge creation. *California Management Review, 40*(3), 133-153.

Walkins, J. (1992). Information systems: The UK retail financial services sector. *Marketing Intelligence & Planning, 10*(6), 13-18.

Wallace, R. (2001). *It's an all-out cyber war as U.S. hackers fight back at China.* Retrieved February 17, 2004, from http://www.foxnews.com/story/0,2933,19337,00.html

Wallach, E. J. (1983). Individuals and organizations: The cultural match. *Training and Development Journal.*

Walther, J. B. (1995). Relational aspects of computer-mediated communication: Experimental observations over time. *Organization Science, 6*(2), 186-203.

Walther, J. B. (1995). Relational aspects of computer-mediated communication: Experimental observations over time. *Organization Science, 6*(2), 186-203.

Walther, J. B. (1996). Computer mediated communication: Impersonal, interpersonal and hyperpersonal interaction. *Communication Research, 22*, 33-43.

Walton, M. (1989). *The Deming management method.* London: Mercury.

Ware, L. (2003). Weighing the benefits of offshore outsourcing. *CIO.com.* Retrieved from http://www.cio.com

Wastell, D. G., White, P., & Kawalek, P. (1994). A methodology for business process redesign: Experiences and issues. *Journal of Strategic Information Systems, 3*(1), 23-40.

Watson, S. (1998, January 26). Getting to "aha!" companies use intranets to turn information and experience into knowledge: And gain a competitive edge. *Computer World.*

Watson-Manheim, M. B., & Belanger, F. (2002). Exploring communication-based work processes in virtual work environments. In *Proceedings of the 36th Hawaii International Conference on Systems Sciences.*

Watson-Manheim, M. B., Chudoba, K. M., & Crowston, K. (2002). Discontinuities and continuities: A new way to understand virtual work. *Information Technology & People, 15*(3), 191-209.

Wayne, H. (2006). Computer security firm uses hacker's instincts, skills. *Knight Ridder/Tribune Business News*, p. 1.

Weaver, R. M. (1967). *A rhetoric and handbook.* New York: Holt, Rinehart, and Winston.

Webster, J., & Trevino, L. K. (1995). Rational and social theories as complementary explanations of communication media choices: Two policy-capturing studies. *Academy of Management Journal, 38*(6), 1544-1573.

Weinberg, G. (1971). *The psychology of computer programming.* New York: Van Nostrand Reinhold Co.

Weis, R., Stamm, K., Smith, C., Nilan, M., Clark, F., Weis, J., et al. (2003). Communities of care and caring: The case of MSWatch.com®. *Journal of Health Psychology, 8*(1), 135-148.

Wenger, E. C., & Snyder, W. M. (2000). Communities of practice: The organizational frontier. *Harvard Business Review*, 139-145.

Wesolowsky, G. O. (2001). Lectures on location theory. *INFOR, 39*(1), 124.

White hat. (2007). *Wikipedia.* Retrieved February 23, 2007, from http://en.wikipedia.org/wiki/White_hat

Whitehead, R. (2001). *Leading a software development team: A developers guide to successfully leading people and projects.* London: Addison-Wesley.

Wiegers, K. (1998). Know your enemy: Software risk management. *Software Development, 6*(10), 38.

Willcocks, L., & Lacity, M. (1998). *Strategic sourcing of information systems: Perspectives and practices.* New York: John Wiley & Sons.

Williams, K. (1997). Has your company outsourced yet? *Management Accounting, 78*(9), 14.

Wills, C. (1993). *The runaway brain: The evolution of human uniqueness*. New York: Basic Books.

Wilson, E. O. (1998). *Consilience: The unity of knowledge*. New York: Alfred A. Knopf.

Wilson, P. (1991). *Computer supported cooperative work*. Oxford, UK: Intellect.

Wright, J., & Schaal, D. (1988). Groupthink: The trap of consensus investing. *Journal of Financial Planning, 1*(1), 41-44.

Wu, Z., Chen, H., & Xu, J. (2003). Knowledge base grid: A generic grid architecture for semantic Web. *Journal of Computer and Technology, 18*(4), 462-473.

Xue, Y., Sankar, C. S., & Mbarika, V. W. A. (2005). Information technology outsourcing and virtual team. *The Journal of Computer Information Systems, 45*(2), 9.

Yin, R. K. (2001). *Case study research: Design and methods* (3rd ed.). CA: Sage Publications.

Yourdon, E. (1993). *Decline and fall of the American programmer*. Upper Saddle River, NJ: Prentice Hall Inc.

Zachary, G. (1994). *Show-stopper! The breakneck race to create Windows NT and the next generation at Microsoft*. New York: The Free Press.

Zaff, J., & Devlin, S. (1998). Sense of community in housing for the elderly. *Journal of Community Psychology, 26*, 381-398.

Zalesny, M. D., Salas, E., & Prince, C. (1995). Conceptual and measurement issues in coordination: Implications for team behavior and performance. *Research in Personnel and Human Resources Management, 13*, 81-115.

Zaltman, G. (1994). Knowledge disavowal in organizations. In R. H. Kilmann, K. W. Thomas, D. P. Slevin, R. Nath, & S. L. Jerrel (Eds.), *Producing useful knowledge for organizations* (pp. 173-187). San Francisco: Jossey-Bass.

Zander, D., & Kogut, B. (1995). Knowledge and the speed of the transfer and imitation of organizational capabilities: An empirical test. *Organization Science, 6*(1), 76-92.

Zanoni, R., & Audy, J. (2004). Project management model: Proposal for performance in a physically distributed software development environment. *Engineering Management Journal, 16*(2), 28.

Zatolyuk, S., & Allgood, B. (2004). Evaluating a country for offshore outsourcing: Software development providers in the Ukraine. *Information System Journal, 28*.

Zeller, T., Jr. (2005, March). Not yet in business school, and already flunking ethics. *New York Times*, p. C3.

Zigurs, I. (2003). Leadership in virtual teams: Oxymoron or opportunity? *Organizational Dynamics, 31*(4), 339-351.

Zigurs, I., & Buckland, B. (1998). A theory of task/technology fit and group support systems effectiveness. *MIS Quarterly, 22*(3), 313-334.

Zigurs, I., Buckland, B. K., Connolly, J. R., & Wilson, E. V. (1999). A test of task-technology fit theory for group support systems. *Database for Advances in Information Systems, 30*(3), 34-50.

Zigurs, I., Evaristo, R., & Katzy, B. (2001). *Collaborative technologies for virtual project management*. Washington, DC: Academy of Management.

Zmud, R. (1983). The effectiveness of external information channels in facilitating innovation within software development groups. *MIS Quarterly, 7*(2), 43-58.

About the Contributors

Ned Kock is an associate professor and founding chair of the Division of International Business and Technology Studies at Texas A&M International University. He holds degrees in electronics engineering (BEE), computer science (MS), and management information systems (MIS; PhD). Ned has authored and edited several books, including the bestselling *Systems Analysis and Design Fundamentals: A Business Process Redesign Approach*. Ned has published his research in a number of high-impact journals including *Communications of the ACM, Decision Support Systems, European Journal of Information Systems, IEEE Transactions, Information & Management, MIS Quarterly*, and *Organization Science*. He is the founding editor in chief of the *International Journal of e-Collaboration*, associate editor of the *Journal of Systems and Information Technology*, and associate editor for information systems of the journal *IEEE Transactions on Professional Communication*. His research interests include action research, ethical and legal issues in technology research and management, e-collaboration, and business process improvement.

* * * * *

Maryam Alavi, PhD, is the John and Lucy Cook chair of information strategy and the former senior associate dean of faculty and research at the Goizueta Business School of Emory University. She also serves as the director of Knowledge@Emory, a Web-based publication of the Goizueta Business School. Dr. Alavi has authored numerous scholarly papers. Her research has been supported by funds and hardware grants from the AT&T Foundation, AT&T Corporation, IBM, and Lucent Technologies. She has served on the editorial boards of several scholarly journals including *MIS Quarterly, Information Systems Research, Journal of MIS*, and *Journal of Strategic Information Systems*. Dr. Alavi was awarded the distinguished Marvin Bower Faculty Fellowship at the Harvard Business School. She also was a recipient of the University of Maryland Distinguished Scholar-Teacher Award, and was elected as the recipient of the prestigious AIS (Association of Information Systems) Fellows Award.

Martha del Alto is currently serving as the software release authority in the Technology Partnerships Division office at NASA Ames Research Center. In this position, she is responsible for the release of software, as well as open-source software, within the center. Additionally, she contributes to the development of partnerships between researchers in the industrial, academic, and governmental sectors for the purpose of advancing the space exploration vision and creating terrestrial benefits. Prior to that she was in the Intelligent Systems Division office, where she cofounded and helped establish successful research document management systems. The first one was the New Millennium Documentation Project, and the second was the Postdoc Project. Her professional interests include issues of software and open source, proprietary data, knowledge management (KM), and the commercialization of space. Her most recent interests lie in contributing to the national vision of space exploration and making it more affordable through partnerships.

Jorge Audy is a professor of computer science and vice president of research and graduate studies at the Pontifical Catholic University of Rio Grande do Sul (PUCRS), Brazil. He holds a doctorate in information systems. He is currently involved in projects related to global software development and project management, with work done with companies like Dell (United States), Hewlett-Packard, SONAE (Portugal), and DB Server (Brazil), and research projects with researchers at HEC and the University of Victoria in Canada, and the Universities of Kentucky and Illinois (Chicago) in the United States. His publications appear in such journals as *Engineering Management Journal*, *Software Process Improvement and Practice Journal*, and *International Journal of e-Collaboration*, and in many conference proceedings.

Bruce J. Avolio, PhD, is a Clifton chair in leadership, College of Business Administration, University of Nebraska-Lincoln; and director of the Gallup Leadership Institute, University of Nebraska-Lincoln. Dr. Avolio has an international reputation as a researcher and practitioner in leadership. He has consulted with public and private organizations in North and South America, Africa, Europe, and Southeast Asia, as well as in Australia, New Zealand, and Israel. His research and consulting includes work with the militaries of the United States of America, Singapore, Sweden, Finland, Israel, and South Africa. Avolio has published five books and more than 95 articles on leadership. His books include *Transformational and Charismatic Leadership: The Road Ahead* (Elsevier Science, 2002), *Full Leadership Development: Building the Vital Forces in Organizations* (Sage Publications, 1999), and *Developing Potential Across a Full Range of Leadership: Cases on Transactional and Transformational Leadership* (Lawrence Erlbaum Associates, 2000). His recent books are *Leadership Development in Balance: Made/Born* (Lawrence Erlbaum Associates, 2005) and *The High Impact Leader: Moments that Matter to Authentic Leadership Development* (McGraw-Hill, 2005). Avolio is a past chair of the OB Division of the Academy of Management. He also serves on the editorial boards of the following journals: *Journal of Applied Psychology*, *Academy of Management Journal*, *Journal of Organizational Behavior*, and *The Leadership Quarterly*.

Richard L. Baskerville is a professor of information systems and past chairman in the Department of Computer Information Systems, Robinson College of Business, Georgia State University. His research specializes in the security of information systems, methods of information systems design and development, and the interaction of information systems and organizations. His interest in methods extends to qualitative research methods. Baskerville is the author of *Designing Information Systems Security* (J. Wiley) and more than 100 articles in scholarly journals, professional magazines, and edited books. He is an editor for *The European Journal of Information Systems* and serves on the editorial boards of *The Information Systems Journal*, *Journal of Information Systems Security*, and the *International Journal of e-Collaboration*. Baskerville holds degrees from the University of Maryland (BS summa cum laude in management) and the London School of Economics, University of London (MS in the analysis, design, and management of information systems; PhD in systems analysis).

Dr. **Irma Becerra-Fernandez** is an associate professor at the Florida International University College of Business Administration. Her research focuses on KM, KM systems, and enterprise systems. She founded the FIU KM Lab, and has obtained funding as principal investigator for over $1.7 million primarily from NASA to develop innovative KM systems. She has published extensively in leading journals including the *Journal of MIS, Decision Sciences, Communications of the ACM,* and *European Journal of Operational Research,* among others. She is an author of the book *Knowledge Management: Challenges, Solutions, and Technologies.* She has delivered many invited presentations at federal sites, universities around the world, and international conferences. She was selected to receive the 2004 FIU Outstanding Faculty Torch Award, and serves as the faculty director for the masters in MIS and on the editorial board of the *International Journal of Knowledge and Learning.* She earned her PhD from FIU in electrical engineering, and her master's and bachelor's in electrical engineering from the University of Miami.

Dr. **Anita Blanchard** is an assistant professor of psychology at the University of North Carolina, Charlotte. She studies the effects of technology on employees' affective attachment. Currently, she has two lines of research. First, she is examining the sense of virtual community in virtual communities, particularly how it develops and what it leads to. Second, she is developing an environmental theory framework to combine the social processes with the technological features that lead to successful virtual group functioning.

Tom Butler is a senior lecturer in business information systems at University College Cork, Ireland. Before lecturing, Tom had an extensive career in the telecommunications industry. Tom has also successfully managed a development project for the United Nations.

James L. Corner is a professor of decision sciences at the Waikato Management School, University of Waikato, in Hamilton, New Zealand. He holds a PhD in operations research from Arizona State University. Before embarking on his academic career, he was a design engineer for Texas Instruments, and held various management positions with U.S. Steel Corp. (now USX). His main areas of research include descriptive decision processes, prescriptive decision modeling (mainly decision analysis), and the interface between theses areas. He has published numerous papers in leading journals, including *Operations Research, Management Science, European Journal of Operations Research, Journal of the Operational Research Society, Decision Analysis, Organizational Behavior and Human Decision Processes, and Journal Multi-Criteria Decision Analysis.* Finally, Jim is the head of the Department of Management Systems and has created degree programs in supply chain management and electronic commerce.

Kevin E. Dow is an assistant professor of accounting in the College of Business, Kent State University. He earned a BS in electrical engineering, and an MBA and PhD in business administration. Dr. Dow's research interests focus on the business value of information technology and the interaction of information technology and managerial accounting. His work has appeared in *Information Systems Research,* the *European Journal of Information Systems,* the *Data Base for Advances in Information Systems,* the *International Journal of Accounting Information Systems,* and others.

Dr. **Roberto Evaristo** is a knowledge management practitioner with 3M in St. Paul, Minnesota. Prior to that, he was on the faculty of the Department of Information and Decision Sciences at the University of Illinois, Chicago. His main research and practical interest is the management of complex environments, with emphasis on distributed projects. He has published in outlets such as *MISQ Executive, Communications of the ACM, International Journal of Project Management, Database, Journal of Engineering and Technology Management, International Journal of Emergency Management, Business Horizons, European Management Journal, Human Systems Management, Journal of Organizational Computing and Electronic Commerce*, and elsewhere. He is an associate editor for the *International Journal of e-Collaboration* and the *Journal of Global Information Management*, and also serves on the editorial board of *Information Technology and People* and the *Journal of Global Information Technology Management.*

Vanessa Garza is a PhD student at the Texas A&M International University College of Business Administration. She holds a bachelor's in business administration (BBA) degree in accounting and management information systems from St. Mary's University, as well as a master's in professional accountancy (MPAcc) from Texas A&M International University.

Surinder S. Kahai is an associate professor of MIS and fellow of the Center for Leadership Studies at SUNY-Binghamton. He has a bachelor's in chemical engineering from the Indian Institute of Technology (IIT), an MS in chemical engineering from Rutgers University, and a PhD in business administration from the University of Michigan. Surinder researches computer-mediated communication, leadership in electronic groups, and IS leadership. His research has been published in journals including *Decision Sciences, Journal of Applied Psychology, Journal of Management Information Systems, Leadership Quarterly*, and *Personnel Psychology*. He is currently serving on the editorial boards of *Group and Organization Management, IEEE-TEM*, and the *International Journal of Electronic Collaboration.* He coedited a special issue of *Organizational Dynamics* on e-leadership and a special issue of the *International Journal of e-Collaboration* on virtual team leadership.

Tim Kayworth, PhD, is an associate professor and department chair of MIS in the Hankamer School of Business at Baylor University. He has prior industry experience in information systems consulting and has also held positions as MIS director and operations manager for private-sector firms. Dr. Kayworth's research interests center on the management of IT in organizations. His recent research projects have included such topics as leadership in global virtual teams, the impact of organizational culture on knowledge management practice, and the role of culture in information systems research. His work has been published in the *European Management Journal*, the *Journal of Management Information Systems, The DATABASE for Advances in Information Systems*, the *Information Resources Management Journal*, as well as in such international conferences as AMCIS, ICIS, and the Strategic Management Society.

Julie E. Kendall, Ph. D., is a professor of management (ecommerce and information technology) in the School of Business-Camden, Rutgers University. Dr. Kendall is the immediate Past Chair of IFIP Working Group 8.2 and was both a Treasurer and a Vice President of the Decision Sciences Institute. She was awarded the Silver Core from IFIP. Professor Kendall has published in MIS Quarterly, Decision Sciences, Information & Management, CAIS, Organization Studies and many other jour¬nals.

Additionally, Dr. Kendall has recently co-authored Systems Analysis and Design, seventh edition. She is also a co-author of Project Planning and Requirements Analysis for IT Systems Development. and co-edited Human, Organizational, and Social Dimensions of Information Systems Development. Dr. Kendall is on the Senior Advisory Board for JITTA and is on the editorial boards of the Journal of Database Management and IRMJ. She also serves on the review board of the Decision Sciences Journal of Innovative Education. Professor Kendall was a functional editor of MIS for Interfaces and an associate editor for MIS Quarterly. She recently served as a Rand Faculty Fellow for the Senator Walter Rand Institute for Public Affairs and was named to the Circle of Compadres of the Ph.D. Project, whose mission is to increase the diversity of business school faculty. Dr. Kendall is researching policy formulation for ICTs in developing countries, and agile methodologies for systems development. She and her co-author (and spouse) Ken are currently examining the strategic uses of Web presence and ecommerce for off-Broadway theatres and other nonprofit organizations in the service sector. They have served as official nominators for the Drama League Awards in New York City. Julie's home page can be accessed at www.thekendalls.org.

Deepak Khazanchi is a professor of information systems and quantitative analysis, and the associate dean for academic affairs at The Peter Kiewit Institute's College of Information Science & Technology at the University of Nebraska in Omaha. He is also the joint coordinator of the Center for the Management of Information Technology (CMIT) in the College of IS&T. His current research and teaching interests are focused on the areas of virtual project management, B2B (business-to-business) assurance services and risk analysis in extended enterprise environments, the impact of IT on internal service quality, and the application of the philosophy of science in the information systems discipline. Deepak received his PhD degree with a specialization in management information systems from Texas Tech University. Prior to this, he completed the MBA degree from Southern Illinois University at Carbondale and obtained an undergraduate degree in civil engineering from the Indian Institute of Technology, Kharagpur, India.

Dorothy E. Leidner, PhD, is the Randall W. and Sandra Ferguson professor of information systems and director of the Center for Knowledge Management at Baylor University. Prior to rejoining the Baylor faculty, she was associate professor at INSEAD and at Texas Christian University. She is a regular visiting professor at the University of Mannheim, Germany, and has been a visiting professor at the Instituto Tecnologico y des Estudios Superiores de Monterrey, Mexico; at the Institut d'Administration des Entreprises at the Université de Caen, France; and at Southern Methodist University. Dr. Leidner received her PhD in information systems from the University of Texas at Austin. Dr. Leidner's research has been published in a variety of journals, such as *MIS Quarterly*, *Information Systems Research*, *Journal of Management Information Systems*, *Decision Sciences*, *Decision Support Systems*, and *Organization Science*. She has received several prestigious best-paper awards, including one in 1993 from the Hawaii International Conference on System Sciences, one in 1995 from *MIS Quarterly*, and one in 1999 from the Academy of Management. She is currently serving as coeditor of the journal *Data Base for Advances in Information Systems*, senior editor for *MIS Quarterly*, and associate editor for the *Decision Sciences Journal* and for the *Decision Support Systems Journal*. She serves on the editorial boards of *MISQ Executive*, the *International Journal of Knowledge Management*, and the *International Journal of Information Technology Education*.

John McAvoy is a lecturer in business information systems at University College Cork, Ireland. Prior to lecturing, John had a variety of roles in the information systems field, ranging from systems administration to managing software development teams.

Rafael Prikladnicki is a professor of computer science at PUCRS, Brazil. He is currently involved in projects related to distributed and global software development, risk management, and software quality, with research collaboration both in industry and academia in Brazil, Canada, the USA, and some countries in Europe. He is also one of the coordinators of the research group on distributed software development (MuNDDoS), and a member of the research group on software quality (QUATES) at his university. He is a member of the working group on global software development, coordinated by the University of Victoria (UVIC). He is working on his PhD at PUCRS and as a visitor student at UVIC. His publications appear, among others, in the *Journal of Software Process Improvement and Practice*, *Communications of the Association for Information Systems*, and in the proceedings of many conferences. He is a member of ACM, AIS, PMI, and the Brazilian Computer Society (SBC).

Dr. **Boris Roussev** is an associate professor of IS at the University of the Virgin Islands. His interests are in the areas of object-oriented and economic-driven software development, requirements engineering, and project management. He has been researching into causes of project failures where the inability of value-neutral principles and practices in software development to deal with the sources of project failures calls attention to the social context in which IT is deployed and produced, and whose due consideration is indispensable to the success of IT management. In addition, Dr. Roussev has experience in software risk management and software process quality factor analysis. All of the above is combined with industry experience in software methods such as domain engineering, product lines, MDA with xUML, generative programming, and aspect-oriented programming. Dr. Roussev has authored numerous journal articles and conference papers. He edited a book entitled *Management of the Object-Oriented Process* published by Idea Group, Hershey, PA, in 2005.

Alexander Serenko is an assistant professor of management information systems with the Faculty of Business Administration, Lakehead University, Canada. He holds an MS in computer science, an MBA in electronic business, and a PhD in management science systems. Dr. Serenko's research interests pertain to user technology adoption, knowledge management, and innovation. Dr. Serenko's articles appeared in various refereed journals, and his papers have received awards at Canadian and international conferences.

Helen Stewart today is the NASA Ames Research Center IT security manager, primarily focused on the Federal Information Security Management Act (FISMA), and information-handling issues dealing with the confidentiality, integrity, and availability of federal information and technology. She founded and managed the New Millennium Documentation Project and the Postdoc Project in the Computational Sciences Division of NASA Ames Research Center, where she has worked since 1987. Her research and development efforts are primarily focused on ensuring compliance with national and NASA policies, IT security, and secure collaborative assistant systems and environments. She has worked on many projects involving secure, advanced federated environments, where primary focus was on secure information sharing for risk management for the Space Shuttle Program and Mission Control Center.

Ms. Stewart continues to do research on policy, procedures, and contractual influences on KM, IT security, and risk management to determine impacts these may have on national space-based assets.

Dr. **Ofir Turel** holds a BS in industrial engineering, an MBA in technology management, and a PhD from the DeGroote School of Business, McMaster University. Before joining the DeGroote School of Business, he held senior positions in the information technology and telecommunications industries. His research interests include human-computer interaction, online collaboration, and mobile services. His work has been published in several peer-reviewed journals and presented at various international conferences.

Dr. **Ganesh Vaidyanathan**, an assistant professor at IU South Bend, School of Business and Economics, has conducted research in the areas of e-commerce, SCM, project management, knowledge management, innovation, and IT value. He has authored over 20 publications in journals including *Communications of the ACM.* He has held executive positions at eReliable Commerce, Honeywell, General Dynamics, Lockheed Martin, and Click Commerce. He launched products that include security, payment processing, procurement, logistics, ERP (enterprise resource planning), SCM, and data warehousing. He has consulted with Fortune 100 companies including United Airlines, Mitsubishi, Motorola, and Honeywell in technology, business, and process reengineering. Dr. Vaidyanathan holds a PhD from Tulane University and an MBA from the University of Chicago.

Professor **Jeffrey Wong** received his PhD from the University of Oregon in 1999 and is a certified public accountant. His research interests focus on understanding how the actions of a firm ultimately map to its financial results, and the interaction between managerial accounting and information systems technology. Professor Wong's research has been presented at national and international conferences and has been published or is forthcoming in journals such as *Database, Behavioral Research in Accounting,* and *Internal Auditing.*

Marcelo Yamaguti is a professor of computer science at PUCRS, Brazil. He holds a doctorate in computer science. He is currently involved in projects related to global software development and software quality with companies like Dell (United States) and SONAE (Portugal). He is also the coordinator of the research group on software quality and testing (QUATES) at his university. His publications appear in many conference proceedings.

Ilze Zigurs is a professor and the Mutual of Omaha distinguished chair of information science and technology in the College of Information Science and Technology at the University of Nebraska at Omaha. She also serves as department chair for the Department of Information Systems and Quantitative Analysis. Her research examines the design, implementation, and use of collaboration technologies, particularly in virtual teams and projects. She has published in such journals as *MIS Quarterly, Journal of Management Information Systems, Journal of Organizational Computing and Electronic Commerce,* and *Group Decision and Negotiation,* among others. She has recently coauthored with Deepak Khazanchi a book published by the Project Management Institute, *Patterns of Effective Management of Virtual Projects.* She serves as editor in chief of *e-Service Journal.*

Index